STUDIES IN PHILOSOPHY

XVII

PHYSIOLOGICAL THEORY AND THE DOCTRINE OF THE MEAN IN PLATO AND ARISTOTLE

by

THEODORE JAMES TRACY, S.J.
Loyola University

1969
MOUTON
THE HAGUE · PARIS

LIBRARY OF CONGRESS CATALOG CARD NUMBER: 68-23201

Printed in The Netherlands by Mouton & Co., Printers, The Hague

Matri Patrique

ACKNOWLEDGEMENTS

To all who have generously supported through their learning, time, and interest, the preparation of this study, I wish to express sincere gratitude; and especially to Professor Whitney J. Oates of Princeton University, under whose patient direction and constant encouragement the work was begun and brought to completion; to Professor Antony E. Raubitschek of Stanford University, who first suggested that some aspect of "the mean" might still be unexplored; to Professor Harold Cherniss of the Institute for Advanced Studies, for a number of valuable insights and suggestions; to Professor Robert Duff Murray of Princeton, for his careful reading of the manuscript; to my Jesuit colleagues, among whom Rev. Lothar C. Nurnberger of Loyola University, Chicago, has been foremost in his interest and assistance; to Dr. Peter de Ridder of Mouton and Company, and Rev. John B. Amberg, S.J. of the Loyola University Press, for their encouragement and support through the uncertainties of publication. *Omnibus maximas vobis gratias.*

I wish also to acknowledge the kindness of several publishers in permitting the use of their translations from the Greek. Wherever possible I have presented the Greek sources through a standard translation in the hope of making the present study useful, and perhaps more acceptable, to a wider group of readers. However, I have not hesitated, after due warning, to depart from the standard translation where this seemed advisable. My gratitude, then, to the Cambridge Philological Society for permission to use W. H. S. Jones' *Medical Writings of Anonymus Londinensis*; to the Clarendon Press, Oxford, for numerous passages from the Oxford translation of Aristotle; to the Harvard University Press for passages from the Loeb Classical Library translations of *Hippocrates* by W. H. S. Jones, *Plato: The Symposium* by W. R. M. Lamb, *Aristotle: Parts of Animals* by A. L. Peck, and *Aristotle: On the Heavens* by W. K. C. Guthrie; to the Johns Hopkins Press for W. H. S. Jones' version of *On Ancient Medicine*, published under the title *Philosophy and Medicine in Ancient Greece*; to Routledge and Kegan Paul Ltd. for the passages from *Plato's Cosmology* by F. M. Cornford.

Loyola University
Chicago
June 1966

T. J. T.

TABLE OF CONTENTS

Acknowledgements 7

Abbreviations and Short Titles 13

I. Introduction 15

II. Physiological Theory in Greek Medicine before Aristotle 22

A. Alcmaeon of Croton 22
B. The Menon Papyrus 25
1. Menecrates 26
2. Petron of Aegina 27
3. Philistion of Locri 28
C. Hippocratic Works 32
1. *Ancient Medicine* 35
2. *The Nature of Man* 45
3. *Airs, Waters, Places* 51
4. *Regimen I* 64
D. Conclusion 67

III. Physiological Theory and "The Mean" in Plato . . . 77

A. The Body of the Cosmos 78
B. The Soul of the Cosmos 90
C. The Human Soul and Body 96
1. The Rational Soul and Its Destiny 96
2. The Physical Organ of the Soul 102
3. The Mortal Soul and Its Organs 106
4. Environment and the Nutritive Process 116
D. Disease and Health in Man 119
1. Disease of the Body 120

2. Disease of the Soul 123
3. The Total Organism and Its Care 136

E. Conclusion: The *Timaeus* and the Medical Writers 142

IV. Physiology and the Doctrine of the Mean in Aristotle . 157

A. Elemental Structure of the Organism in Health and Disease 157
1. The Nature of Health and Other Bodily Excellence 158
2. Elemental Structure of the Organism 163
3. Environment and Its Influence 174

B. Proportion and the Mean in the Larger Aspects of the Organism 178
1. Ultimate Basis of the Vital Processes and Organs 178
2. The Principal Organs and Their Functions . . . 182
3. Summary 194

C. Proportion and the Mean in Sense Perception . . 197
1. Body, Soul, and the Living Composite 197
2. The Problem and Conditions of Sense Perception 201
3. The Special Senses and the Mean 205
4. The Central Sense Power and the Mean 212
5. Summary 220

D. Moral Excellence and the Mean 222
1. The Method of the *Ethics* 223
2. General Comparison of Bodily and Moral Excellence 227
3. Bodily and Moral Virtue and the Equilibrium of Opposites 231
4. Further Characteristics of Bodily and Moral Virtue 237
5. The Psycho-Physical Aspect of Emotion and Moral Excellence 247
6. Moral Excellence and the Intellect 261
7. Moral Excellence and Happiness 277

E. The Political Organism and the Mean 283
1. The Principle of Natural Subordination 286

2. The Principle of Order among Equal and Opposite Parts: Μεσότης 295
3. Unnatural or Deviation Forms; the Relative Mean 301
F. Conclusion: The Expert and His Norms of Value 311
1. Objectives of Physician, Moral Guide, and Statesman 311
2. Norms and Values; the Absolute Good 326

V. Conclusion 334

Appendix I. On the Notion of ΜΕΣΟΤΗΣ in Greek Mathematics 344
Appendix II. Concerning the Theory of Francois Nuyens on the Development of Aristotle's Psychology 347
Appendix III. The Heart as Source and Center of External Action 354

Bibliography 360

Greek Words, Phrases, Sentences 369

General Index 369

Index of Passages Cited 377

ABBREVIATIONS AND SHORT TITLES

AJP *American Journal of Philology.*

CJ *Classical Journal.*

CMG *Corpus Medicorum Graecorum* ediderunt Academiae Berlinensis Havniensis Lipiensis. Berolini in aedibus Academiae Litterarum, 1927 —.

Commentaria in Aristotelem Graeca The Berlin Academy's edition of Greek Commentators on Aristotle, published in 23 volumes (Berlin, 1882-1909) with the *Supplementum Aristotelicum* in 3 volumes (1882-1903). For individual commentators and editors, see the bibliography.

Diels-Kranz *Die Fragmente der Vorsokratiker*, edited by H. Diels with additions by W. Kranz. The 6th and 7th editions (Berlin, 1951-1952 and 1954) are photographic reprints of the 5th (1934) with Nachträge by Kranz.

Doxogr. Graec. *Doxographi Graeci*, collegit recensuit prolegomenis indicibusque instruxit H. Diels. 3rd ed. (photographic reprint of 1879 ed.), Berlin, 1958.

JHS *Journal of Hellenic Studies.*

Kühn *Claudi Galeni Opera Omnia* (Greek-Latin edition of Galen's works), edited by C. G. Kühn. 20 volumes in 22. Leipzig, 1821-1833.

Oxford Translation *The Works of Aristotle*, translated into English under the editorship of (J. A. Smith and) W. D. Ross. Oxford, 1908-1931. (For translators of individual works, see entry in bibliography under "Ross").

RE *Real-Encyclopädie der classischen Altertumswissenschaft*, the work of A. Pauly continued by G. Wissowa, W. Kroll *et alii*. Stuttgart, 1893—.

ARISTOTLE'S WORKS

Cat.	*Categories.*
De An.	*De Anima.*
De Gen. An.	*De Generatione Animalium*
De Gen. et Cor.	*De Generatione et Corruptione.*
De Motu An.	*De Motu Animalium.*
De Part. An.	*De Partibus Animalium.*

E.E.	*Eudemian Ethics.*
Hist. An.	*Historia Animalium.*
Met.	*Metaphysics.*
Meteor.	*Meteorology.*
N.E.	*Nicomachean Ethics.*
Parva Nat.	*Parva Naturalia*, i.e.
	De Sensu et Sensibilibus (463a1-449b4).
	De Memoria et Reminiscentia (449b4-453b11).
	De Somno (453b11-458a32).
	De Somniis (458a33-462b11).
	De Divinatione per Somnum (462b12-464b18).
	De Longitudine et Brevitate Vitae (464b19-467b9).
	De Vita et Morte (467b10-470b5).
	De Respiratione (470b6-480b30).
Post. Anal.	*Posterior Analytics.*
Pr. Anal.	*Prior Analytics.*
Rhet.	*Rhetoric.*

I

INTRODUCTION

The term ἡ μεσότης, generally translated into English as "the mean", appears throughout the works of Aristotle, though most commonly associated with the *Nicomachean Ethics*, where it assumes critical importance in the famous definition of moral excellence as a lasting disposition (ἕξις) of man's affective faculties specified by being constituted ἐν μεσότητι (1106b 36-1107a1). Consequently every serious student of Aristotle has at some time been concerned with the meaning of the term, so that the bibliography on the subject, ranging from brief comment to book-length study, assumes forbidding proportions. It may seem an impertinence, then, to add another item to the list. However, the importance of the term for an understanding of Aristotle's ethics alone will, I trust, justify one in pursuing any well-founded hope of seeing it in a new perspective, of gaining some fresh insight into its meaning.

In order to reveal the basis of this hope, though without attempting anything like a complete survey and evaluation of the literature, we may cite a number of studies pertinent to the problem, indicating how they suggest, and leave room for, the investigation undertaken here. At the risk of some oversimplification, these discussions may be grouped roughly into three classes according to their general approach.

1. Among studies with a predominantly *historical approach* can be cited, in chronological order, those of Kalchreuter (*Die Mesotes bei und vor Aristoteles*, 1911),[1] Souilhé (*La Notion platonicienne*

[1] Full data on this and subsequent publications mentioned in the text will be found in the Bibliography.

d'intermédiaire dans la philosophie des dialogues, 1919), Jaeger (*Paideia*, III, ch. I: "Greek Medicine as Paideia", 1944), and Wehrli ("Ethik und Medizin: Zur Vorgeschichte der aristotelischen Mesonlehre", *Museum Helveticum* VIII, 1951). These specifically investigate the historical roots of the doctrine of "the mean" as it appears in Aristotle. The earlier studies trace the ultimate origins of the notion from Homer and the Delphic maxim (μηδὲν ἄγαν) through the lyric poets and Presocratic philosophers. All indicate contemporary *medical* theory as an important immediate source, establishing the general similarity of terminology and concept in the works of Aristotle and certain of the Hippocratic writings. None that I know, however, attempts a detailed study of the basic notion and its implications in both the medical writers and Aristotle.

Less pertinent, but stressing the close relationship between Greek medical writers and philosophers, are the studies of Poschenrieder (*Die naturwissenschaftlichen Schriften des Aristoteles in ihrem Verhältnis zu den Büchern der hippokratischen Sammlung*, 1887), Keus (*Über philosophische Begriffe und Theorien in den hippokratischen Schriften*, 1914), Klippel (*La Médecine grecque dans ses rapports avec la philosophie*, 1937), Miller ("Philosophy and Medicine", *CJ*, XLIV, 1949), Diller ("Hippokratische Medizin und attische Philosophie", *Hermes*, LXXX, 1952), Edelstein ("The Relation of Ancient Philosophy to Medicine", *Bulletin of the History of Medicine*, 1952), Longrigg ("Philosophy and Medicine: Some Early Interactions", *Harvard Studies in Classical Philology*, 1963), and Schumacher (*Antike Medizin: Die naturphilosophischen Grundlagen der Medizin in der griechischen Antike*, 2nd ed., 1963, with bibliography).

2. Next, there is the *expository* or *analytical* approach, which is most common in discussions of "the mean". This is usually employed, first of all, in the more or less summary treatments of the doctrine in general works on ethics, on the whole range of Aristotle's work, on his ethical and political theory, and in commentaries on the *Ethics* and *Politics*. Such would be the approach found, for example, in the broad studies of Zeller (*Aristotle and the Earlier Peripatetics*, 1897), T. Gomperz (*Greek Thinkers*, IV, 1912), and Überweg-Praechter (*Grundriss der Geschichte der Philo-*

sophie, I, 1920) among the older authors, or of Ross (*Aristotle*, 1923), Mure (*Aristotle*, 1932), Allan (*The Philosophy of Aristotle*, 1952), and Randall (*Aristotle*, 1960) among the more recent; in studies of Aristotle's ethical and political theory like those of Barker (*The Political Thought of Plato and Aristotle*, 1906) and Gauthier (*La Morale d'Aristote*, 1958); in the classical commentaries of Grant (*The Ethics of Aristotle*, 1874), Newman (*The Politics of Aristotle*, 1887-1902), Stewart (*Notes on the Nichomachean Ethics*, 1892), Burnet (*The Ethics of Aristotle*, 1900); and in the more recent commentaries of Joachim (*Aristotle: The Nichomachean Ethics*, 1951), Dirlmeier (*Aristoteles: Nikomachische Ethik*, 1956), and Gauthier — Jolif (*L'Éthique à Nicomaque*, 1958-1959).

Because they are concerned with larger objectives these works, however excellent, must necessarily restrict their treatment of any one doctrine, and must often present summary conclusions rather than cite evidence at length. And even where the discussion is somewhat protracted, as in Zeller, the doctrine of "the mean" is generally considered only within the context of the *Nichomachean Ethics*.

There are, however, a number of analytical studies devoted specifically to the doctrine of "the mean" in Aristotle, and to this extent their objective coincides more closely with that of the present investigation. Among these, in chronological order, are the studies of Stocks ("ΛΟΓΟΣ and ΜΕΣΟΤΗΣ in the *De Anima* of Aristotle", *Journal of Philology*, XXXIII, 1910), Schilling (*Das Ethos der Mesotes*, 1930), Joseph ("Aristotle's Definition of Moral Virtue and Plato's Account of Justice in the Soul", *Philosophy*, IX, 1934), Oates ("The Doctrine of the Mean", *Philosophical Review*, XLV, 1936), Ohmsted ("The 'Moral Sense' Aspect of Aristotle's Ethical Theory", *AJP*, LXIX, 1948), Van der Meulen (*Aristoteles: die Mitte im seinem Denken*, 1951), and Godfrey ("The Idea of the Mean", *Hermathena*, LXXXI, 1953).

These studies afford close and valuable analyses of Aristotle's doctrine of "the mean". Most of them, however, concentrate upon the notion as it is found in the *Nicomachean Ethics*, with few references to other works except perhaps the *De Anima*. Van der Meulen, whose study reaches book length, comments briefly on a large collection of texts from Aristotle's logic and analytics, metaphysics, psychology, and ethics, but leaves the biological works and the *Politics* almost untouched. Moreover, he abstracts completely from the historical context of Aristotle's thought, from the immediate environment established by Plato and the medical writers, with the light they cast upon the origin of the notion and its implications for Aristotle. If the general conclusion of the historical studies is correct, however, it can be hoped that greater attention to the notion as it appears in the writings of the Hippocratic corpus, in Plato's *Timaeus*, and in the biological works of Aristotle may help toward a new perspective and fuller understanding of the doctrine in the *Ethics* and *Politics*.

3. There are, finally, some commentators who in fact recommend a *biological* approach to Aristotle, among them D'Arcy Thompson (Oxford Translation, IV, 1910, p. vii), Hanz (*The Biological Motivation in Aristotle*, 1939), Le Blond (*Logique et methode chez Aristote*, 1939, pp. 346-370; *Aristote, philosophe de la vie*, 1945, pp. 7-14), Peck (Loeb edition of *De Generatione Animalium*, 1953, p. viii), and especially Marjorie Grene (*A Portrait of Aristotle*, 1963). These scholars emphasize the influence of Aristotle's biological interests upon the formation of his philosophical notions. None of them, however, has attempted to treat in detail the possible biological or physiological background of the doctrine of "the mean", though this is suggested by some.[2]

In summary, then: The *historical* studies suggest that a proximate source of Aristotle's notion of μεσότης is the medico-biological thought of his contemporaries. However, none of the *analytical* studies, as far as I know, has attempted to explore the notion in the light of this background. And those writers who advocate a *biological approach* have not yet employed it to study the doctrine of "the mean". There would seem to be room, then, for a study which attempts to examine this notion and its application in the medico-biological tradition contemporary with Aristotle and in the biological writings of Aristotle himself, with the hope of gaining

[2] See A. L. Peck, *op. cit.*, p. 480, note a; J. P. Anton, *Aristotle's Theory of Contrariety* (London, 1957), p. 129, note 52.

fresh insight into the implications of μεσότης as it appears in the *Ethics* and *Politics*.

The link between Aristotle's μεσότης of moral excellence and his biological or physiological theory is in fact supplied by the *Nicomachean Ethics* itself. For throughout that work Aristotle constantly employs *health of body* as the prime analogue in explaining and illustrating the nature of moral excellence, just as he constantly compares the moral guide or statesman to the physician, and vice to disease. This has been noted and abundantly documented in a study by Werner Jaeger, who maintains that "the medical example, far from being a casual analogy, is present to the philosopher's mind throughout. It belongs to the very foundation of his ethical science, at least in the form it has taken in the *Nicomachean Ethics*".[3] There seems to be solid hope, then, that inquiry into Aristotle's notion of *health*, with all its implications, may prove useful in casting light upon his notion of moral excellence, both of them being constituted ἐν μεσότητι.

Our procedure, then, will be to determine first of all what the notion of health implied for Aristotle, and how the term μεσότης is involved. To determine this we shall look first into the scientific tradition to which Aristotle belonged, examining the notion of health and allied physiological concepts in certain medical writers, and their development in Plato's *Timaeus*. Next we shall trace these physiological notions as they appear in Aristotle's own scientific writings, exploring the implications of the term μεσότης in this context, and its application on the psychological level. We shall then be in a position to study, in the light of his physiological analogies, Aristotle's use of μεσότης in connection with moral excellence in the individual, and its application to the corporate organism of the state. Thus it is hoped that the detailed study of the notion of health and other important physiological concepts used by Plato and Aristotle in describing the psychological and ethical aspects of man may contribute to a better understanding of their thought, and particularly of the doctrine of "the mean".

Certain qualifications should be noted. First, the primary ob-

[3] "Aristotle's Use of Medicine as a Model of Method in His Ethics", *JHS*, LXXVII (1957), 56-57.

jective of the investigation will be to *elucidate* the notion of μεσότης as it appears in Aristotle's *Ethics*, not to establish its *origins*. Consequently it is not necessary to agree with previous studies either that the pivotal ideas of Aristotle's philosophy are based upon his biological investigations, or that the notion of "the mean" is drawn immediately from contemporary medical literature. Such claims could not be made, however, if there were not a close association and similarity between Aristotle's biological and strictly philosophical thought; and this is borne out by his frequent use of physiological analogy in the *Ethics* and *Politics*. Granted this, a study of "the mean" in one sphere should help to elucidate its analogue in the other.

Second, the investigation will not attempt to settle problems of *chronology* in Aristotle, e.g., to establish that his biological works as we have them are earlier than the *Nicomachean Ethics* or *Politics*, or any part of them. It is enough to know that Aristotle's biological interests and investigations began early,[4] regardless of when the present treatises were composed. In particular instances, however, exception must be taken to certain aspects of the developmental hypothesis as proposed by Jaeger and Nuyens. These will be discussed as they arise.

Thirdly, although Plato is mentioned in the title, the proposed study does not pretend to a comprehensive coverage of his works, but concentrates upon the *Timaeus* particularly as illustrating his physiological theory and his broad application of the doctrine of "the mean".

A final word should be said regarding method. In treating each group of writings, this study will attempt first to gather, organize, and present the evidence at some length and in context, then to draw conclusions. This involves the abundant use of direct quotation and paraphrase, and a certain amount of repetition where different aspects of the same passage or subject must be emphasized. At times this method may strike the reader as unnecessary and tedious. However, it was adopted, first, out of a sincere desire to present the evidence adequately, allowing the ancient authors

[4] See H. D. P. Lee, "Place Names and the Date of Aristotle's Biological Works", *Classical Quarterly*, XLII (1948), 61-67.

to speak for themselves wherever possible. Secondly, the writer is deeply convinced that *context* is of extreme importance to the understanding of a particular statement or passage, not only the immediate context but the broader contexts of the whole work, of the author's thought as known from other works, and of the intellectual tradition to which he belongs. Reconstruction of context often involves repetition of materials otherwise well known. We hope that such repetition will to some extent be justified by a sharper understanding and deeper appreciation of the significance of particular passages when seen in relation to the whole. Summary and conclusions will be found at the end of each chapter.[5]

[5] Two important recent works became available to me only after the present study was in process of publication: Ingemar Düring, *Aristoteles: Darstellung und Interpretation seines Denkens* (Heidelberg, 1966) and Hans Krämer, *Arete bei Platon und Aristoteles* (Heidelberg, 1959). The latter presents an extensive metaphysical analysis of the ethical mean in Plato and Aristotle.

II

PHYSIOLOGICAL THEORY IN GREEK MEDICINE BEFORE ARISTOTLE

A. ALCMAEON OF CROTON

The general theory of health and disease common to many of Aristotle's predecessors, both physicians and philosophers, may be illustrated by the well known passage of Aetius (V, 30, 1) concerning Alcmaeon of Croton. Aristotle reports that Alcmaeon conceived the human composite in terms of opposites, which he grouped into an indefinite number of pairs — white-black, sweet-bitter, good-bad, large-small, etc.[1] Aetius ascribes to him a theory of health based upon this concept, as follows:

Constitutive of health is a state of equality (ἰσονομία) among the powers (δυνάμεις), moist-dry, cold-hot, bitter-sweet, etc. On the other hand, a state of domination (μοναρχία) among them is productive of disease. For the domination of one member of a pair is destructive of the other. Disease occurs through the immediate agency of an excess (ὑπερβολή) of heat or cold, for example, occasioned by taking too much or too little nourishment, and localized in the blood, marrow, or brain. But it is also

[1] *Met.* 986a27-b1; and cf. Diogenes Laertius, VIII. 83. For a full discussion of Alcmaeon see J. Wachtler, *De Alcmaeone Crotoniate* (Leipzig, 1896); L. Stella, *Importanza di Alcmeone nella storia del pensiero greco* (Rome, 1939). W. Heidel, *Hippocratic Medicine* (N.Y., 1941), pp. 42-49, offers a brief appraisal and calls attention to the fundamental difficulties of chronology and authenticity; on these, see L. Edelstein's review of Stella's work in *AJP*, LXIII (1942), 371-372. It should be remarked that it is not the aim of the present discussion to enter into the involved historical problem of the development of medical theory in its relations with philosophy, but rather to examine some of the passages *illustrative* of the general theory of health in question, as it appears in medical writings commonly regarded as earlier than, or at least contemporary with, Aristotle. Hence, whether the passage from Aetius represents a theory proposed by Alcmaeon *ca.* 500 B.C. or, as Edelstein suggests, a Pythagorean view contemporary with Socrates, it is still useful as a summary of one theory of health current before and during the time of Aristotle.

engendered in these at times from external causes — waters ... the locality ... and the like. Health, however, is the proportionate blending (σύμμετρος κρᾶσις) of the qualities.[2]

An analysis of the statement reveals the following features:

(1) The bodily organism is conceived as a complex of elemental "powers" (δυνάμεις),[3] moist-dry, hot-cold, bitter-sweet, etc., each of which is *opposed* to another by reason of being what it is. This implies that while combined in the organism the powers never completely lose their peculiar character, but each tends to work within the whole the effect that is proper to it, though this runs directly counter to the tendency of the opposite power.

(2) The *health* or well-being of the bodily organism is thought to depend upon the relationship of the opposing powers. The body is healthy as long as the powers remain in a state of *equality* (ἰσονομία),[4] i.e., as long as each is counterbalanced by its opposite so

[2] The text is that of Diels-Kranz, 24B4; the translation is mine.

[3] On the technical signification of this difficult term in early Greek medical writing cf. A. Keus, *Über philosophische Begriffe und Theorien in den hippokratischen Schriften* (Cologne, 1914), pp. 46-58; J. Souilhé, *Etude sur le terme ΔΥΝΑΜΙΣ* (Paris, 1919), pp. 31-57. W. H. S. Jones comments on the term in his *Philosophy and Medicine in Ancient Greece* (Baltimore, 1946), pp. 93-96, and *The Medical Writings of Anonymus Londinensis* (Cambridge, Engl., 1947), where he remarks (p. 9): "The meaning of δύναμις in Greek science is very different from the Aristotelian 'potentiality'. Cornford (*The Cosmology of Plato*, p. 53) calls δυνάμεις the properties of bodies considered as having the power to act and be acted upon, e.g., hotness is the property of fire that is manifest when fire makes something hot or causes in sentient things a sensation of heat. 'Property', 'quality', or 'characteristic' are possible but partial equivalents in many cases; what is lacking in these renderings is that a δύναμις always belongs to a thing's real nature. It is, in fact, the active side of a φύσις. 'Essence' or 'essential characteristic' are thus also possible equivalents, but even these fail to suggest the idea of force or action associated with the word".

[4] The term ἰσονομία, as Liddell-Scott-Jones (9th ed., 1940) indicates, has a political flavor, meaning "equality of political rights" among the elements of the social group. Cf. Herodotus III, 80, 142. In the present context it is contrasted to another obviously political term, μοναρχία. The implied comparison between political and physiological structure is interesting. Note that ἰσονομία implies a *balanced distribution of power*, a notion that is lost in the inadequate English translation "equality". A. E. Taylor (*A Commentary on Plato's Timaeus* [Oxford, 1928], p. 448) translates: "Health is kept up by the *balance of powers* in the body, the moist, the dry, etc. ... The *monarchy* of any of these causes disease, because the monarchy of one destroys its opposites." See also G. Vlastos, "Isonomia", *AJP*, LXXIV (1953), pp. 337-366.

that they maintain an *equilibrium*. Health thus conceived is not something static, but a *dynamic* equilibrium maintained through the active interplay of opposing forces within the living organism.

(3) *Disease*, on the other hand, occurs when one or the other of the powers grows disproportionately strong, so that the equality among them, and consequently their equilibrium, is destroyed. The resulting state of μοναρχία, in which the stronger power dominates the others and reigns supreme, is disease.

(4) This condition is immediately caused by the *excess* (ὑπερβολή) of the power dominating the others *within* the organism. However, this excess may be traced to factors *external* to the organism which have a direct influence upon it, for example, the nature and amount of food and drink, and the character of the locality or environment. These external factors are regarded as bearing directly upon the powers within the body and affecting the internal relationship. They are, therefore, considered of prime importance in explaining health and disease.

Finally, the notion of health is summed up in the phrase τὴν σύμμετρον τῶν ποιῶν κρᾶσιν. Health exists when the "qualities" or powers are *blended* (κρᾶσις) in proper *proportion* (σύμμετρος). The technical term κρᾶσις must be explored for a full understanding of the formula. For the present it is enough to remark that κρᾶσις implies a close and relatively stable union of the opposing powers within the organism, whereas the important qualifying σύμμετρος indicates that the powers thus united must be properly proportioned both among themselves and with respect to the external factors mentioned, to constitute a state of ἰσονομία, equality or equilibrium, without excess or deficiency. This proportionate blending of the powers of the organism is considered the essence of health — τὴν δὲ ὑγιείαν τὴν σύμμετρον τῶν ποιῶν κρᾶσιν.

This concept of health, then, as a dynamic equilibrium of opposing constituents in proportionate blend is attributed by the doxographer to Alcmaeon. There is reason to believe that this same theory was current in the late fifth and fourth centuries, and that it was accepted by the predecessors and contemporaries of Aristotle.

B. THE MENON PAPYRUS

The currency of this general theory among medical men and philosophers immediately before Aristotle and its connection with Aristotle are indicated in the medical papyrus known as *Anonymus Londinensis*.[5] This document is of particular interest for us because the second section (roughly chapters IV-XX), concerned with the etiology of disease, offers a summary of opinions of some twenty medical authorities which scholars believe to be based upon the Ἰατρικά of Aristotle's pupil Menon,[6] and which therefore represents medical theory contemporary with Aristotle at the latest.

The twenty authorities are unevenly divided into two groups according to their general approach to the source of disease.[7] The fourteen authorities comprising the *first* group are characterized as holding that "diseases arise because of the residues from nutriment" (IV, 26) and as "those who attribute the causes of diseases to residues and secretions" (XIV, 7). Contrasted to them are the members of the *second* group, "those who hold that our bodies are composed of a combination of elements" (XIV, 9) and who explain diseases as "due to the elements" (IV, 28). Now it is in the accounts of this second group that the general theory of health and disease outlined above is most easily discernible, as we shall see. And, significantly, the first authority cited as representative of this group is Plato.

Menon's outline of Plato's physiological theory, drawn from

[5] The *editio princeps* is that of H. Diels, *Anonymi Londinensis ex Aristotelis Iatricis Menoniis et aliis Medicis Eclogae* (Berlin, 1893). Diels' text, with a few changes, is reprinted with a helpful introduction, translation, and notes by W. H. S. Jones, *The Medical Writings of Anonymus Londinensis*. Passages will be cited according to Jones' edition, and in his translation.

[6] The name actually given several times by the author as his source for this section is *Aristotle*. However, scholars generally believe that he refers to a collection of medical opinions compiled by Menon, since Galen, in his commentary on Περὶ Φύσιος Ἀνθρώπου (Kühn, XV, 25), mentions such a collection as attributed by some to Aristotle but really written by his pupil Menon. See H. Diels, "Über die Excerpte von Menons *Iatrika*", *Hermes*, XXVIII (1893), 407 ff.; Jones, pp. 1-8.

[7] Diels, *Hermes* XXVIII, 415, believes that this division of authorities according to opinion goes back to the original doxography of Menon. For the basis of the division, see Diels, *ibid.*, and Jones, p. 3.

the *Timaeus*, is not only the first but also the longest account in this group. However, for the present we shall postpone the discussion of Plato, that we may give him the separate consideration his importance deserves, and that we may first examine the general theory of health and disease as it is found in writers whose primary interest is medicine. Hence we shall begin by tracing the theory and its variations briefly through the accounts of three other members of this group — Menecrates, Petron of Aegina, and Philistion. Having thus established its outline in greater detail, we shall turn to representative works of the Hippocratic collection for a more complete understanding of the theory as it was developed by the medical writers. We shall then be in a better position to discuss the theory as it appears in Plato and in Aristotle.

1. *Menecrates*[8]

The account of Menecrates begins as follows (XIX, 18-33):

> Menecrates surnamed Zeus when setting forth in his book *Medicine* an explanation of bodies, and when about to trace the causes of affections, first goes at some length into the qualities of bodies, saying that bodies are composed out of four elements, two of which are hot, and two cold, the hot being blood and bile, the cold, breath and phlegm. And when these do not disagree, but are in a state of harmony, the body is healthy; when they are not in harmony, it is diseased (τούτων ... μὴ στασιαζόντων, ἀλλ' εὐκρότως διακειμένων ὑγιαίνειν τὸ σῶμα, δυσκρότως δὲ ἐχόντων νοσεῖν). For then there are expelled from our bodies phlegms, boils and the like. Catarrhs also of various kinds arise from the excess (ὑπερβολή) of phlegm ...

There follows some further explanation of specific diseases, how the excess if carried to the hips brings on lumbago, to the lungs, pneumonia, to the ribs, pleurisy, etc. However, it is clear that Menecrates' general theory of health and disease is essentially the same as that attributed to Alcmaeon. Health is a state of *harmony among opposing constituents*, here reduced to four, opposed on

[8] A Syracusan traditionally associated with Philip of Macedon. See H. Raeder, "Menekrates", *RE*, XV, Pt. I, col. 802, #29; Diels, *Hermes*, XXVIII, 416-417.

the basis of hot and cold.[9] Στάσις occurs, the harmony is disrupted, when one of the constituents is in *excess*. Unless this excess is expelled it develops into one specific disease or another depending upon the part of the body in which it settles. In other words, health is a state of dynamic equilibrium among opposing forces in the body, and remains as long as those forces are properly proportioned. When one force grows excessive or disproportionate the result is στάσις, loss of harmony or equilibrium, disease.

2. *Petron of Aegina*[10]

The same basic theory is attributed to Petron of Aegina, as follows (XX, 1-14):

> Petron of Aegina says that our bodies are composed of a pair of elements, the cold and the hot, and to each of these he assigns a partner, to the hot the dry and to the cold the moist, and out of these are our bodies composed. And he says that diseases may arise simply through the residues of nutriment. Whenever the belly taking in nutriment not commensurate (σύμμετρα) with it, but overmuch (πλείω), cannot digest it, the result is that diseases occur. He also derives diseases from the aforesaid elements, when they are disproportionate (ἀπὸ τῶν στοιχείων ... ὅταν ἀνώμαλα ᾖ).

Here again the organism is conceived as a combination of opposites, hot, cold, moist, dry. Disease is assigned to two sources: First, it may arise from a purely internal condition, i.e., whenever the opposing elements become ἀνώμαλος (ἀν-ὁμαλός) — "uneven", "unequal", "disproportionate". Inequality implies excess of one element, which destroys the balance or equilibrium among

[9] Menecrates' pairs of hot and cold elements may have been further distinguished, as in other medical writers, on the basis of the opposites moist and dry, i.e., blood as hot and moist, bile as hot and dry, phlegm as cold and moist, breath as cold and dry. Cf., for example, [Hippocrates], *On the Nature of Man*, 7, where this pattern is assumed, though black bile, rather than breath, is the cold-dry element. On the traditional opposition of hot and cold, see C. Fredrich, *Hippokratische Untersuchungen*, Philologische Untersuchungen, Vol. XV (Berlin, 1899), pp. 134-136.

[10] On Petron or Petronas of Aegina, contemporary of Hippocrates, see K. Deichgräber, "Petronas", *RE*, XIX, Pt. I, col. 1191; Fredrich, pp. 135-136 and 138.

them.[11] This is the central notion of the Alcmaeon passage. Secondly, disease may arise when food is taken that is not *proportionate* (σύμμετρος) to the stomach's capacity for digestion, but excessive (πλείω), so that a harmful residue (περίττωσις)[12] remains.

As the Alcmaeon passage also indicates, the Greek physician placed great importance upon the exact proportioning of food to the individual organism, believing that failure in this, either by way of excess or defect, threatened the balance of the opposing elements within the organism. Consequently the diet must be carefully balanced so as not to destroy the proportion of elements in the healthy individual or to help restore the proper proportion in the sick. The same principle of balance and proportion that distinguishes the disposition of elements *within* the healthy organism is applied to the assimilation of new elements from outside, demanding that diet, to be "healthy", be balanced or proportioned to suit the internal disposition of the individual. Further illustration from the medical literature will make this clear.

3. *Philistion of Locri*[13]

Following Petron of Aegina, the Menon papyrus offers an account of Philistion of Locri (XX, 25-49). It begins thus:

[11] Plato's use of the word, and his formulation of the principle (*Timaeus* 57E), provide an excellent commentary here: οὕτω δὴ στάσιν μὲν ἐν ὁμαλότητι, κίνησιν δὲ εἰς ἀνωμαλότητα ἀεὶ τιθῶμεν· αἰτία δὲ ἀνισότης αὖ τῆς ἀνωμάλου φύσεως. Thus, as long as "even-ness" or equality among the opposing elements of the body is maintained, the relative stability of health exists. When inequality arises the balance is broken and the elements tend to shift, causing disturbance and pain in the body.

[12] On the concept of περίσσωμα (= περίττωσις) cf. Jones, pp. 11-12. Aristotle (*Hist. An.* 511b10) lists the περισσώματα as dung, phlegm, black and yellow bile. Menon remarks that Petron considered bile the result of disease (XX, 16-23). Plato (*Timaeus* 82E) considers bile, and phlegm, as arising from disease. This is contrary to the medical theory that looked upon blood, phlegm, and bile as normal constituents of the body. Cf. Menecrates above, and [Hippocrates], *On the Nature of Man*, 7.

[13] Philistion is grouped with Empedocles as representative of the Sicilian school of medicine by Galen, *De Methodo Medendi I*, 1 (Kühn X, 5). For the problems connected with his date, life, and influence, see H. Diller, "Philistion", *RE*, XIX, Pt. II, cols. 2405-2408. M. Wellmann collects the fragments in *Die Fragmente der sikelischen Ärzte* (Berlin, 1901), I, 109ff.

Philistion thinks that we are composed of four "forms", that is, of four elements — fire, air, water, earth. Each of these too has its own power (δύναμις); of fire the power is the hot, of air it is the cold, of water the moist, and of earth the dry.

Philistion conceived the body as a combination of the four elements of Empedocles,[14] each of which is identified with one of the traditional opposites as its characteristic "power". For him also health evidently consists in a proper proportion or equilibrium of these opposing forces, since the first general source of disease ascribed to him is an excess or deficiency of one or other element (XX, 33-37): "The elements cause disease when the hot and the moist are in excess, or when the hot becomes less and weak ..." (ἐπειδὰν πλεονάσῃ τὸ θερμὸν καὶ τὸ ὑγρόν, ἢ ἐπειδὰν μεῖον γενῆται καὶ ἀ ... υ ... μαυρὸν τὸ θερμόν).

According to *Anonymus*, Philistion taught that a second general source of disease was *external* influences (παρὰ τὰ ἐκτός), which he divided into: 1) wounds and sores; 2) excess of heat, cold, etc., or change of heat to cold or cold to heat; 3) change of nutriment to the unsuitable and corrupt. It is clear why Philistion would consider a wound, with loss of blood, or a festering sore, accompanied by fever, as a source of disease originating outside the or-

[14] The doctrines of Empedocles and their influence on medical theory are discussed by J. Burnet, *Early Greek Philosophy*, 3rd ed. (London, 1920), pp. 200-250. See especially pp. 200-202: Empedocles founder of the Sicilian school; its influence on Plato and Aristotle; pp. 228-232: the four "roots" and the opposites hot-cold, moist-dry; the unique function of fire (hot) in cosmology and biology; p. 240: organic substances a combination of the "roots" in proper proportion; pp. 244-249: physiology — respiration through pores of the whole body; blood, especially that near the heart, the seat of sensation. On Empedocles' doctrine of the elements among medical men and minor philosophers contemporary with Socrates, see A.-J. Festugière, *Hippocrate, l'Ancienne Médicine* (Paris, 1948), p. 58, n. 69. One significant aspect of Empedocles' teaching is his use of the notion of *proportion* or *measure* to explain the formation of things. He holds that the four elements are *equal*, though they differ in nature (Diels-Kranz 31 B17.27-29); from their mixture or blend (κρῆσις) all things are formed (B 21.9-14); as by a painter who mixes his pigments in due proportion (ἁρμονίη), using more of some, less of others (B 23); proportion (ἁρμονίη) binds earth, water, and fire together in fixed amounts to form bone (B 96), while the four in nearly equal proportion form blood and flesh (B 98). See also Aetius V. 22. 1.; Aristotle, *Met.* 993 a17-23, and other passages cited by Diels-Kranz 31A78; and H. Cherniss, *Aristotle's Criticism of Presocratic Philosophy* (Baltimore, 1935), pp. 233-234.

ganism. And the notions explained above regarding the close relationship between health and diet account in general for his mentioning nutriment as an external source of disease. But the causes described as excess or changes of heat, cold, etc., are not clear until we realize the supreme importance placed by many Greek physicians upon the external *environment* as a factor in health and disease.

The individual human organism was viewed against the background of the particular locality and climate in which it lived. The same elements, e.g., hot, cold, moist, dry, which composed the living body also composed its surroundings, and this affinity was the basis of constant interplay between them. Hence a climatic condition of extreme heat or cold, for example, was regarded as capable of upsetting the balance of hot or cold elements *within* the organism by producing an excess of one or the other and thus bringing on disease. Furthermore, *sudden* changes in temperature, even though not between extremes, were thought to disturb the internal equilibrium since they gave the organism no time for a slow and gradual adjustment, which was regarded as the "natural" mode of reaction. For this reason both extremes and sudden changes of temperature were considered external causes of disease. This will become evident when we examine the medical writings.

Finally, Philistion is said to have named as a third general source of disease "the condition of the body" (παρὰ τὴν τῶν σωμάτων διάθεσιν), which he explains as follows (XX, 42-49):

The condition of the body is a cause of disease in the following way. When ... the whole body breathes well and the breath passes through unhindered, health is the result. For breathing takes place not only by way of mouth and nostrils, but also over all the body. When the body does not breathe well, diseases occur, and in different ways ...

At this point the papyrus is badly mutilated, so that the explanation is lost. However, some indication of the theory may be inferred from the doctrines attributed to Philistion's master Empedocles, and from the portions of Plato's *Timaeus* generally supposed to have been influenced by the Sicilian school.[15] Respiration

[15] Empedocles on respiration: Aetius IV. 22. 1 (Diels-Kranz 31A74); O. Gilbert, *Meteorologische Theorien des Griechischen Altertums* (Leipzig, 1907), pp.

served to cool the innate heat residing in the heart, and the regular cycle of breathing was carried on through the pores as well as through the lungs. If the pores are closed off and the body does not breathe well, the balance of hot and cold is disturbed, air is captured within the body and gathers to excess, while the parts not refreshed by constant renewal of air tend to disintegrate. Through the constant inhaling of cool air and exhaling of warm the equilibrium of elements within the organism must be maintained, and interference with this process results in disease. Some such theory may have been in Philistion's mind when he mentioned as the last general source of disease "the condition of the body".

In the accounts of these three medical authorities — Menecrates, Petron, and Philistion — it is possible to detect the theory summarized in the Alcmaeon passage, centred about the notion that health is an equilibrium of opposing constituents in proper proportion, and that disease is basically a disturbance of that equilibrium through the excess or deficiency of one of the constituents. Opinion may differ on the nature of the elementary constituents, e.g., whether they be the "humors" or earth, air, fire, and water. Authorities may differ in emphasizing the importance of one or other source of disturbance and disease, diet or climate or respiration. But the basic theory remains the same. And the fact that it appears in the accounts of these authorities indicates that it was current in medical circles of the late fifth and early fourth centuries.

To study in detail the basic features of the theory of health and disease outlined in the Alcmaeon passage and in the accounts of the doxographer, we should like to be able to examine the writings of these men themselves. However, none of the works of the individual medical authorities cited by the Menon papyrus survives, with the happy exception of Plato's *Timaeus*.[16] For this reason

339, 343-344. Plato on respiration: *Timaeus* 70C-D, 79A-E, 84C-E, 86A; Taylor, *Commentary on Plato's Timaeus*, pp. 503-504. Cf. Philistion, fragment 6 (Wellmann, *Fragmente* I, 81); Diller, *RE*, XIX, Pt. II, col. 2407; W. Jaeger, *Diokles von Karystos* (Berlin, 1938), pp. 61-63, 211ff., esp. 216-218.

[16] Menon's account of Plato's medical views is a fairly accurate summary of matter from *Timaeus* 69E-86A. Menon does not follow the order of the *Timaeus*, however. Jones, *Anonymus Londinensis*, p. 3, lists the parts of the *Timaeus* summarized, in the order in which Menon treats them.

the *Timaeus* constitutes a most valuable source of medical theory, one which we may suppose to have had great influence on the formation of Aristotle's notion of health and disease, and which we must therefore study closely. However, in order first to obtain a broad view of physiological theory contemporary with Plato and a basis of comparison with the doctrines of the *Timaeus*, we should examine the purely medical writings of the period. And these survive only in the anonymous works of the Hippocratic collection.[17]

C. HIPPOCRATIC WORKS

The Hippocratic Corpus, a large collection of every type of medical writing from formal treatises to note-book jottings, is thought to date, as a whole, from the latter part of the fifth century and the first half of the fourth.[18] Many of the works are purely technical — practical handbooks of medicine, physiology, or anatomy. Others are concerned with basic theory, touching at times upon natural philosophy. From this group we may select a few representative works for detailed consideration, and the purpose will be well served by four which have received more than ordinary attention from scholars, the treatises *Ancient Medicine*; *Nature of Man*; *Airs, Waters, Places*; and *Regimen I*.[19]

[17] On the Hippocratic Corpus and the problem of authorship, see L. Edelstein, *RE*, Suppl. VI, cols. 1290-1343; W. H. S. Jones, "'Hippocrates' and the *Corpus Hippocraticum*", *Proceedings of the British Academy* XXI (1945), 103-125. No considerable portion of Greek medical writing can be assigned with confidence to a particular physician earlier than Diocles of Carystos (Jaeger, *Diokles*, p. vi). In regard to Diocles, it would be tempting to use his fragments as illustrative of medical theory contemporary with Aristotle if the controversy over his dates and influence did not make this inadvisable. Jaeger regards him as post-Aristotelian, 340-270 B.C. (*Sitzungsberichte der Preussischen Akademie der Wissenschaften* [Berlin], Jahrgang 1938, p. 237, no. 3). Edelstein accepts as probable 375-295 B.C. (in his review of Jaeger's work, *AJP* LXI [1940], 483-489).

[18] "Some works of the Hippocratic corpus may be pre-Hippocratic and a few of the third century B.C., but most were written between 430 and 350, i.e., roughly within the lifetime of Hippocrates himself". Jones, *Proceedings*, p. 113. So also Festugière, *L'Ancienne Médecine*, p. viii, n. 3.

[19] The first of these four treatises has been edited separately by A.-J. Festugière, *Hippocrate, l'Ancienne Médecine* (Paris, 1948), and W. H. S. Jones,

Ancient Medicine[20] is a polemic written by a physician of the "old school" against those who build their theory and practice upon philosophical hypotheses. The author favors the old empirical approach to medicine, which he practically identifies with dietetics. The treatise is of special interest as illustrating the insistence of Greek medicine upon correct diet, and as representing the author's conception of pure *medical* theory, free from the speculations of philosophy.

The *Nature of Man*[21] is a short treatise stressing the elementary

---

Philosophy and Medicine in Ancient Greece (Baltimore, 1946), both with intro duction, notes, and translation. Jones's text and translation will be used in discussing *Ancient Medicine*, and citations given according to page and line number of his Greek text (e.g., ch. 3, p. 51. 22-24). For text and translation of the other three treatises, the Loeb Classical Library edition will be used, with references to the Greek cited according to the line numbering in that edition: *Hippocrates*, with an English translation by W. H. S. Jones (London, 1923, Vols. I-II; 1931, Vol. IV). The individual treatises, and the corresponding texts in the older editions of E. Littré, *Œuvres completes d'Hippocrate* (Paris, 1839) and H. Kühlewein, *Hippocratis Opera* (Leipzig, 1894), are located as follows: *Ancient Medicine*: Jones *supra*, or his earlier edition in the Loeb I, 1-64; Littré I, 558-632; Kühlewein I, 1-30. *Airs, Waters, Places*: Jones in Loeb I, 65-137; Littré II, 1-93; Kühlewein I, 31-71. *Nature of Man*: Jones in Loeb IV, 1-41; Littré VI, 28-69. *Regimen I*: Jones in Loeb IV, 224-295; Littré VI, 462-525.

[20] Jones dates this work between 430 and 400 B.C. (*Philosophy and Medicine*, p. 47), Festugière, between 440 and 420 B.C. (p. 60, n. 69, *ad finem*). Jaeger (*Diokles*, p. 170) thinks it is not later than 400 B.C. However, H. Diller, in "Hippokratische Medizin und attische Philosophie", *Hermes* LXXX (1952), 385-409, maintains that *Ancient Medicine* is written to justify the empirical approach to medicine against the *Platonic* concept of medicine as ἐπιστήμη, that it is therefore post-Platonic, and that it had considerable influence upon Aristotle. This would make the work especially significant as illustrating the purely *medical* theory of health and disease contemporary with Aristotle. For a summary of the problem see G. E. R. Lloyd, "Who Is Attacked in *On Ancient Medicine*?" *Phronesis* VIII (1963), 108-126. Lloyd sees Philolaus as the object of attack, and dates the work in the late fifth or early fourth century.

[21] Jones (Loeb IV, xxvi) and Festugière (p. xxi, n. 1) consider as constituting the *Nature of Man* only chapters 1-8 of the MS. text; chapters 9-15 do not belong to the same treatise, while chapters 16ff. form the essay known as *Regimen in Health*. Jones dates chapters 1-8 not earlier than 440 and probably not later than 400 B.C. (*ibid.*). The Menon papyrus (XIX. 2-18) ascribes the doctrines of chapters 2-4 to Polybus, the son-in-law of Hippocrates and one of those grouped with Plato, Menecrates, Petron, and Philistion. Aristotle (*Hist. An.* 512b13ff.) quotes part of chapter 11 as the work of Polybus also. The Menon papyrus, on the other hand, attributes some of the doctrine of chapter 9 to

nature and the importance of the four humors, and indicating principles of diagnosis and treatment on the basis of this concept of the human composite.

Airs, Waters, Places[22] is a somewhat longer essay, or pair of essays, dealing with the influence of external environment on health and disease. The first part (chs. 1-11) lays down general principles on the effect of various seasons, waters, and locations, while the second (chs. 12-24) discusses the relative merits of Europe, Asia, and Egypt and their impact on human physiology and psychology.

Regimen I[23] is the first part of a long treatise emphasizing the supreme importance of balanced regimen — diet and exercise — as the key to health. This introductory section is strongly philosophical in character, apparently the work of an independent eclectic, combining elements of Heraclitus, Empedocles, Anaxagoras, and others into a coherent physiology.

Hippocrates himself (VII. 15-24), and Galen thinks the whole of chapters 1-8 should be assigned to Hippocrates (*CMG* V.9.1). See Jaeger, *Paideia* III, 294, n. 18. K. Deichgräber, *Die Epidemien und das Corpus Hippocraticum* (Berlin, 1933), pp. 163, 169-170, considers the *Nature of Man* and *Airs, Waters, Places* as closely related to the genuine Hippocratic works of the Corpus.

[22] In 1923 Jones wrote (Loeb I, 66): "No ancient critic appears to have doubted the authenticity of this work, and only Haller among the moderns has rejected it." Edelstein, however, in his *ΠΕΡΙ ΑΕΡΩΝ und die Sammlung der hippokratischen Schriften* (Berlin, 1933), p. 59, divides the work into two essays, one prognostic (chs. 1-11), the other ethnological (chs. 12-24); he suggests that they are by different authors, both unknown (p. 181). The authenticity of the work has been reasserted by M. Pohlens, *Hippokrates* (Berlin, 1938), pp. 79-80 (but cf. Edelstein's review in *AJP*, LXI, 1940, 221-229) and W. Nestle, "Hippocratica", *Hermes*, LXXIII (1938), 35. Wellmann thinks it probable that the work was known to Diocles of Carystos (*Fragmente* I, 64). F. Heinimann, *Nomos und Physis* (Basel, 1945), p. 209, fixes the date at shortly before 430 B.C. See his note 109 for other opinions.

[23] The conventional date for this treatise is *ca.* 400 B.C. (see Jones, Loeb IV, xlvi). Fredrich (*Untersuchungen*, pp. 171 ff. and 196 ff.) and Wellmann (*Fragmente* I, 61) hold that it was known to Diocles of Carystos; Jaeger (*Paideia* III, 36-37) places it in the generation of Diocles' father, Archidamus, about the mid-fourth century. Edelstein's dating of Diocles (375-295 B.C.) would put it a quarter century earlier. On the other hand, G. S. Kirk, *Heraclitus: The Cosmic Fragments* (Cambridge, Eng., 1954), pp. 27-28, argues that the work shows Peripatetic influence (*contra* Jaeger, *Paideia* III, 37) and is therefore comparatively late. Kirk, however, is anxious to maintain that the πάντα ῥεῖ interpretation of Heraclitus, apparent in *Regimen I*, stems from Plato and Aristotle.

These works, then, will serve to illustrate in detail the basic features of the theory of health we have seen sketched in the Alcmaeon passage and the accounts of *Anonymus Londinensis*. Where it seems advisable material from some of the other Hippocratic treatises will be introduced to fill out the picture.

1. *Ancient Medicine*

The author of *Ancient Medicine* takes a firm stand against those who introduce into medical theory and practice hypotheses borrowed from philosophy, i.e., those who attempt to explain man and all else through a few causes like heat, cold, moisture, and dryness (ch. 1). There is no need, he maintains, for a medical man to know the ultimate causes of the universe in order to practice. Natural science should be left to philosophers like Empedocles. Or rather, clear knowledge of nature can come only from medicine, and the physician's first duty is to know man in relation to what he eats and drinks, and to his habits generally (ch. 20). The author sees medicine as a development of dietetics, and explains the constitution of the human organism in terms of its relations and reactions to food and drink. He applies the notions of δύναμις, μῖξις, πέψις and κρᾶσις not only to the nature and preparation of suitable nourishment for the body, but adopts them as the basis of his physiology in explaining the constituents and function of the body itself.

Dietetics and medicine are, according to *Ancient Medicine*, old and long established arts, independent of later speculation (ch. 2). Their principles were discovered first when man learned to *proportion* his food to his nature, in the following way (ch. 3): In his primitive state man ate the fruit, wood, and grass which the other animals found nourishing and painless. But he suffered terribly from this "strong and brutish regimen" (ὑπὸ ἰσχυρῆς τε καὶ θηριώδεος διαίτης), consisting as it did of foods that were "crude, unblended, and possessed of strong powers" (ὠμά τε καὶ ἄκρητα καὶ μεγάλας δυνάμιας ἔχοντα). This diet caused many, then as it would today, to fall into violent pains and diseases, quickly followed by death.

It was for this reason that primitive men too ... sought for food that suited their constitution (τροφὴν ἁρμόζουσαν τῇ φύσει) and found that which we now use. So from the grain by steeping, winnowing, grinding, sifting, kneading, baking they finally produced a loaf, and from barley a cake. Experimenting with these materials they mixed and boiled and baked, and compounding the strong and unblended with weaker substances they molded all to suit the constitution and the power of man, thinking that if foods were taken which were strong the constitution would not be able to digest (κρατέειν) them, and they would of themselves cause pains, diseases, and death; while from food that can be digested would come nutrition, growth, and health (ch. 3, p. 51. 26-36).

Noteworthy in this account of the discovery of dietetics are the following principles: 1) The "powers" (δυνάμιες) of *raw* food are *too strong* for men: hence unsuitable, indigestible, the cause of pain, disease, and death. As the author remarks later (ch. 6, p. 53. 4-6): "All the causes of pain can be reduced to one: the strongest nutriment (τὰ ἰσχυρότατα) hurts a man most, and most obviously, whether he be well or ill".[24] 2) Man needs nourishment that is in *harmony* with his constitution (τροφὴ ἁρμόζουσα τῇ φύσει) and *proportioned* to his strength (πρὸς τὴν τοῦ ἀνθρώπου ... δύναμιν). This notion of harmony and proportion of food in relation to the organism is the essence of dietetics. 3) The strong powers of raw food can be modified and proportioned to the human constitution by mixing (μῖξις) with a weaker substance (e.g., water) and *cooking* (πέψις) until they form a *blend* (κρῆσις).[25] Food properly proportioned through this process can be assimilated by man and brings nourishment, growth, and health.

[24] Jones (*Philosophy and Medicine*, p. 70, note to ch. 6) calls attention to the fact that the author of *Ancient Medicine* draws the distinction between an extreme of *quality* (τὸ ἰσχυρότατον) and mere excess of *quantity*, a distinction important for Plato's theory of pleasure in the *Philebus* and for Aristotle's theory of "the mean". The same distinction appears even more clearly in the passage from chapter 5 quoted next, *infra*.

[25] It is important to note the basic process of preparing food described in the passage, and the terminology used: The extremes of strong food, which are *unblended* (ἄκρητα), modified by *mixing* (μῖξις) and by *cooking* (πέψις) to form a *blend* (κρῆσις). The same process is thought of as taking place in the digestion (πέψις) and assimilation of food *within* the body. And both process and terminology are employed by Plato and Aristotle in explaining the formation of the body, as we shall see.

The discovery of these principles in regard to the generality of men in *normal* condition was, according to *Ancient Medicine*, the beginning of the dietetic art; their application to the *sick* is described as the beginning of medicine (ch. 5, p. 52. 19-31):

Those who sought and found the medical art ... at first ... merely lessened the quantity (πλῆθος) of food, leaving it unchanged, and instead of more (πλείονα) they made it very small in quantity (ὀλίγιστα). Finding, however, that this was sufficient for some patients, who were manifestly helped by it, but not for all, since others were in such condition that they could not assimilate food in the smallest quantity but seemed to require something weaker (i.e., in quality), they invented gruels by mixing (μίξαντες) a small quantity of strong food with much water and taking away the strength by blending (κρῆσις) and boiling (ἕψησις). When some patients were unable to assimilate even the gruel, they took this away also and hit upon potions, watching carefully that these by their blend and quantity should be proportionate, administering nothing more, nor more unblended, nor more deficient than it should be (ταῦτα τῇσι τε κρήσεσι καὶ τῷ πλήθει διαφυλάσσοντες ὡς μετρίως ἔχοι, μήτε πλείω τῶν δεόντων μήτε ἀκρητέστερα προσφερόμενοι μηδὲ ἐνδεέστερα).[26]

Here is the basic notion of the "ancient medicine" as the author conceived it — the art of establishing a diet *proportionate* to the strength, or weakness, of the patient. To accomplish this the physician must modify the food substances not only *quantitatively*, by regulating the amount, but also *qualitatively*, by mixing the right quantity of "strong" food (e.g., raw barley grains) with the right quantity of a "weaker" element (e.g., water) and through the application of heat (ἕψησις) producing a proper *blend* (κρῆσις).[27] The result of κρῆσις is a compound which is 1) *homogeneous*, in that the "strong" and "weak" substances do not exist separately but are evenly distributed and blended throughout the mixture, and 2) *qualitatively modified*, being weaker than the "strong" sub-

[26] The writer is responsible for the translation here.

[27] The term κρᾶσις (Ionic κρῆσις) in its various contexts and implications has been studied by W. DenDulk, ΚΡΑΣΙΣ, *Bijdrage tot de Grieksche Lexicographie* (Leiden, 1934). See especially his discussion of the term as used in physiology, pp. 67-95. See also Festugière's excellent note (*L'Ancienne Médecine*, p. 37, n. 25) reviewing the use of κρῆσις in *Ancient Medicine* and other medical works.

stance but stronger than the "weak", so that it *constitutes a mean between them.* Conceived thus, κρῆσις implies a sort of *proportion*, forming an intermediate substance which links the two extremes, being identified with neither of them as such, yet having something common with both.

This compound, whether it take the form of gruel or be thinned out to a potion, must, according to *Ancient Medicine*, be *measured* (ὡς μετρίως ἔχοι) so as to be neither excessive in quantity or in blending (quality), nor weaker than it should be (deficient). The object to which the nutriment must be measured or proportioned is, obviously the condition (διάθεσις) of the *individual* human organism.[28] The author makes this clear (ch. 6 and 8) by setting up the proportion of various types of diet to various conditions of the body, thus: solid food / normal or healthy body = gruel / moderately ill = potion / very ill. Mistakes in diet are discussed on the basis of this proportion in terms of the relative gravity of the pain they cause. Thus it is said that gruel does not help the very sick but increases their pain and fever, while even a very little solid food hurts them ten times more, because of its strength in relation to their condition (ch. 6, p. 53. 1-2; διὰ τὴν ἰσχὺν ... πρὸς τὴν διάθεσιν). Hence a proper proportion or balance must be struck between the nutriment and the nature or disposition of the *individual.* This emphasis on the individual is, in fact, the distinguishing note of the discoverer of *medicine*, in contrast to dietetics, in that he "sought to remove whatever cannot be digested in the temporary condition of *each individual sufferer*" (ch. 7, p. 53. 14-15). Interest in the individual does not prevent the author from recognizing the validity of general principles, however. Rather he looks upon the condition of the individual organism as determining the general principles to be applied here and now.

Ancient Medicine points out that the difficulty of striking a balance between the nutriment and the individual organism is intensified by the fact that *deficiency* in diet can be as harmful and

[28] On this point see W. Muri, "Περὶ 'Αρχαίης 'Ιητρικῆς IX", *Hermes*, LXXI (1936), 467-469.

dangerous as excess (ch. 9).[29] The greatest possible accuracy is necessary if nutriment is to be exactly proportioned to the disposition of the individual without excess or defect. Hence there is need of a *standard* — μέτρον — a norm which will mark off the right quantity and blend of nutriment from the excessive or deficient.[30] The author asserts, however, that in this there is "no measure, either number or weight, reference to which would give accurate knowledge, other than the *response* (αἴσθησις — sensitivity, reaction) of the body. Therefore it is difficult to acquire knowledge (of the case) so accurate that we miss the mark only by a little one way or the other (by excess or defect)."[31]

The author is acutely conscious of the difficulty of fitting general principles to individual cases. He asserts the validity of the gen-

[29] The harmful effects of both excess and deficiency in diet are illustrated concretely in chapters 10 and 11.

[30] Ch. 9, p. 54. 6-8: πολλὰ δὲ καὶ ἄλλα κακὰ ἑτεροῖα μὲν τῶν ἀπὸ πληρώσιος, οὐχ ἧσσον δὲ δεινὰ καὶ ἀπὸ κενώσιος· διότι πολλὸν ποικιλώτερά τε καὶ διὰ πλέιονος ἀκριβείης ἐστί· δεῖ γὰρ μέτρου τινὸς στοχάσασθαι. The terminology of the last clause is very like that which Aristotle uses to express much the same idea. Cf. *N.E.* 1109 a 20-24.

[31] Ch. 9, p. 54. 8-11: K. Deichgräber's attempt to emend αἴσθησιν to διάθεσιν ("Zu Hippokrates Περὶ Ἀρχαίης Ἰητρικῆς", *Hermes*, LXVIII [1933], 356-358) has been rejected by Muri (*Hermes*, LXXI, 467-469), Festugière (*L'Ancienne Médecine*, p. 34), Jones (*Philosophy and Medicine*, p. 73), and Diller (*Hermes* LXXX, 400). Jones suggests (*ibid.*) that "αἴσθησις could mean either (a) the sensitiveness or reaction of the body to δυνάμεις, or (b) the "feel" of the body to the touch of the physician". It is rather the first; cf. chapter 15, where wood and leather are said to be less *sensitive* or *responsive* than man (ἀνθρώπου ἀναισθητότερα). Thus Muri (*ibid.*), who translates αἴσθησις "der Empfindlichkeit des Körpers", Festugière (*ibid.*, p. 7), who translates "la sensibilité du corps", and Diller (*ibid.*), interpreting it as the *reaction* of the body. That the genitive τοῦ σώματος can only be *possessive* (or "subjective"), see Deichgräber and Muri, *ibid.*

It might be noted that just as the term αἴσθησις is ambiguous in its common derived sense of "perception", standing for both the *capacity* to perceive and the *operation* of that capacity, so here in its more fundamental meaning the word seems to imply both the capacity to respond or react ("responsiveness", or "reactivity") and the operation of that capacity ("response", or "reaction").

Festugière (p. 59, n. 69) shows that the author of *Ancient Medicine*, in setting up the "responsiveness" or the "response" of the individual body as a μέτρον, is *not* adopting the relativity of Protagoras, but regards medicine as a τέχνη with generally valid laws based upon observation and experience.

eral principles, but looks for some standard in selecting and applying them to a particular patient. He finds the only concrete norm to be the response or reaction of the individual organism, a norm which may reduce the margin of error but does not guarantee complete accuracy. He does not explain *how* the reaction of the body constitutes a norm for the compounding of proportionate nutriment, but we have seen him attribute the discovery of proper diet to observation of the painful effect of strong food upon the body and of the reaction of growth, strength, and a sense of well-being that followed the reduction of this strong diet through cooking. We may suppose, then, that he expects the physician to determine the diet proportionate to each individual, first, by knowing the habitual reactions of the body in varying conditions of health and disease to the different gradations of diet, and then by observing closely the *reaction* of the patient to the particular diet prescribed. A reaction of pain, distress, and weakening will mean that the diet fails to strike a balance or proportion with the organism and must be excessive or deficient, whereas a general response of strengthening and freedom from pain will indicate that the diet is at least very close to being exactly proportionate. The reactions of *pleasure* and *pain*, therefore, are of considerable importance as an index of properly proportionate diet. Conversely, the type of diet which causes pleasure or pain is an index of the condition of the organism, i.e., if ordinary solid food brings a sense of well-being and not pain, the body is normal and healthy, while if it causes pain, the body is diseased and the gravity of the disease may be gauged by the severity of the pain. This then is the measure (μέτρον) at which the physician must aim (στοχάζεσθαι) in determining the proper quantity and blend of nutriment.

Another important consideration in the difficult art of harmonizing nutriment and organism is *timing* — καιρός (chs. 10-12). Breaking the regular routine of meals, taking food when the stomach is not prepared for it or abstaining from food when it is, causes pain and disturbance even in those who are healthy. The weaker a man is by constitution the more quickly and seriously is he disturbed by mistakes in this matter. Hence the sick are much more liable to suffer when the exact καιρός is missed (ch. 12, p. 55. 31-

33). The physician's full responsibility is to administer the proper measure (μέτρον) of nutriment at the proper time (καιρός). In this way μέτρον and καιρός are linked in the medical writings, and one suggests the other.

Ancient Medicine, as we have mentioned, rejects the "recent" hypothesis of the hot, cold, moist, and dry as the basis of physiology and medicine. In place of this it holds a theory of physiology based on dietetics, which it traces to the founders of medicine, whose discoveries in dietetics led them to conclusions on the nature of man (ch. 14). According to *Ancient Medicine*, the earliest physicians found, for example, that the constituents of bread and barley cake, when prepared in various ways, have corresponding effects upon the body; it made a difference to the organism whether or not flour was sifted, wheat winnowed, whether it was kneaded with much water or little, well baked or underbaked, etc. For "the powers of each are great, and no one is like any other" (ch. 14, p. 56. 33). They discovered that the *strength* (τὸ ἰσχυρόν) of each quality caused trouble when it was too great for the constitution of man and could not be assimilated, and so sought to modify the "strongest", that is, the extreme (ἀκμή), of each quality — the sweetest of sweetness, the bitterest of bitterness, the most acid of acidity, etc. For they assumed that these δυνάμιες were also *present in man as constituents*, and that the extremes were what caused him injury (ch. 14, p. 57. 14-31):

For the constituents of man are salt, bitter, sweet, acid, astringent, insipid, and countless other things, possessing powers of all kinds in number and in strength. These when mixed and blended neither are manifest nor cause a man pain; but when one of them has been isolated and set apart by itself, then it both is manifest and causes a man pain. Again, such foods as are unsuitable for us and injurious to a man when taken, are in each case either too bitter or salt or acid or something else unblended and strong, and therefore we are disordered by them, as we are also by the secretions that are isolated in our bodies. But all that a man eats or drinks are foods plainly free entirely from such an unblended and overpowering flavor, e.g., bread, barley-cake ... which men are accustomed to use habitually in great quantity ... When these are taken in great quantity by a man no disorder arises in the least degree, nor isolation of the powers of the body, but a very high degree of strength, growth, and nourishment, simply because the food is well blended, with nothing unblended or strong, forming a single, simple whole.[32]

The passage reveals the main features of the author's physiology and its connection with the dietetic theory previously discussed:

(1) The body is conceived as a complex of "the salt, bitter, sweet, acid, astringent, insipid, and countless others" each having its own peculiar "power". Later these constituents are referred to as χυμοί — "juices" or "humors" (cf. chs. 14, 18, 19), and a "power" is defined as the "extreme" and the "strength" of each of them (ch. 22, p. 62. 24-25: δύναμιν μὲν εἶναι τῶν χυμῶν τὰς ἀκρότητάς τε καὶ ἰσχύν).

(2) The author seems to think of these innumerable χυμοί as *opposites*, at least in the sense that the "power" of one in the body can be counteracted by that of another. Thus, in discussing the relative unimportance of heat and cold in comparison with other δυνάμιες like insipid and astringent, he says (ch. 15, p. 58. 9-12): "I am convinced that each member of these two pairs produces exactly the opposite result of that produced by the other, not only in man, but in a vessel of leather or wood and in many other things less sensitive (ἀναισθητότερα) than man". And again (ch. 16, p. 58. 17-20): "So long as the hot and cold are combined together in the body they cause no pain; by blending cold moderates the heat, and heat moderates the cold" (ὃν μὲν ἂν δήπου χρόνον μεμειγμένα αὐτὰ ἑωυτοῖσιν ἅμα τὸ θερμόν τε καὶ ψυχρὸν ἐνέῃ, οὐ λυπέει· κρῆσις γὰρ καὶ μετριότης τῷ μὲν θερμῷ γίνεται ἀπὸ τοῦ ψυχροῦ, τῷ δὲ ψυχρῷ ἀπὸ τοῦ θερμοῦ).

(3) Κρῆσις and μετριότης of all the opposing components in the body is the essence of health. "These when mixed and blended neither are manifest nor cause a man pain" (ch. 14, p. 57. 16-17). We have seen that in the preparation of food "extreme" or "strong" powers were moderated by being mixed with their opposites and cooked until κρῆσις took place. The same process, including "cooking" (πέψις), is thought of as taking place among the powers

[32] This is perhaps the most important single passage in *Ancient Medicine*, and is of special significance because its physiological theory, and terminology, represent for the author the pure *medical* tradition, as distinct from natural science based upon philosophy, which he rejects.

that constitute the body, and man is said to be "in the best possible condition when there is complete coction and peace,[33] with no particular power manifesting itself plainly" (ch. 19, p. 61. 5-7). This is the condition of the powers *within* the body which corresponds to, and is promoted by, "healthy" diet, described above as "well blended, with nothing unblended or strong, forming a single, simple whole". Κρῆσις, as we have seen, implies a certain equality, proportion, equilibrium, or balance among opposing forces, a blend by which they are united and brought into harmony, without excess or defect.[34] This relationship of the opposing constituents of the organism constitutes health. In the phrase of the Alcmaeon passage: τὴν δὲ ὑγιείαν τὴν σύμμετρον τῶν ποιῶν κρᾶσιν.

(4) *Disease*, on the other hand, occurs when one of the constituents is separated from the blend, isolated so that it is no longer tempered by its opposite. It becomes, therefore, excessively strong in the body, makes its presence felt, causes pain. This relationship of the constituents of the body corresponds to that of the constituents of food that is unsuitable and injurious to man, being "either too bitter or salt or acid or something else unblended and strong" (ch. 14, p. 57. 20-21). And as unsuitable food can only be made "healthy" by cooking and blending, so the excessive element in the body must be reduced to proportion by purging, "coction", and blending if the body is to be restored to health.

This becomes clear from the specific cases introduced by the author in subsequent chapters to prove that it is the excess of some

[33] Jones (*Philosophy and Medicine*, p. 83) translates ἡσυχίη here as "rest", which obscures the essential *dynamic* notion of health as a "balance of powers". In a state of κρῆσις no one power is manifest, but this is not because the various powers cease to operate, as "rest" implies, but rather because they *cooperate* to maintain an equilibrium, the influence of one being tempered or balanced by that of its opposite so that they are at *peace*. If the elements ceased to operate and exert their influence, each according to its nature, they would cease to be "powers".

[34] Festugière, in his discussion of κρῆσις, remarks: "Comme on le voit, la κρῆσις est immediatement liée a la μετριότης, c'est-à-dire a l'idée de proportion, d'équilibre ..." (*L'Ancienne Médecine*, p. 38). He also cites Jaeger (*Paideia* III, 27), who says: "Closely connected with the conception of proportion and symmetry is that of mixture, which really means a sort of equal balance between the various forces ruling the organism".

humor, and not of heat or cold alone, which causes disease. Thus the common cold is accompanied by an unusually acrid discharge which causes swelling and inflammation of the nasal passages, and which ceases when the discharge "flows thicker and less acrid, being mellowed (πέπον) and more mixed (μεμιγμένον)" (ch. 18, p. 59. 31-32). Likewise diseases of the eye are caused by strong and acrid humors and last "until the discharges have been concocted and thickened ... this coction being the result of mixing, blending, and boiling together with one another" (ch. 19, p. 60. 10-12).[35]

Furthermore, all other complaints to which man is liable proceed in every case from powers. For instance, when there has been an efflux of the bitter humor which we call yellow bile, what nausea, burning and weakness grip the patients! But on getting rid of it sometimes by purging, with or without medicine, should the particular remedy be seasonable (ἐν καιρῷ), they plainly get rid both of the pains and of the heat. So long, however, as the noxious fluid is unsettled, without coction and blending, by no means can the patients be free of the pains and fever (ch. 19, p. 60. 24-31).

Thus any single constituent, existing apart in the body, unblended with, and untempered by, its opposite, is *stronger* in comparison with the others, excessive, upsetting the proportion, the balance and equilibrium among them. This is the nature of disease. Recovery can come only if this excess is somehow overcome by purging, mixing, and "coction" until κρᾶσις of all the constituents is restored. οὐ παύεταί τι τούτου πρότερον πρὶν ἢ ἀποκαθαρθῇ τε καὶ καταστορεσθῇ καὶ μιχθῇ τοῖσιν ἄλλοισιν (ch. 19, p. 60. 33-34).

The treatise on *Ancient Medicine*, then, representing what the author considers to be pure *medical* theory, emphasizes the close connection between health and diet. Health is the proportionate blend (κρῆσις) of opposing powers within the body; disease is the isolation and excessive strength of one of them. This balance or

[35] Note that the terms used to describe this process taking place *within* the body are the same as those describing the preparation of food: *Coction* or cooking (πεφθῆναι) of the elements disturbing the body results from *mixing* (μιχθῆναι), *blending* (κρηθῆναι), and *boiling together* (συνεψηθῆναι). It is a basic assumption in Greek physiology that the "vital heat" within the body must constantly "cook" the elemental constituents if the body is to retain its "blend" or integrity.

loss of balance among the constituents of the body depends upon diet for the most part,[36] and medicine is the art of maintaining or restoring the balance of health by *proportioning* the diet to the *individual* organism.[37] The physician's norm, in diagnosing the condition of the individual patient and in proportioning diet to that condition, is the response of the body, its growing strength and sense of well-being, or weakness and feelings of pain. Keeping his eye upon this broad and rather inexact criterion he must seek to administer, without excess or deficiency, the right amount and blend of nutriment at exactly the right time to maintain or produce that proportionate κρῆσις of the bodily elements which constitutes health.

2. *The Nature of Man*

The author of this short treatise sets himself apart, first of all, from those who hold that man is constituted of a *single* element like air, fire, water, or earth (ch. 1). Secondly, he rejects the theory of certain medical thinkers who try to reduce man to some other single constituent, either blood or bile or phlegm, which they claim can change its nature and power (τὴν ἰδέην καὶ τὴν δύναμιν) under the influence of heat and cold so as to become either bitter or sweet, white or black, etc. Rather, the author holds 1) that the constituents of man are *manifold*, and 2) that they remain the same κατὰ νόμον καὶ φύσιν throughout every age and season, though the *quantity* of each constituent may increase or decrease within the body (ch. 2).

In arguing for a plurality of constituents in man (ch. 3) the author assumes that both man and the rest of the universe, both microcosm and macrocosm, follow the same basic laws and are ultimately explicable in terms of hot, cold, moist, and dry. The nature of all things, including man, is such that *generation* will not take

[36] The author also recognizes other factors in disease, e.g., the harmful effects of fatigue and baths taken ἀκαίρως (ch. 21) and the various shapes of the bodily organs (ch. 22-23).

[37] In addition to what has been said on the necessity of individual treatment, cf. ch. 20, which points out the contrary effect of the same food (cheese) upon different individuals, and ch. 23, which recognizes the importance of individual bodily structure.

place unless the hot, cold, moist, and dry are disposed toward each other *proportionately* and *equally* (μετρίως καὶ ἴσως) and unless they are *well-blended* with each other (ἢν μὴ τύχῃ καλῶς ἔχοντα τῆς κρήσιος τῆς πρὸς ἄλληλα). Man, like all else, is begotten through the combination of several elements, and these retain in his body the powers which each contributed to his generation. Likewise, man and all other things dissolve in death when the components return each to its own nature, hot to hot, cold to cold, moist to moist, dry to dry. "All things are born in a like way and all things die in a like way. For the nature of them is composed of all those I have mentioned, and each thing ... ends in that from which it was composed" (ch. 3. 24-28).

Having set man against the background of the universe, the author now outlines his physiological theory. "The body of man has in it blood, phlegm, yellow bile, and black bile; these make up the nature of his body, and through these he feels pain or enjoys health" (ch. 4. 1-4). In subsequent chapters (5, 6, 7) these four substances are shown to be distinct, elemental, and present in man at all times. Not only their conventional names but their particular natures (ἰδέη) are always distinct, for they differ in appearance and to the sense of touch they are not equally warm, cold, moist, or dry; hence they are not one, but each is elemental in the body (ch. 5. 1-17). However, (ch. 7) the nature of each bodily component is conceived in terms of the four basic opposites which compose the whole of nature; thus phlegm is cold and moist, blood, warm and moist, yellow bile, warm and dry, black bile, cold and dry. It is clear that, thus conceived, the four humors are regarded as *opposing* powers within the body they compose.

Like the other medical writings we have seen, the *Nature of Man* describes health as a relationship of *proportion* or *equilibrium* among the components of man's body (ch. 4. 4-9).

Now man enjoys the most perfect health when these elements are duly proportioned in their blend with one another, i.e., in power and quantity, and when perfectly mixed. Pain is felt when one of these elements is in defect or excess, or is isolated in the body without being blended with all the others.[38]

[38] The translation represents a slight modification of Jones, Loeb IV, 11.

Thus health exists when the opposing elements are in *proportion* (μετρίως ἔχει) with regard to *power* (δύναμις) and *quantity* (πλῆθος), and *blended* (κρῆσις), i.e., mixed in proper distribution and union, so that none is isolated from the rest.

The connection between pain and the isolation, excess, or deficiency of any one element is then explained (ch. 4. 10-20). When a component is isolated and stands alone, the place in the body from which it came must be diseased (as lacking a necessary element) and the place where it stands in excess, being overfilled, gives pain and distress. If a component flows out of the body in greater quantity than is needed to relieve excess, the emptiness causes pain. If, on the other hand, emptying and shifting and separation take place entirely within the body, there occurs a double pain, one from the place left empty, the other from the place where the component is gathered in excess. Pain, then, is the result of excess, deficiency, and isolation of one of the components, and so an indication of malfunction within the organism.

The *Nature of Man* introduces an aspect of Greek medicine disregarded by *Ancient Medicine*, for it conceives man as part of the universe, explaining both through the opposites hot, cold, moist, dry, and tracing the influence of the changing *environment* upon the human organism (ch. 7). Each of the seasons in the cycle of the year is seen as a period in which one combination of opposites dominates the others. Winter is cold and moist, spring, warm and moist, summer, warm and dry, autumn, cold and dry. The change from one to the other through the advancing year is thought to bring a corresponding change within the body of man, so that in each season the corresponding bodily element tends to increase excessively, to dominate the organism, and to bring on specific diseases. Thus winter is thought to increase the cold-moist element, phlegm; hence in that season especially the head is congested, swellings turn white, and diseases are generally phlegmatic. As spring comes on with its warm rains the warm-moist element, blood, tends to increase, and growing excessive brings on dysenteries, hemorrhages, etc. Summer induces an increase of yellow bile, the warm-dry element, while its opposite, phlegm, is weakest at this time. Characteristic diseases are fevers and bilious attacks. This is true for autumn also, when black bile, the cold-dry element predominates.

All these elements then are always comprised in the body of a man, but as the year goes round they become now greater and now less, each in turn and according to its nature. For just as every year participates in every element, the hot, the cold, the dry, and the moist — none in fact of these elements would last for a moment without all the things that exist in this universe, but if one were to fail all would disappear, for by reason of the same necessity all things are constructed and nourished by one another — even so, if any of these congenital elements were to fail, the man could not live. In the year sometimes the winter is most powerful, sometimes the spring, sometimes the summer, and sometimes the autumn. So too in man sometimes phlegm is powerful, sometimes blood, sometimes bile, first yellow, and then what is called black bile (ch. 7. 48-66).

Thus the microcosm is subject to a series of changes corresponding to, and under the influence of, the seasonal changes of the macrocosm. Each season brings its own peculiar threat to the equilibrium of elements within the body, favoring the domination of one over the others, so that the physician must reckon with this in the prevention and cure of disease. He can expect that diseases brought on by one season will be helped by the coming of another (ch. 8). The phlegmatic complaints of winter and spring, arising from excess of cold and moisture, should disappear when the balance is restored by the warmth and dryness of summer. "The physician ... must treat diseases with the conviction that each of them is powerful in the body according to the season which is most conformable to it" (ch. 8. 11-14).

What is the general method of treatment? After diagnosing the cause of disease as an excess of one element or other, and the consequent deficiency of its opposite, the physician must use every means at his disposal to help nature restore the equilibrium of health. He must know that "diseases due to repletion (πλησμονή) are cured by evacuation (κένωσις), and those due to evacuation are cured by repletion; those due to exercise are cured by rest, and those due to idleness are cured by exercise" (ch. 9. 2-6). In prescribing treatment for an individual the physician must take into account a whole complex of factors which affect the organism. "To know the whole matter, the physician must set himself against the established character of diseases, of physical constitutions, of season, and of ages" (ch. 9. 6-8). If the disease be due, not to a general epidemic, but to faulty regimen, then "this the physician should learn and change, and carry out treatment only after examination of the patient's constitution, age, physique, the season of the year and the fashion of the disease, sometimes taking away and sometimes adding ... and so making changes in drugging or in regimen to suit the several conditions of age, season, physique, and disease" (ch. 9. 36-44). The same general principle is recommended for the preservation of health by the treatise *Regimen in Health* (ch. 2. 14-18): "In fixing regimen pay attention to age, season, habit, locality, and physique, and counteract the prevailing heat or cold. For in this way will the best health be enjoyed."

The difficulty of the physician's task becomes apparent when we consider the various factors which he was to reckon with in attempting to prescribe exactly the diet, exercise, and medication which, administered in proper measure (μέτρον) and timing (καιρός), would preserve or restore the equilibrium of elements within the body. First, he has to take into account the influence of the season, as we have seen. Secondly, the age of the individual had to be considered, since each period of man's life was thought to favor the predominance of one element in the body. Thus infancy and childhood, being warm and moist like spring, were liable to an excess of blood; youth, warm and dry like summer, favored the increase of yellow bile; maturity, cold and dry like autumn, black bile; old age, cold and moist like winter, phlegm.[39] Thirdly, the sex and physique of the individual had to be considered. Males were thought to be normally dry and warm; females, moist and cold.[40] Fleshy, soft physiques were moist by nature; the lean and sinewy, dry.[41] Finally, of course, the physician must

[39] Cf. *Nature of Man*, ch. 12, 34-44; ch. 15, 30-37; *Regimen in Health*, ch. 2, 8-14; *Regimen I*, 33.
[40] Cf. *Regimen in Health*, ch. 6, 7-11; *Regimen I*, 25, 34.
[41] Cf. *Regimen in Health*, ch. 2, 1-8 and ch. 3, 4.

recognize the specific character of the disease and know its source. To take an uncomplicated example, he must recognize, for instance, that his patient is suffering from quartan fever, and know that this is caused by an excess of black bile (cold-dry).[42] He must then take steps to draw off the excess of black bile or increase the opposite element, blood (warm-moist), and so help to restore the equilibrium of health.

Some of the techniques used by the physician in "setting himself against the established character of diseases, of physical constitutions, of seasons, and of ages" may be gathered from the short companion piece to the *Nature of Man* known as *Regimen in Health*. First, diet should be regulated by selecting the quantity and quality of food and drink which will produce effects in the organism directly opposite to those produced by the changing seasons (ch. 1). In winter (cold-moist), eat as much and drink as little as possible. Food should be bread and roasted meats (dry); drink, undiluted wine (warm) — "for so will the body be most dry and hot". In spring (warm-moist) gradually increase and dilute wine; cut quantity of food, which should be softer, with boiled meat. In summer (warm-dry) take soft foods, boiled meats, drink wine dilute and copious. "For one must use these, when it is summer, that the body may become cold and soft. For the season is hot and dry, and makes bodies burning and parched" (ch. 1. 22-26). According to the same principle, in autumn (cold-dry) food must be gradually increased, made drier, with less wine, and that less diluted, "so that winter may be healthy". Diet must also be regulated according to physical types (ch. 2), the fleshy and soft (moist) keeping to a rather dry regimen all year, while the lean and sinewy (dry) should have a more liquid diet. Likewise, young people, being dry and firm, need a softer, more moist diet, while the old, being moist, soft, and cold, need a drier. And as diet is regulated according to these various factors, so different exercise, bathing habits, clothing, and bedding must be prescribed for each individual (ch. 3, 4) to counteract the element which tends to grow excessive and to maintain a balance within the organism against the influence of environment and the propensities of age and physique.

[42] Cf. *Nature of Man*, ch. 15, 21-37.

To summarize, then: In these two short essays, the *Nature of Man* and *Regimen in Health,* we are introduced to the medical theory which explains man and all else in terms of the hot, cold, moist, and dry, equating the components of the human body, the four humors, each with a particular combination of these fundamental opposites. Here, as in the other medical writings we have seen, health is a proper proportion and equilibrium of these elements, while disease is the excess and domination of one of them.

The treatises are of special interest as illustrating the integrated view of man and the universe held by Greek physicians, and as explaining the strong emphasis which they placed upon environmental factors, particularly the variations in climate, in accounting for health and disease. The assumption is that man normally reproduces the cycle of changing seasons within his own body, both within the period of each recurrent year and over his entire lifetime from infancy to old age. The microcosm follows the pattern of the macrocosm. Furthermore, the normal balance of elements within the body varies with individuals, not only at different seasons and times of life but also according to sex and physique; each has its tendency toward excess in one direction or the other, and a consequent propensity for certain diseases. The physician must consider all these factors in using his resources of medication, diet, exercise, etc., to restore the normal balance when it is lost in disease; and even in health he must bolster the organism against the impact of hostile environmental influences, bringing it as close as possible to the ideal condition of perfect equilibrium.

3. *Airs, Waters, Places*

The importance of environmental factors and the extent of their influence upon the human organism are illustrated most clearly by the essay *Airs, Waters, Places.* Here the physician is told (ch. 1) that he cannot pursue his profession properly unless he considers the following: 1) The seasons (ὧραι), which differ in the effects they produce. 2) The winds (πνεύματα), whether they be hot or cold. 3) The waters (ὕδατα), which differ in taste and weight, each having its own power. 4) The position or situation (θέσις — "aspect") of the town with regard to the prevailing winds and the rise or setting of the sun. 5) The soil and topography (γῆ), since rocky land is dry; wooded, moist; high, cold; low, warm. 6) The regimen or way of life that pleases the natives, i.e., their habits of eating, drinking, and exercise.

The author of this treatise maintains that it is of the greatest advantage to the physician to know the medical significance of these factors (ch. 2). Equipped with such knowledge, upon entering a new town the physician will be able to determine the nature of the diseases common to the locality even before treating individual cases. As time passes he will be able to foretell which diseases are likely to attack the town as a whole in winter or in summer, and which threaten each individual from a change in regimen. Observing the changes of seasons and the rise and setting of the stars in the particular locality, he can tell beforehand what kind of year it will be, i.e., whether or not the seasons will change at the normal time and follow their normal character. Applying this knowledge, he will be able best to foretell the development of disease in each individual case and to "hit upon" health in the greatest number possible (προγινώσκων τοὺς καιροὺς μάλιστ' ἂν εἰδείη περὶ ἑκάστου καὶ τὰ πλεῖστα τυγχάνοι τῆς ὑγιείης). To one who objects that this is more meteorology than medicine the author replies that astronomy is of the greatest value to the physician, since men's diseases and digestive systems change with the seasons.

Airs, Waters, Places, like the *Nature of Man*, assumes that man and all other things are ultimately explicable in terms of hot, cold, moist, and dry; that health is an equilibrium of these opposites; and that because of the basic affinity between man and his environment, the equilibrium within man is subject to influence by seasonal changes, the nature of the locality, etc. On the basis of these assumptions, environmental factors are thought to exert a strong influence upon the inception and the course of disease. For example, a patient afflicted during the winter with an overabundance of phlegm (cold-moist) will be differently affected if spring be early or late, cold or warm, wet or dry. Consequently, the physician should be able to forecast the time and character of the coming season so as to regulate accordingly the medication, diet, and exercise of his patient.[43] Hence some knowledge of meteorology is important for him, as well as any other science that explains the nature of the environment in its relationship to man. Having made this point, *Airs, Waters, Places* attempts a systematic summary of these environmental factors and their effect upon the human organism, both physical and psychological.

The *situation* or "aspect" of a city, its location with regard to the prevailing winds and the rise or setting of the sun, is thought to determine the physical, and even the psychological, character of the inhabitants (ch. 3-6). Thus a city with a *southerly* aspect (facing ESE to WSW)[44] will be exposed to the hot winds and sheltered from the cold. Accordingly, its waters will be plentiful and brackish, and the moisture and heat characteristic of such a locale will produce in man a flabby physique and a relaxed, listless character. Consequently the heads of the natives here are moist, and

[43] On the aim and importance of *prognosis* in Hippocratic medicine, cf. Edelstein, *ΠΕΡΙ ΑΕΡΩΝ und die Sammlung der hippokratischer Schriften*, pp. 60-88. The basic objective of *prognosis* and its necessity even in health are well summarized thus (pp. 66-67): "Ausser dieser Prognose der Krankheiten und der Kranken kennt die antike Medizin aber auch eine Prognoselehre der Gesunden. Denn vielen Ärzten gilt die Gesundheit als ein Gleichgewicht, sei es der Säfte im Körper, sei es der von aussen zugeführten Stoffe, der menschlichen Handlungen oder beider, die Krankheit aber als eine Störung des Gleichgewichts, also als ein Missverhältnis der Teile. Notwendigerweise kann dann das Gleichgewicht der Gesundheit nie dauernd sondern immer nur für Augenblicke erreicht werden. Denn das Verhaltnis der Teile ist durch die immer neue Nahrungsaufnahme, immer neue Handlungen, automatisches Steigen und Verringern der Säfte in den Jahreszeiten oder ihr Fliessen im Körper je nach der Einwirkung von Hitze und Kälte fortwährenden Veränderungen unterworfen. Darum muss jeder Mensch in jedem Augenblick das Missverhältnis, das entsteht, auszugleichen versuchen. So ist die Diätetik von grosser Bedeutung, und der Gesunde untersteht ebenso wie der Kranke dem Arzt. Wie man bei der Behandlung von Krankheiten Prognosen stellen kann, so kann man es auch für den gesunden Menschen. Der Körper verändert sich durch die allgemeinen Bedingungen der Jahreszeiten und der Winde, ja durch die besonderen Formen des Lebens. Aus dem Vorherwissen aber ergibt sich überhaupt erst die Möglichkeit des Vorbeugens und der Hilfe."

[44] See Jones's note on periods of the year and directions in the Hippocratic writings, Loeb I, 67-69.

they are subject to phlegmatic complaints, intestinal disturbances, running inflammations, and other diseases linked with an excess of moisture (ch. 3).

On the other hand a city with a *northerly* aspect (facing ENE to WNW), exposed to the cold winds and protected from the hot, will have opposite characteristics. Its waters will be hard and cold for the most part. The cold and dryness result in physiques that are sinewy and spare, and characters that are more fierce than mild (τά τε ἤθεα ἀγριώτερα ἢ ἡμερώτερα). The natives here eat much, drink little, are more bilious than phlegmatic. Their general tenseness and dryness easily produce internal rupture, pleurisy, ulcers. They are generally healthier than natives of the southerly cities, but when disease occurs it tends to be acute and violent (ch. 4).

Both northern and southern exposures are said to be detrimental to childbearing. Many women of the northern cities are barren because of the nature of the water supply, while many of the southern are so because of disease; the former find childbirth difficult because of their tension and dryness; the latter often deliver prematurely because of their moist, flabby physiques.

However, the author maintains that the city which faces the *east*, being situated midway between north and south so that heat and cold *temper* each other, enjoys the advantages of the other two aspects with none of the disadvantages (ch. 5. 6-28):

Those that lie towards the risings of the sun are likely to be healthier than those facing the north and those exposed to the hot winds, even though they be but a furlong apart. In the first place, the heat and the cold are more moderate (μετριώτερον). Then the waters that face the risings of the sun must be clear, sweet-smelling, soft and delightful in such a city. For the sun, shining down upon them when it rises, purifies them. The persons of the inhabitants are of better complexion and more blooming than elsewhere, unless some disease prevents this. They are clear-voiced, and with better temper and intelligence than those who are exposed to the north, just as all things growing there are better. A city so situated is just like spring, because the heat and cold are tempered (κατὰ τὴν μετριότητα τοῦ θερμοῦ καὶ τοῦ ψυχροῦ); the diseases, while resembling those which we said occur in cities facing the hot winds, are both fewer and less severe. The women there very readily conceive and have easy deliveries.

This passage is particularly significant as illustrating not only the decisive influence which environment was thought to exert upon man, but also the notion that *balance* between opposites determines the *best* environment for man. We have seen that the medical writers conceive health — the best condition of the body — as depending upon a balance and blend of opposing constituents which remain operative in the body, but in such a way that they temper each other so that no one — whether it be bitter or sweet, hot or cold — dominates the organism. Again, we have seen the same notion applied to diet, especially in *Ancient Medicine*, where the best nutriment is conceived as reproducing the balance and blend of constituents found in the healthy body. Now, in *Airs, Waters, Places* (ch. 5), we find the eastern aspect described as best because here the extreme of heat, which prevails in the southern aspect, meets the cold, which dominates the northern, and the two *balance* or *moderate* each other. This balanced climate constitutes the best situation for a city because it tends to form man correspondingly well balanced in body and soul. Unlike the fleshy, listless citizen of the hot southern city and the sinewy, fierce citizen of the cold northern city, each threatened with characteristic disease, those who dwell in this moderate climate enjoy the bloom of health, well balanced emotions, and superior intelligence.

Finally, it is instructive to note that as the eastern aspect is best for health, so its opposite, the western, is the most unhealthy of all (ch. 6). Apparently it is not dominated by one extreme to the exclusion of the other, but rather is subject to all extremes of heat and cold, moisture and dryness, in such a way that they do not blend and temper each other, but alternate in quick succession. The city that faces west is cut off from the rays of the morning sun and exposed to the cold winds from the north, the hot winds from the south, and the rainy winds from the west. In the morning the western city, being sunless, is chilled by cool breezes and heavy dew; there the sun does not rise gradually but appears suddenly at its height and scorches the inhabitants throughout the afternoon. The effect of these extreme changes can only be a violent disturbance of the equilibrium within the body, the suddenness of their succession leaving no time for the organism to adjust. Accordingly, the inhabitants of the westerly city are described as pale and sickly, subject to all the diseases found in the other cities so that no one disease is peculiar to them. Western exposure combines the disadvantages of the northern and southern cities with none of their advantages, and represents the complete antithesis of the well balanced eastern aspect.

In discussing the various types of water and their properties in relation to health (ch. 7-9) the author applies the same general principle — that excellence lies in a balance of opposites. Thus the spring water that is most unhealthy (causing dropsy, dysentery, etc.), comes from low, soft, marshy ground; while the next worst is that which springs from rock or from ground with high mineral content. "For all these result from the violence of heat. So from such earth good waters cannot come, but hard, heating waters, difficult to pass and causing constipation" (ch. 7. 53-57). The best water flows not from one of these extremes of soft marshland or hard, rocky terrain but from a combination of the two, from "elevated ground and earthy hills" (ἐκ μετεώρων χωρίων ῥεῖ καὶ λόφων γεηρῶν). Coming from this *middle* type ground, not too hard and not too soft, the water flows sweet and clear, warm in winter and cool in summer.

Connected with this is an interesting note on the use of various waters (ch. 7. 79-90). First, it is said that a "man in health and strength can drink any water that is at hand without distinction" (ch. 7. 80-82). Apparently health and strength imply not only an equilibrium of bodily constituents but a certain *stability* of equilibrium which can resist or adjust to external influences without breaking easily. In line with this notion is the remark of *Ancient Medicine* (ch. 12, p. 55. 30-33) that such constitutions (φύσιες) as "feel quickly and powerfully the effects of their mistakes in diet are weaker than others. The weak man is very near to being a sick man, but the sick man is yet weaker than he", being more likely to suffer from any mistake at all (ὅ τι ἂν τοῦ καιροῦ ἀποτυγχάνῃ). Presumably, then, a man of strong, healthy constitution would not "feel quickly and powerfully" the effects of mistakes in diet but would adjust to them painlessly, as he would to water from any source, though it might be excessively hard or turbid. This aspect of stability helps to the understanding of health as an abiding disposition or ἕξις[45] of a body which tends to retain its internal equilibrium despite disturbing influences from the outside.

On the other hand, *Airs, Waters, Places* advises (ch. 7. 82-93) that a sick man seeking the most suitable drink can best "hit upon" health (μάλιστα τυγχάνοι τῆς ὑγιείης) by drinking the sweetest, lightest waters if his digestive organs are hard and easily heated, while if they are soft, moist, phlegmatic he must take water that is hard, harsh, and saline. Here again the physician attempts to restore the balance by setting himself against the excessive element which dominates the diseased organism.

Following the long passage on waters and their effect upon health (chs. 7-9), *Airs, Waters, Places* offers the physician some principles regarding the cycle of the seasons, by which he may determine whether the year will prove healthy or unhealthy. As we have already seen (p. 47 *supra*), winter was thought of as cold-moist, spring as warm-moist, summer as warm-dry, and autumn as cold-dry, each combination of opposites being dominant in turn and the whole cycle constituting a sort of equilibrium in which winter is balanced by summer and spring by autumn. It now appears that this cycle was considered normal and any deviation was likely to have unhealthy repercussions upon the human organism, disrupting the normal cycle *within* the body and causing disease. Thus the physician is told (chs. 10, 11) that he may expect a healthy year if "the signs prove normal (κατὰ λόγον) when the stars set and rise", and if the extreme of each season be somewhat tempered, e.g., if the dryness of autumn be relieved by some rain, or if "the winter be moderate (μέτριος), neither too mild nor unseasonably cold" (ὑπερβάλλων τὸν καιρὸν τῷ ψύχει). However, should this normal pattern be broken, the year will see an outbreak of diseases of various types, depending upon the nature of the

[45] The term ἕξις is applied to health in the Hippocratic Corpus itself, e.g., *Nutriment*, 34: ἕξις ὑγιεινὴ κρείσσων ἐν πᾶσιν; *Regimen I*, ch. 32, 3: ... σώματι ὑγιεινοτάτην ἕξιν ἀποδεικνύει ... The author of the latter essay evidently regards health as a relatively stable disposition when he says (ch. 2, 61-67): "For diseases do not arise among men all at once; they gather themselves together gradually before appearing with a sudden spring. So I have discovered the symptoms shown in a patient before health is mastered by disease ... "

season which deviates from the normal and the violence of the climatic changes involved. The author discusses a number of deviations from the regular "healthy" cycle of seasons, and indicates the specific diseases resulting from each.[46] Considered as bad are those years in which: 1) winter is dry and cold, spring wet and warm; 2) winter is warm and wet, spring cold and dry; 3) summer and autumn are wet and warm; 4) summer is dry and cold, autumn wet and warm.

The passage illustrates not only the influence of climate upon health and disease in general, but the particular importance attached to *regularity* and *violent change* in climate. We have noticed the value placed upon regularity of eating habits in *Ancient Medicine* (cf. pp. 40-41 *supra*). And in regard to change, *Airs, Waters, Places* has a final word at the close of the first part (ch. 11), warning the physician against undertaking any strong treatment like purging, cautery, or surgery within ten days of the summer and winter solstice and the spring and autumn equinox, since at these times seasonal changes are liable to be most violent, and so most dangerous.

The second part of *Airs, Waters, Places* (chs. 12-24) presents in concrete form the effects of climate and topography upon the body, and the soul, of man. The author compares Europe and Asia, and traces the differences in the physique and character of the inhabitants for the most part to environmental factors. Here again the environment most favorable to physical growth and health is said to be that which combines opposites so that they temper each other and no one dominates the whole. Remarkable, too, is the influence which climate is thought to exert upon national character (ch. 12. 7-13):

I hold that Asia differs very widely from Europe in the nature of all its inhabitants and of all its vegetation. For everything in Asia grows to far greater beauty and size; the one region is less wild than the other,

[46] The assumption that irregular seasons cause specific diseases may also be illustrated by *Epidemics* I and III, where the four καταστάσεις record seasonal variations notably divergent from the normal and then present accounts of symptoms and diseases evidently regarded as *effects* of these abnormal variations. The fourth κατάστασις (III, 15) says, for example: "In all the cases described spring was the worst enemy, and killed the greatest number."

the character of the inhabitants is milder and more gentle (τὰ ἤθεα τῶν ἀνθρώπων ἠπιώτερα καὶ εὐοργητότερα).

The reason for this superiority of Asia in physical growth and beauty, and for the temperate character of the natives, as well as the general physiological principle involved, is given as follows (ch. 12. 13-19):

> The cause of this is the temperate climate (ἡ κρῆσις τῶν ὡρέων), because it lies towards the east midway (ἐν τῷ μέσῳ) between the risings of the sun, and farther away than is Europe from the cold. Growth and freedom from wildness are most fostered when nothing is forcibly predominant (μηδὲν ᾖ ἐπικρατέον βιαίως), but equality in every respect prevails (παντὸς ἰσομοιρίη δυναστεύῃ).

This applies, of course, not to the whole of Asia, but to the region situated *midway* between the heat and cold (ἐν μέσῳ ... τοῦ θερμοῦ καὶ τοῦ ψυχροῦ) where the land is neither scorched with heat and dried out by drought nor oppressed by cold and by excessive rain and snow. There the country is well wooded, fruitful, mild, and supplied with excellent water; the harvests are plentiful, cattle flourish, the men are "well nourished, of very fine physique and very tall, differing from one another but little either in physique or stature" (ch. 12. 35-38). In its natural constitution (κατὰ τὴν φύσιν) and the moderation of its climate (κατὰ ... τὴν μετριότητα τῶν ὡρέων) the region is most like spring.

The terminology of the passage should be noted. This "healthiest" region is situated ἐν μέσῳ τοῦ θερμοῦ καὶ τοῦ ψυχροῦ, where there is κρῆσις τῶν ὡρέων, i.e., μετριότης τῶν ὡρέων, where μηδὲν ᾖ ἐπικρατέον βιαίως ἀλλὰ παντὸς ἰσομοιρίη δυναστεύῃ. The expressions used here to describe the healthy relationship of the opposing *climatic* elements recall those of the Alcmaeon passage describing that of the opposing *bodily* elements (τὴν σύμμετρον τῶν ποιῶν κρᾶσιν; μοναρχία; ἰσονομία), as well as the κρῆσις καὶ μετριότης applied by *Ancient Medicine* (cf. p. 42 *supra*) to both the constituents of healthy nutriment and of the healthy body. Furthermore — and this is especially significant — the point of κρῆσις and μετριότης is said to be ἐν μέσῳ τοῦ θερμοῦ καὶ τοῦ ψυχροῦ, i.e., to lie in a *mean* between the extremes of heat and cold, in this case a geographical position and its corresponding climatic condition in which there is neither too much nor too little of either extreme.

The very qualities which make this region ideal for physical growth and health, however, discourage the development of the sterner virtues among the inhabitants. "Courage, endurance, industry, and high spirits could not arise in such conditions ... but pleasure must prevail" (ch. 12. 40-44). The uniformly perfect blend of climatic conditions, continuing without violent change, brings health and pleasure without struggle or effort, so that the natives never develop the strong physical and moral qualities born of conflict, and in this they are inferior to the inhabitants of Europe.

For the inhabitants, like the topography of the land itself, mirror the climate of the region (ch. 13. 8-24):

The land is affected by them [changes of climate] exactly as human beings in general are affected. For where the seasons experience the most violent and the most frequent changes, the land too is very wild and very uneven; you will find there many wooded mountains, plains, and meadows. But where the seasons do not alter much, the land is very even. So it is too with the inhabitants, if you will examine the matter. Some physiques resemble wooded, well-watered mountains, others light, dry land, others marshy meadows, others a plain of bare, parched earth. For the seasons which modify a physical frame differ; if the differences be great, the more too are the differences in the shapes.

Detailed illustration of this general principle is found in the discussion of the dwellers on the Phasis (ch. 15), the Scythians (chs. 18-20), and others. But in addition to climate and topography there is another factor which is said to be of considerable importance in accounting for the differences among men, and that is νόμος. One striking example of the power of νόμος is that of the so-called "Longheads" (ch. 14) whose custom of shaping the heads of their babies eventually led to the acquisition of the trait by heredity, so that what was once νόμος eventually became φύσις, the natural structure of the body. This belief in the inheritance of acquired characteristics may help to explain the aristocratic bias of many ancient thinkers, and the great interest of political philosophers like Plato and Aristotle in regulating the training of youth.

The differences in character between Asiatics and Europeans

are explained by the author of *Airs, Waters, Places* on the basis of these factors, climate, topography, and νόμος (ch. 16. 3-14):

With regard to the lack of spirit and of courage among the inhabitants, the chief reason why Asiatics are less warlike and more gentle in character than Europeans is the uniformity of the seasons, which show no violent changes either towards heat or towards cold, but are equable. For there occur no mental shocks nor violent physical change, which are more likely to steel the temper and impart to it a fierce passion than is a monotonous sameness. For it is changes of all things that rouse the temper of man and prevent stagnation. For these reasons, I think, Asiatics are feeble.

The same principle is invoked later to explain the uniformity and lack of vigor among the Scythians (ch. 19. 33-35): "For neither bodily nor mental strenuousness is possible where the changes are not severe." It is therefore not so much mildness of climate that is thought to produce this lack of mental and physical energy, since the Scythians live in constant cold and moisture, but rather *sameness* of climate which does not vary much from one season to another.

In another passage (ch. 23) the variation in seasons is made responsible for the greater variety of physical types in Europe, in that climatic conditions are said to affect the nature of the seed at the time of generation, since it is "not the same for the same seed in summer as in winter, nor in rain as in drought" (ch. 23. 9-12). And the author continues (*ibid.*, 19-30):

The same reasoning [principle — λόγος] applies also to character. In such a climate arise wildness (τὸ ἄγριον), unsociability (τὸ ἄμεικτον), and spirit (τὸ θυμοειδές). For the frequent shocks to the mind impart wildness, destroying tameness and gentleness. For this reason, I think, Europeans are also more courageous (εὐψυχότεροι) than Asiatics. For uniformity engenders slackness, while variation fosters endurance (αἱ ταλαιπωρίαι) in both body and soul; rest and slackness are food for cowardice, endurance and exertion for bravery.

Climate and topography alone, however, do not account for the character of a people but, as we have mentioned, laws (νόμοι) and institutions are regarded as an important formative factor. The lack of strong qualities among Asiatics is attributed not only to their natural environment, but to the fact that most of them are governed by kings (ch. 16. 14 ff.). Where men are not their own masters (αὐτόνομοι) but ruled by despots they avoid warlike virtues, since then they are forced to fight and endure all for their masters, and bravery only profits the rulers, not themselves. Hence even when a man is by nature courageous and stouthearted, this disposition is changed by the νόμοι of despotism. In proof of this the author cites the example of those Asiatics, whether Greek or barbarian, who are independent and autonomous. These, he claims, prove to be the most warlike of all. And in a later passage he summarizes his contention thus (ch. 23. 30-41):

> Wherefore [because of climatic variation] Europeans are more warlike, and also because of their institutions (νόμοι), not being under kings as are Asiatics. For ... where there are kings, there must be the greatest cowards. For men's souls are enslaved, and refuse to run risks readily and recklessly to increase the power of somebody else. But independent people, taking risks on their own behalf and not on behalf of others, are willing and eager to go into danger, for they themselves enjoy the prize of victory. So institutions contribute a great deal to the formation of courageousness (οἱ νόμοι οὐχ ἥκιστα τὴν εὐψυχίην ἐργάζονται).

Finally, the determining influence of environment is illustrated in detail in the description of the peoples of Europe (ch. 24), where the same reasons are given to account for the differences of various tribes in size, physique, and moral qualities. Where the land is mountainous, rugged, high, well watered, exposed to sharp contrasts of climate the natives are likely to be large in frame, strenuous, courageous, rather wild and fierce. But in the low-lying, meadowy regions where the climate is stifling and the winds and waters warm, the natives will be, not tall and straight, but squat, fleshy, dark rather than fair. The same bravery and vigor are not natural to them, but law can produce these qualities in them. If the terrain be high and level, windy and well watered, the inhabitants will be uniformly large, but somewhat tame and unmanly in disposition; while thin, dry, bare soil in a region where seasonal changes are not well blended will produce natives who are hard and sinewy in physique, stubborn and independent in character and temper. In general, where climatic changes are most frequent and sharply contrasted, there will be found the greatest differences of physique, character, and constitution (εἶδος, ἦθος, φύσις).

Climate, topography, and waters shape the physiques and *mores* (τρόποι) of men. All things that grow conform to the land they grow in.

Besides affording a striking illustration of the extent to which environment was thought capable of determining the physical and psychological constitution of man, these final chapters of *Airs, Waters, Places* introduce some interesting notions regarding man's reaction to, and control of, environmental influence.

First, from the comparison of Asiatic and European climate and the men each produces it becomes clear that the organism may most easily maintain the equilibrium of health, growth, and beauty in the perfectly tempered climate, but it can never develop strong qualities, physical and psychological, unless it is forced to exercise itself in an effort to keep that equilibrium against the opposition of *changing* seasons. Where the organism is beset by hostile factors in the environment it is constantly stimulated and in the struggle to adjust develops energy, strength, and endurance, i.e., the ability to face and overcome opposition. Psychologically the result is alertness, courage, and independence. These qualities can never be developed in a land of perpetual springtime where the environment maintains the organism in good condition without exertion. Hence the organism, physical and psychological, may be well balanced without being strong, simply through the influence of favorable environment; to be strong it must be exercised through opposition.

Secondly, we learn that, just as a man's bodily constitution and the influence of environment can be counteracted by regimen and the medical art, so his psychological disposition (γνώμη) can be modified by νόμος. The author mentions the Asiatics (ch. 16. 31-33) who are born courageous and high spirited but are changed by their νόμοι, and the Europeans (ch. 24. 19-22), natives of hot, wet valleys, in whom bravery and endurance are not natural characteristics (ἐν τῇ ψυχῇ φύσει) but "the imposition of law can produce them artificially" (νόμος δὲ προσγενόμενος ἀπεργάζοιτ' ἄν). This implies that the νομοθέτης, the statesman, can do for the soul what the physician does for the body, and that their problem is basically the same — the development of a well balanced organism, body and soul. The analogous function of physician and statesman is made explicit in Plato and Aristotle.

4. *Regimen I*

The author of *Regimen I* conceives of man, and all other things, as constituted of two basic elements, fire and water (ch. 3). Fire, which is hot and dry (ch. 4), is the active, moving element, while water, being moist and cold, is passive and must be moved. Each is necessary for the other, since water nourishes fire and fire imparts motion to water, and each can share its characteristic quality of moistness or dryness with the other. Both fire and water exist in varying degrees of condensation and rarefaction, from the finest and rarest to the strongest and most dense (ch. 32). From these opposites in varying proportions, then, all things are formed.

That the author considers health to be the proportionate blending of these opposing elements is clear from his description of the healthiest type of body (ch. 32). That body, he says, exhibits the healthiest disposition (ὑγιεινοτάτην ἕξιν) which is composed of the finest (λεπτότατον) water and rarest (ἀραιότατον) fire. The reason given is that since they exist in their rarest and finest form it is not likely that either element will become excessively dense or strong under the influence of varying seasons, times of life, and diet, so as to upset the balance between them and cause disease. Rather, being most rare and fine, they are said to form the most perfect blend, like "the softest and rarest bronze that admits of the most thorough blending and becomes most beautiful" (πλείστην κρῆσιν δέχεται καὶ γίνεται κάλλιστος). Bodies of this type enjoy good health at least until forty, often into extreme old age.

This passage illustrates once more the notion that health depends upon: 1) a proper proportioning of the elements, constituting a balance or equilibrium (μετριότης) between them so that they temper each other and neither is excessive or deficient, and 2) a wide distribution and mingling of the one with the other, so that they blend (κρῆσις). The excellence of the state of health (ἕξις ὑγιείνη) is apparently measured by its stability, the most stable being that in which the elements are most perfectly proportioned and most perfectly blended.

Regimen I explains not only the various physical types but also differences of psychological endowment on the basis of the opposing elements and their proportions in combination (ch. 35). Intelligence (φρόνησις)[47] depends upon the particular blend of elements present in the individual (περὶ μὲν οὖν φρονίμου καὶ ἄφρονος ψυχῆς ἡ σύγκρησις αὕτη αἰτίη ἐστίν ...). Thus, the highest degree of intelligence is said to result when the moistest fire and the driest water are blended, since each of the elements in this state has the most of the characteristic quality of the other, and so is most self-sufficient (αὐταρκέστατον) in itself, independent of the other, free from necessity and force. In this state its moisture lends the greatest constancy and steadiness to fire, whereas its dryness imparts the greatest self-movement to water. The elements thus being in themselves as nearly balanced and self-sufficient as their natures permit, upon being combined form the most perfect and self-sufficient blend. The soul formed by such a blend is most intelligent and has the best memory.

Here again, in the psychological sphere, we find the perfection of operation associated with the equilibrium of opposing elements blended in perfect proportion so as to constitute a mean between extremes, each modifying and balancing the other without excess or defect. It is interesting to note also that, according to the same author (ch. 35), when something occurs to disturb the proportion and blend between the elements in this perfect state, so that one becomes excessive or deficient, the results are most devastating, since each element in this state is most capable of independent activity. "But if by the influence of some addition one or the other of these grow or diminish, there will result something most unintelligent ..." (ch. 35. 15-17). *Corruptio optimi pessima.*

In *Regimen I*, as in the other treatises we have examined, the task of the physician is to restore, maintain, and improve the equi-

[47] The term φρόνησις here seems to be broader than the English "intelligence", and to involve a number of psychological operations, i.e., quickness of perception, good memory, cool and deliberate judgment, the power of concentration, alertness.

librium of body and soul. This he is to do by regulating the diet and exercise of the patient (ch. 2). Diet and exercise are considered correlative, both necessary for health, though having opposite functions, since by nature exercise uses up material, while food and drink fill up deficiencies. So the physician must know the properties and effects of all types of diet and exercise, the proper proportions (συμμετρίαι) of one to the other and of both to the various ages and types of men, seasons of the year, changes of wind, localities, and climatic cycles.

However, even when he has mastered the principles of his science and has learned to forecast changes in the external universe by the movements of the stars, the physician is still faced with the difficulty of applying his general principles to the individual human organism. And in this there can be no absolute accuracy (ch. 2. 41-48):[48]

> If indeed ... it were possible to discover for the constitution of each individual a due proportion of food to exercise, with no inaccuracy either of excess or of defect, an exact discovery of health for men would have been made. But as it is ... this last discovery cannot be made.

However, despite the acknowledged impossibility of absolute accuracy in the practice of medicine, *Regimen I* proceeds to describe the basic constituents of all things, the development of the individual, medicine and the other arts as imitating the functions of nature, the factors involved in conception and in determining sex, various physical types, ages, and psychological endowment. This discussion forms the basis of the rest of the work (*Regimen II-IV*), which is mainly occupied with a description of various foods, drink, and exercise in terms of the heating, cooling, moistening, or drying effects they have upon the human constitution. The work is designed to offer a proximate guide in helping the physician, or the patient himself, to restore or increase the balance of these qualities

[48] The terminology should be noted: εἰ μὲν γὰρ ἦν εὑρετὸν ἐπὶ τούτοισι πρὸς ἑκάστου φύσιν σίτου μέτρον καὶ πόνων ἀριθμὸς σύμμετρος μὴ ἔχων ὑπερβολὴν μήτε ἐπὶ τὸ πλέον μήτε ἐπὶ τὸ ἔλασσον, εὕρητο ἂν ὑγιείη τοῖσιν ἀνθρώποισιν ἀκριβῶς. The impossibility of complete accuracy in prescribing for the particular individual is reiterated in *Regimen III*, 67. Later in the same book (ch. 69) the author outlines the regimen which, he believes, "approximates to the truth as closely as possible".

and thereby to improve not only bodily health but psychological acumen as well.[49]

D. CONCLUSION

Having traced the theory of health outlined in the Alcmaeon fragment through Menon's accounts of some of the physicians with whom Plato is associated, and having studied that theory, with its variations and ramifications, as it appears in several representative works of the Hippocratic collection, we may now venture some general remarks by way of synopsis.

First, it has, we hope, become abundantly clear that all the medical writers examined conceive the human body as a *complex of opposing factors*, though these are variously named (ποιά, δυνάμιες, χυμοί, στοιχεῖα) and designated differently, as follows:

Alcmaeon — hot, cold; bitter, sweet; moist, dry; etc.
Menecrates — blood and bile (hot), breath and phlegm (cold)
Petron — the hot (dry), the cold (moist)
Philistion — fire (hot), air (cold), water (moist), earth (dry)
Ancient Medicine — bitter, sweet; acid, astringent; salt, insipid; hot, cold; etc.
Nature of Man — phlegm (moist-cold), blood (moist-hot), yellow bile (dry-hot), black bile (dry-cold)
Airs, Waters, Places — explains man and all else in terms of hot, cold, moist, dry
Regimen I — fire (hot-dry), water (cold-moist)

These opposing factors unite to form man's body, just as they are generally thought to form the rest of the universe. In compound the elements never lose their natural "power" or characteristic activity, though the power of one may be balanced or tempered by the power of its opposite. Hence the human body is regarded as a *dynamic* union of opposing forces.

[49] The author maintains that regimen can improve mental capacity or harm it. For example, in speaking of the type of individual whose soul is blended of slightly more water than fire, he remarks (ch. 35. 25-27): "This kind of man under right regimen may become more intelligent (φρονιμώτερος) and sharper (ὀξύτερος) than natural endowment warrants." For further illustration of the dependence of φρόνησις upon the condition of the body, cf. *Breaths* 14, *Sacred Disease* 17.

When properly *proportioned* (σύμμετρον) these opposites counteract or temper each other in the compound so that a balance or dynamic *equilibrium* (μετριότης) is established. When this condition prevails the body is in health, functioning properly and smoothly, and with a sense of well-being. However, if something occurs to shift the proportion in the direction of one or other element, so that it grows excessive or strong, while its opposite becomes correspondingly deficient or weak, then the equilibrium is broken, resulting in malfunction and pain, and the body is diseased.

This centıal notion of health as an equilibrium between opposites, and of disease as loss of proportion through excess and defect, is common to all the medical writers we have seen, though expressed in various ways. In the Alcmaeon fragment, for example, health is ἰσονομία τῶν δυνάμεων, a σύμμετρος κρᾶσις τῶν ποιῶν, whereas disease is μοναρχία, through ὑπερβολή of one or other power. For Menecrates, health means that the constituents are disposed εὐκρότως, μὴ στασιάζοντα, whereas in disease they exist δυσκρότως, and in a state of στάσις. For Petron disease occurs when the constituents become ἀνώμαλα, while in health they are ὁμολά. For Philistion disease means that one element is excessive (πλεονάζει) or deficient (μεῖον) so that for him also health implies equality or proportion among them. And thus the manner of expression differs while the underlying notion remains basically the same — that of equality, proportion, harmony, equilibrium for health, and inequality, disproportion, disharmony, excess, and defect for disease.

However, the medical writings themselves, especially *Ancient Medicine* (cf. pp. 41-45 *supra*) and the *Nature of Man* (cf. p. 47 *supra*), make it clear that health involves not merely quantitative equality or proportion of opposing elements but also their proper distribution and blend (κρῆσις). For the opposing powers must be blended to counteract and temper each other, and none must be left in isolated strength to dominate the mass. *Ancient Medicine* explains (cf. p. 42 *supra*) that, as in the preparation of food, so *within* the body this blending is the result of combining opposites in proper proportion, mixing (μῖξις), and cooking or "coction" (πέψις), and the body is said to be "in the best possible condition when there is complete coction and peace."

Now opposing elements which have been thoroughly blended tend to remain in that condition and are not so easily separated again into their "raw" or unblended state. Hence κρῆσις is thought to induce a certain stability (cf. pp. 56-57, 64-65 *supra*) so that the equilibrium (μετριότης) of elements is constituted in an abiding state or condition (ἕξις). Only then can the body be said to be truly healthy when its dynamic equilibrium tends to remain constant, habitual, and not likely to be disrupted at any moment. Generally, the more perfect the proportion and blend of its opposing elements, the healthier the body, since then it approaches more closely the state of perfect equilibrium in its most permanent form.

However, no matter how perfect the blend of elements within the human body, their union is never a static one, but requires continual adjustment. Since his body is constituted of the same opposing elements that form the rest of the universe, man's internal equilibrium is constantly subject to the powerful influence of his regimen and environment (cf. esp. *Ancient Medicine* and *Airs, Waters, Places*). Diet and exercise, clothing and baths, the varying seasons, the situation of one's native city, its "aspect", soil, waters, and prevailing winds, all exert a continual play of opposing forces upon the balance of elements within the body. The organism must constantly adjust to these influences by responding proportionately, compensating for the changes and strength of diet and environment by shifting its internal proportion correspondingly in the opposite direction, thus tempering the forces exerted upon it and maintaining its equilibrium. As long as it makes this *proportionate response* the body remains healthy; when it fails, the impact of diet or environment causes an excess of one element or another, the internal equilibrium is disrupted, and disease ensues.

Since this is so, it becomes clear why diet and environment are considered "healthiest" when the elements which compose them are in a state corresponding to the μετριότης and κρῆσις of the healthy body. Thus, for example, *Ancient Medicine* (cf. p. 43 *supra*) describes diet as healthy when it is "well-blended, with nothing unblended or strong, forming a single, simple whole", and *Airs, Waters, Places* (cf. p. 59 *supra*) considers that climate most favorable for growth and beauty where there is ἡ κρῆσις τῶν ὡρέων ... ἐν μέσῳ τοῦ θερμοῦ καὶ τοῦ ψυχροῦ, where "nothing is forcibly predominant, but equality in every respect prevails". Under such diet and climate the internal equilibrium is easily maintained, without need of any violent, and dangerous, adjustment. Conversely, in "unhealthy" diet and environment one or other extreme prevails, or sudden and violent changes occur, so that the body maintains its internal equilibrium only with the greatest difficulty, if at all. All the details of diet and environment are evaluated on this basis — specific foods and beverages, winds, waters, temperature, topography, "aspect", seasons, and the climatic cycle of the year — and those judged "healthiest" which display μετριότης and κρῆσις. The microcosm is studied as existing in, and molded by, the macrocosm, and the same principles are seen to govern both.[50]

Once we understand the complex interplay of internal and external forces under which the equilibrium of health was supposed to be maintained, we can begin to appreciate the difficulty of the objective which the Greek medical man set himself — the restoration of that equilibrium when lost in disease, or its preservation and perfection in health. This required not only that he understand the present state of health or disease, but that he foresee the changes that would take place within the individual and in his environment, and that he reckon with both (prognosis). His task was further complicated by the belief that each time of life, each sex and type of physique, had its own tendency toward one or other extreme (cf. pp. 49-50 *supra*). As physician he was expected to recognize each specific disease throughout its various phases,

[50] The principles of μετριότης and κρῆσις are thought to govern not only the normal functioning of man and the universe, but also the *generation* of all things, man included; thus *Nature of Man*, 3 (for man, cf. also *Regimen I*, 8 and 9); *Aphorisms* V, 62: ὁκόσαι δὲ ἐξ ἀμφοτέρων τὴν κρᾶσιν σύμμετρον ἔχουσιν, αἱ τοιαῦται ἐπίτεκνοι γίνονται. The close affinity between microcosm and macrocosm is further illustrated by *Regimen I*, 10, which describes man as a miniature copy of the universe (ἀπομίμησις τοῦ ὅλου); and *Nature of Man* 7, which traces the same yearly cycle of elements in man and the seasons.

and to know its basic cause in terms of excess or deficiency of the elements. Then he must know, in terms of the same elements, the effects of different foods and potions, medicines and exercises, baths, applications, anointings, purging, bleeding, cauterizing, etc. But, most difficult and important of all, he must administer these in just the right measure (μέτρον) and at the precise time (καιρός)[51] necessary to draw off the excessive element, supply the deficient,[52] and thus help nature to restore the normal equilibrium in *this particular individual*, with his own peculiar make-up and circumstances.

The Greek physician recognized both the general principles of his science and the individuality of each case (cf. pp. 38-39, 49-50 *supra*). He felt keenly the difficulty of understanding completely the complex of general laws operating here and now in this particular concrete situation, and of applying precisely the right general principles in precisely the right measure to this individual patient. And so in the practice of his art he spoke of "aiming at" the right measure (στοχάζεσθαι τοῦ μέτρου) and of "hitting upon" health (τυγχάνειν τῆς ὑγιείης), and even denied the possibility of absolute accuracy in the matter (cf. pp. 39, 66). For the μέτρον he sought was not completely determined by any absolute scale, but was dependent upon the nature and condition of this particular individual under these particular circumstances.[53] And the individual is always, to a certain extent, incomprehensible.

[51] These two terms are so closely associated in the medical literature that they come to imply each other, and are used synonymously. Cf. *Airs, Waters, Places*, 10. 7, where μέτριος = μήτε ὑπερβάλλων τὸν καιρόν. Cf. also *Regimen II*, 66. 61-62; *Regimen I*, 7. 18.

[52] This is the "allopathic" method, described briefly in *Breaths*, 1. 3-7: "To sum up in a single sentence, opposites are cures for opposites. Medicine in fact is subtraction and addition, subtraction of what is in excess, addition of what is wanting. He who performs these acts best is the best physician ..." Cf. also *Nature of Man*, 9.

[53] Cf. *Epidemics I*, 23: "The following were the circumstances attending the diseases, from which I framed my judgments, learning from the common nature of all (ἐκ τῆς κοινῆς φύσιος ἁπάντων) and the particular nature of the individual (καὶ τῆς ἰδίης ἑκάστου), from the disease, the patient, the regimen prescribed and the prescriber ... from the constitution ... of the weather and of each region, from the custom, mode of life, practices and ages of each patient, from talk, manner, silence, thoughts, etc. ... " See also *Regimen III*, 67. 1-13, and W. Muri, *Arzt und Patient bei Hippokrates* (Bern, 1936), ch. VIII "Individualität", pp. 41-47.

However, the Greek physician was not without norms in dealing with the individual patient. According to *Ancient Medicine* (cf. pp. 39-40 *supra*) the physician could judge the effectiveness of his treatment in each case by the reaction or response (αἴσθησις) of the body. Here pleasure and a sense of well-being, with their opposites, pain and distress, are important, since pain is considered the natural result and concomitant of malfunctioning and disease, as pleasure seems to be the sign of proper functioning and a return to health.[54] Hence if the physician's treatment results in further pain he must take it as a warning, whereas if it brings a sense of well-being he may be fairly sure he has struck the right measure or proportion for this particular patient.[55]

Moreover, the reaction of pleasure or pain gives the physician some indication of the relative strength or weakness of his patient by comparison with the reactions of the normal, healthy body (cf. pp. 38-40 *supra*). The healthy man, for example, will find pleasure in solid food, while the sick will be distressed by it, and the very sick cannot stand even a small portion. Thus the reactions of pleasure and pain constitute a relative norm for the condition of any particular individual, while the functions and reactions of the healthy man constitute, as it were, an absolute norm for all.[56]

[54] For *pain* as a symptom of malfunction, cf. *Ancient Medicine* 14, p. 57. 16-23; 19, p. 60. 24-34; *Nature of Man* 4; *Places in Man* 42ff. The medical writers generally do not speak of *pleasure* as a sign of recovery and health; rather they speak of *relief* or *freedom from pain*. But cf. *Airs, Waters, Places* 12. 40ff., where we are told that μετριότης of climate produces a people in whom "pleasure must prevail".

[55] For pain as indicating excessive or deficient *diet*, cf. *Ancient Medicine* 6 and 10; excessive *exercise*, *Regimen II*, 66. 61ff. In both cases there is question of adapting regimen to the *individual*.

[56] Cf. *Prognostic* 2: "In acute diseases the physician must conduct his inquiries in the following way. First, he must examine the face of the patient, and see whether it is like the faces of healthy people, and especially whether it is like its usual self". *Ibid.* 3: "The patient ought to be found ... reclining on his right or left side, with his arms, neck, and legs slightly bent ... for so also recline the majority of men when in health, and the best postures to recline in are most similar to those of men in health". Also *Aphorisms* VII, 66, 67. Muri, *Arzt und Patient*, p. 55, points out the double norm in *Prognostic* 2, i.e., the faces of healthy people in general and the face of *this* individual when in health, of which the second is the important proximate norm. The first would be absolute in that it holds for all, the second relative to the individual. Muri also calls attention (pp. 10-14) to a distinction made by the treatise *On Joints*

Furthermore, in the practice of his art the Greek physician-trainer was guided by a system of values implicit in his analysis of nature. For him, as a medical man, the highest good was health, and he regarded as good all that constituted or promoted health. If he explained health as "due to the elements", this would include, as we have seen: 1) μετριότης — equilibrium of opposing factors or tendencies, lying in a mid-point or "mean" between them, where each moderates the strength of the other and neither is excessive or deficient but both of equal power (cf. pp. 42, 47 *supra*); 2) κρῆσις — blending of perfectly proportioned elements, by which the compound becomes, as far as possible, "a single, simple whole" (cf. p. 42 *supra*), approaching simplicity or oneness, and bringing with it the greatest possible a) stability, permanence, resistance to change, and b) self-sufficiency, independence, freedom from necessity (by uniting in itself an equal share of existing components, without excess or deficiency to require the addition of others; cf. pp. 64-65 *supra*). Thus, though he may never have formulated a theory of values, the physician holding this theory of health and disease associates the good with equilibrium, the mean between extremes, the measure (μέτρον, καιρός) between excess and defect, moderation, proportion, equality, coction, and blending.[57] He evaluates compounds as "better" or "best" insofar as they approach simplicity, permanence, and self-sufficiency. Moreover, these standards of excellence apply not only to man's body but, by analogy, to other compounds as well — to prepared food (pp. 37-38 *supra*), natural waters (p. 56 *supra*), temperature (pp. 58-59 *supra*), "aspect" (p. 55 *supra*), the climatic cycle of the year (pp. 57-58 *supra*),

---

between the norm of *complete* health, to which faulty structure or function is compared (e.g., ch. 57), and the relative norm of *limited* health (ἱκανῶς ὑγιηροί) which cripples of a certain type may enjoy (ch. 56). Thus the medical writers seem to recognize an *absolute* norm of perfect health and a *relative* norm of health attainable by the individual, given his particular constitution and circumstances.

[57] Cf. for example the use of καλός with whatever favors health, with καιρός (μέτρον) and "coction" in *Epidemics I*, 2, 11, 14, 17, 19. That this basic theory of health and the values implied were established in purely *medical* thought seems clear from the fact that they are incorporated in *Ancient Medicine*, which eschews the philosophic speculations of the "meteorological" school of medicine. Cf. Heinimann, *Nomos und Physis*, p. 175.

and even to the soul (pp. 64-65). The good is, in this sense, one for both microcosm and macrocosm.

On the other hand, all that threatens or destroys the equilibrium of health is regarded as "bad" or "harmful". This includes: 1) excess and deficiency, lack of proportion and moderation, inequality, etc.; 2) sudden or violent action and change (cf. pp. 30, 58);[58] 3) the imposition of force, interfering with the regular order or functioning of nature and bringing instability, distortion, pain.[59] And we have seen in two instances at least ("aspect", p. 55 *supra*; intelligence, p. 65 *supra*) that what was considered "worst" was somehow a perversion of the "best". Such seem to be the values assumed by those medical authorities who analyzed nature in terms of opposing elements.

The principles discovered in man's body and the physical universe were also used by the medical writers to explain differences in man's ethical character and intellect. We need only recall those chapters in *Airs, Waters, Places* (cf. pp. 53-55; 59-63 *supra*) where the physical environment is made responsible not only for a particular type of physique (εἶδος), but also for character (ἦθος, τρόποι) and intelligence (σύνεσις). The citizens of a city that faces the moist warmth of the south, for example, will be fleshy and listless; those facing the dry cold of the north, sinewy and fierce; while those enjoying an eastern exposure are most healthy, well-balanced emotionally, and of superior intelligence κατὰ τὴν μετριότητα τοῦ θερμοῦ καὶ τοῦ ψυχροῦ (cf. pp. 54-55 *supra*). Equable climate reacts favorably not only upon the body of man but upon his soul, both emotions and intellect.

Moreover, the principle is formulated that severe changes of climate and rugged terrain are needed to develop physical and mental vigor and the hardy virtues like courage and perseverance.

[58] Cf. *Humors* 15: "It is changes that are chiefly responsible for diseases, especially the greatest changes, the violent alterations both in the seasons and in other things. But seasons which come on gradually are the safest, as are gradual changes of regimen and temperature, and gradual changes from one period of life to another".

[59] Thus *Airs, Waters, Places* 12. 16-18: "Growth and freedom from wildness are most fostered when nothing is forcibly predominant ..." (μηδὲν ᾖ ἐπικρατέον βιαίως). And *Regimen I*, 7. 18: "But if untimely violence be applied there is no success" (ἢν δὲ βιῆται παρὰ καιρόν, παντὸς ἀποτεύξεται).

"For it is changes of all things that rouse the temper (γνώμην) of man" (*Airs, Waters, Places*, ch. 16. 12-14). "For uniformity engenders slackness, while variation fosters endurance in both body and soul; rest and slackness are food for cowardice, endurance and exertion for bravery" (*ibid.*, ch. 23. 25-30). This supposes a very close connection between the quality and functions of body and soul, and invokes the same principle as governing both.[60]

Finally, the concluding chapters of *Airs, Waters, Places* (cf. pp. 60-63 *supra*) indicate that the natural character of man, determined by his physical constitution and environment, may be modified by the laws and institutions under which he lives. We have seen how the physician must control the natural disposition (φύσις) of the body and soul of the individual. Now we learn that the natural character, the disposition of soul, in a whole nation or community can be modified by νόμος.[61] Implied is the notion that the νομοθέτης, the statesman, should so frame the laws as to correct the natural weaknesses and to develop the necessary virtues in the body politic. And this conception of the parallel roles of physician and statesman, common enough in the late fifth and fourth centuries, would seem to make these medical theories highly pertinent to the ethical-political thought of the period.[62]

It is perhaps not surprising, then, that the two outstanding philosophers of the period, Plato and Aristotle, with their strong

[60] The close connection between body and soul, between the brain and φρόνησις, is illustrated by *Sacred Disease* 17, which traces all psychological disturbances to the brain "when it is not healthy, but becomes abnormally hot, cold, moist, or dry ... But all the time the brain is still (ἀτρεμήσῃ) a man is intelligent" (φρονεῖ). Here and in *Regimen I*, 35, intelligence is made dependent upon the proportion and stability of the constituents of man, and *Airs, Waters, Places* 5 teaches that an equable, moderate climate favors intelligence. Yet in this last work (ch. 12, 16, 23) just the opposite — mental shocks (ἐκπλήξιες τῆς γνώμης) and severe physical changes (μετάστασις ἰσχυρὴ τοῦ σώματος) — are recommended to develop fierce, rugged, energetic, courageous character. Here, in the medical writers, we seem to have a physical basis for the dichotomy between the "contemplative" and the "active" life.

[61] Cf. Heinimann, *Nomos und Physis* (p. 168 esp.) on the meaning of these concepts in *Airs, Waters, Places* and the literature of the fifth century B.C.

[62] In connection with the influence of medical theory upon political thinkers, cf. K. Weidauer, *Thukydides und die Hippokratischen Schriften* (Heidelberg, 1954). For indications of medical theory in Herodotus and Euripides, cf. Heinimann, pp. 206-207.

interest in ethics and politics, should also devote themselves to the study of medical theory to such an extent that both became, in their own right, authorities in the field. Against the background of what we have found in the medical writings we may now look briefly at the theory as it appears in Plato.

III

PHYSIOLOGICAL THEORY AND "THE MEAN" IN PLATO

It may strike us as strange that Plato, the philosopher of the immutable and eternal Idea, should be regarded as an authority in the very practical field of medicine. Yet, as we have remarked, the Menon papyrus places him at the head of one group of medical thinkers. And the influence which he enjoyed with Galen[1] assures him a significant role in the history of medicine for all time.

The chief source of Plato's medical theory is the *Timaeus*. The fact that the Menon papyrus presents a comparatively long summary of part of this dialogue (69 E-86A) and that Galen composed a commentary upon it indicates the seriousness with which it was considered as a contribution to medicine in ancient times. We must turn to the *Timaeus*, then, to discover whether Plato's general theory of health and disease agrees substantially with that traced in the medical writers we have seen. If it does, we may indicate further the extent of its application in the *Timaeus* and the modification which it finds there, either by way of elaboration or omission.

The *Timaeus* is constructed upon a comparison between man and the visible universe, just as the *Republic* compares man and the state. Plato, early in the dialogue, tells us[2] that the cosmos is

[1] L. Edelstein, in his account of Galen in the *Oxford Classical Dictionary* (Oxford, 1949), *s.v.*, p. 377, remarks: "Plato and Hippocrates were his gods; Aristotle he held in sincere respect". For a defence of Plato as a serious and competent medical theorist see H. W. Miller, "The Aetiology of Disease in Plato's *Timaeus*", *Transactions of the American Philological Assn.* XCIII (1962), 175-187.

[2] I am assuming, with most commentators, ancient and modern, that the doctrine in the *Timaeus* is Plato's and not, as A. E. Taylor maintains in his

a *living being* with body and rational soul (30B), formed by the Demiurge on the model of the ideal Living Being (30C). "For the god, wishing to make this world most nearly like that intelligible thing which is best and in every way complete, fashioned it as a single visible living creature ..." (30D).[3] That the cosmos is the most perfect visible expression of its model is clear from the words with which Plato concludes his account (92C): "Having received in full its complement of living creatures, mortal and immortal, this world (κόσμος) has thus become a visible living creature embracing all that are visible and an image of the intelligible, a perceptible god, supreme in greatness and excellence, in beauty and perfection, this Heaven single in its kind and one." Now as the Demiurge formed the body and soul of the universe on the ideal model, so the created gods form man on the model of the universe (41C; 42E; 44D; 69C; etc.). Man's body and soul are as close a copy of the body and soul of the universe as his mortality will allow. Hence we may consider as pertinent to Plato's notion of health and disease not only those passages dealing directly with *man's* body, but also those concerning the body of the *universe*, which represents the structure and function of living body in its most perfect and typical state.

A. THE BODY OF THE COSMOS

When the "creation" myth of the *Timaeus* begins describing the formation of the cosmic body it becomes immediately evident that Plato, like the medical writers we have seen, conceives the

---

otherwise excellent *Commentary on Plato's Timaeus* (Oxford, 1928), that it belongs to the fifth century Pythagorean Timaeus. See Cornford's criticism of Taylor's thesis, *Plato's Cosmology* (London, 1937), pp. viii-xi. These two works have been most helpful and I have relied upon them almost exclusively in the present discussion. They will be cited henceforth simply as "Taylor" and "Cornford".

[3] Cornford's translation will be used throughout, and any significant departure from it will be noted. The Greek text will be that of J. Burnet, *Platonis Opera* (Oxford, 1905), Vol. IV.

living organism as a compound of opposing elements mixed in proportion:

Now that which comes to be must be bodily, and so visible and tangible; and nothing can be visible without fire, or tangible without something solid, and nothing is solid without earth. Hence the god, when he began to put together the body of the universe, set about making it of fire and earth. But two things alone cannot be satisfactorily united without a third; for there must be some bond between them drawing them together (δεσμὸν ... ἐν μέσῳ ... ἀμφοῖν συναγωγόν). And of all bonds the best is that which makes itself and the terms it connects a unity in the fullest sense; and it is of the nature of *proportion* (ἀναλογία) to effect this most perfectly (31 B, C).

Thus Plato begins with the two elements of Empedocles which he considers most radically opposed (cf. 55D-56A)[4] and invokes the notion of proportion, basic in the medical writers, to unite them. Plato may have geometric proportion especially in mind here, though he does not specify.[5] He goes on to explain why proportion must constitute the most perfect bond, in the physical world as in mathematics:[6]

[4] Thus Proclus, II, 17.17ff., characterizing fire and earth as τὰ παντάπασιν ἐναντία. References to Proclus will be cited according to volume, page, and line in the Teubner edition of E. Diehl, *Procli Diodochi in Platonis Timaeum Commentaria* (Leipzig, 1903).

[5] Cornford, p. 44, translates ἀναλογία as "continual geometric proportion", a qualification which I have omitted. Cf. *Gorgias* 508A, where ἡ ἰσότης ἡ γεωμετρική is said to be the key to understanding the bonds which hold together "heaven and earth and gods and men" and make the universe a *cosmos*. Cornford also cites (p. 45, n. 1) an "ancient tradition" which made geometric proportion the proportion *par excellence*. However, Taylor (p. 96) holds that: "The word ἀναλογία was subsequently extended to *all* proportion ... It is in this sense that the word is being used at 31 C 3". He agrees with Proclus, who concedes that geometric proportion is proportion κυρίως (II, 20.26) but maintains that what is said here applies to arithmetic and harmonic proportion as well (II, 18-20).

[6] Cornford, p. 44, restricts the principle enunciated here to *mathematics*, translating the terms ὄγκοι and δυνάμεις in the first clause as follows: "For whenever, of three numbers, the middle one between any two that are either solids (cubes?) or squares, is such that, etc. ... " In this he follows Sir Thomas Heath (cf. Cornford, pp. 46-47), and agrees with Archer-Hind (*The Timaeus of Plato* [London, 1888], p. 97) and Bury (in his translation of the *Timaeus*, Loeb Classical Library [London, 1942], p. 59). On the other hand, Proclus (II, 23.9ff.) holds that the terms are meant in a *physical* sense as well, and Taylor (p. 99) agrees that "the explanation of Proclus is pretty clearly right.

For whenever of three numbers, or masses, or "powers" of any kind, there is a middle one (τὸ μέσον) such that, as the first is to it, so it is to the last, and conversely as the last is to the middle, so is the middle to the first, then since the middle becomes first and last, and again the last and first become middle, in that way all will necessarily come to play the same part towards one another, and by so doing they will all make a unity (32A).

He takes ὄγκοι to mean, as it usually does, 'bulks', 'volumes' ". However, when he comes to δυνάμεις Taylor interprets Proclus in keeping with his own central thesis: "What the δυνάμεις mean Proclus explains by saying that the 'high' (ὀξύ), and the 'low' (βαρύ), in musical pitch are an instance of such a δύναμις ... Thus Timaeus will be illustrating his general proposition about 'means' by an example taken from each of three Pythagorean studies, arithmetic (ἀριθμῶν), geometry (ὄγκων), music (δυνάμεων)."

While agreeing that "the explanation of Proclus is pretty clearly right", it seems to me that when Proclus introduces the identification of δυνάμεις with *tonal* qualities like "high" and "low" (cf. II, 21.14ff.) he is giving not his own interpretation so much as the interpretation of *others* who held that Plato spoke of ἀριθμοί, ὄγκοι, and δυνάμεις in the sense of tonal qualities to suggest thereby the types of proportion characteristic of each, namely, arithmetic, geometric, and harmonic. Actually, Proclus *rejects* this interpretation, maintaining that all three types of proportion can be formed in all three classes, i.e., with numbers, volumes, or tonal qualities. Rather, Proclus protests against an *exclusively mathematical* interpretation of this passage, such as even Taylor would give it, and insists that the terms must also be understood in their *physical* sense (II, 23.9ff.) — φυσικὸς γὰρ ὁ διάλογος. In this sense he identifies δυνάμεις as αἱ συνεκτικαὶ καὶ εἰδοποιοὶ τῶν σωμάτων (II, 25.4-8) and gives as examples θερμότης ἢ ψύξις ἢ ὑγρότης ἢ ἄλλη τις τοιάυτη (II, 25.13; and cf. his illustration of analogy among these physical elements at II, 26.29-27.7).

As to the reasons offered in support of the purely mathematical interpretation by Heath and Cornford (pp. 46-47), the following may be said:

1) With regard to δύναμις, even though we concede that the word is used regularly in mathematics to mean "square", as well as "square root" and "multiple" (cf. *Theaetetus* 147 D ff.), the problem still remains whether it has that meaning here, and whether ὄγκος can mean "regular solid" or "cube". On this latter point the evidence from the ancient sources presented by Cornford (p. 50, n. 1) seems very tenuous, and Heath arrives at the conclusion only by inference (cf. Cornford, pp. 46-47): on the basis that it is "*a priori* probable (if not perhaps quite certain) that Plato πυθαγορίζει" in this passage, δύναμις is given its mathematical meaning "square"; ὄγκος must then be interpreted mathematically as "solid", and "if δυνάμεις are squares, it is reasonable to suppose that the ὄγκοι are also equilateral, i.e., the 'solids' are cubes." The only argument I find that δυνάμεις *must* be interpreted mathematically here is Cornford's rather impatient reference to Proclus' interpretation as involving "the false notion that δυνάμεις in our passage has a physical sense, and means the sensible qualities elsewhere called 'powers' ... Such qualities (pairs of opposites) form, in Plato's view, an ἄπειρον, and could not possibly stand as

Plato here introduces a feature which is characteristic of the *Timaeus* — the conscious effort to formulate in mathematical terms "scientific" data and hypotheses such as we have seen in the medical writers, and to push back to a more ultimate and systematic explanation of these through the exploitation of their mathematical aspect. It is a pioneer effort toward a mathematical phys-

---

terms in a numerical proportion" (p. 45, n. 1). However, this seems to me to beg the question, since it assumes that Plato is speaking here of proportion in an exclusively mathematical sense, i.e., of *numerical* proportion only. *De facto* Plato does speak of the "mean" (τὸ μέσον) of just such an ἄπειρον or pair of opposites, e.g., *Critias* 112C: "the mean between luxurious display and meanness" (τὸ μέσον ὑπερηφανίας καὶ ἀνελευθερίας). Moreover, throughout the *Timaeus* proportions are established by the Demiurge among δυνάμεις in the *physical* sense (cf. 56C); and δύναμις is used in this sense almost exclusively in the dialogue (an exception will be found in 54B: τριπλῆν κατὰ δύναμιν).

2) ὄγκοι normally means physical "masses" or "bulks", and this makes very good sense in the present passage, which is concerned with proportion established among the cosmic masses of earth, air, fire, and water. Cornford himself, in closing his discussion of the passage, writes (p. 51): "It may be conjectured that the quantities in question are the total volumes of the four primary bodies ... Plato would not imagine that anyone could know what the actual quantities were. He is only convinced that they must be linked in some definite proportion ..." And he quotes with approval from Hirzel (*Themis*, p. 309): "Gleichheit der elementaren Masses ahnte schon das älteste Denken in der Welt." The context, then, is concerned with proportion among the four cosmic masses, each with its own peculiar "powers". The present sentence explains how proportion unites them, as well as numbers.

This passage, which introduces the notion of proportion as the unifying principle of earth, air, fire, and water, should be interpreted, I believe, in the light of the passage which closes the discussion of their formation. There (56C) we are told that the god joined these together by proportion (συνηρμόσθαι ταῦτα ἀνὰ λόγον) by effecting proportions as accurate as possible with regard to their *masses* (πλήθη; cf. 57C), their *motions* (κινήσεις), and *other* "powers" (τὰς ἄλλας δυνάμεις). On this Cornford remarks (p. 223): "The last paragraph adds to the determination of the shapes the adjustment in due proportion of the numbers, motions, and powers of the primary bodies. 'Numbers' or 'quantities' (πλήθη) probably means the total quantities of the several kinds, as at 57C the same word means the main mass of each kind."

On the force of εἴτε ... εἴτε, see Taylor, p. 99. Cf. also Aristotle's use of ὄγκος and δύναμις in the same sentence in *N. E.* 1178a1. For a discussion of the purely mathematical *vs.* the mathematical-physical interpretation of Plato, and of this passage in particular, see P. Grenet, *Les Origines de l'analogie philosophique dans les dialogues de Platon* (Paris, 1948), pp. 141-142 and 147-163 (especially 154-163).

ics.[7] The first step in this process is the explanation of why there must be four elements in the proportion which unites fire and earth:

Now if it had been required that the body of the universe should be a plane surface with no depth, a single mean (μεσότης) would have been enough to connect its companions and itself; but in fact the world was to be solid in form, and solids are always conjoined, not by one, but by two means (μεσότητες). Accordingly the god set water and air between fire and earth, and made them, so far as was possible, proportional (ἀνὰ τὸν αὐτὸν λόγον) to one another, so that as fire is to air, so air is to water, and as air is to water, so is water to earth, and thus he bound together the frame of a world visible and tangible. For these reasons and from such constituents, four in number, the body of the universe was brought into being, coming into concord by means of proportions (δι' ἀναλογίας ὁμολογῆσαν), and from these it acquired Amity (φιλία), so that coming into unity with itself it became indissoluble by any other save him who bound it together (32A-C).

The theory of "plane" and "solid" numbers involved in this explanation has been adequately discussed by the commentators. It is enough here to underscore, first, the fact that Plato's cosmic body is made up of the four elements of Empedocles; secondly, that it is constituted as a *cosmos* by the imposition of proportion, the all-pervading and all-important principle of organic unity; thirdly, that the *mathematical* term μεσότης, carrying the notion of extremes linked by a mean, is introduced to express, and explain this proportionate relationship of the *physical* elements.[8] Thus

[7] Taylor (p. 98), in keeping with his central thesis, considers this "simply a play of mathematical fancy, such as would naturally occur to a Pythagorean interested in bringing his arithmetic into connexion with the medicine of the Sicilian school originated by Empedocles". That it is an application of mathematics to medical theory seems clear; but that it is a mere "play of mathematical fancy", and not a serious attempt to fix the ultimate explanation of visible phenomena in the realm of the pure intelligibles, seems inconsistent with the carefully elaborated theory of the geometric structure of the elements basic to the cosmic physiology of the *Timaeus*.

[8] This is the first time we have come upon the key term μεσότης in a passage from an ancient author, and Plato here obviously borrows it from the language of contemporary mathematics. It is important, then, to examine the mathematical significance of the term, and this will be done briefly in Appendix I: *On the Notion of* Μεσότης *in Greek Mathematics*.

The mathematical term, however, is used here by Plato to express a mean or proportion among *physical* realities, the cosmic masses of fire, air, water, and earth. Μεσότης and τὸ μέσον will have for him, as for his successors like

Plato conceives the cosmic body as a compound of opposing elements in proper proportion, which he analyzes in mathematical terms.

Now since, as we have seen, the cosmos is the most perfect reproduction of the intelligible Living Being, this internal proportion of its elements must be most perfect, so that it will be in the best possible condition.[9] Moreover, to imitate its eternal, unchanging model and preserve this condition, the cosmic body must not lose the perfect balance of its elements through the impact of hostile elements from outside. Plato provides for this (32C) by including in the cosmos *all* existing materials of earth, air, fire, and water, "leaving no part (μέρος) or power (δύναμις) of any one of them outside." He gives as his reasons (32D-33A), first, that the cosmos "might be in the fullest measure a living being whole and complete, of complete parts; next, that it might be single, nothing being left over, out of which another might come into being; and moreover that it might be *free from age and sickness*" (ἵν' ἀγήρων καὶ ἄνοσον ᾖ). To explain this last reason, Plato enunciates a principle which recalls the teaching of the medical writers on the influence of environment:

For he [the Demiurge] perceived that, if a body be composite, when hot things and cold and all that have strong powers (δυνάμεις ἰσχυραί)

Aristotle, more than a merely *mathematical* content, but will suggest proportion, balance, equilibrium among factors on any level, like the μετριότης of the medical writers. Philoponus, for example, analyzes the proportions of this passage in purely physical terms as follows (Commentary on Nicomachus' *Introduction to Arithmetic*, edited by R. Hoche, Bk. II [Berlin, 1867], p. 28, XXIV, 11.4-9): "For this reason, then, the god placed air and water between fire and earth, that as air is to fire and water, this water would be to air and earth. For air, being hot and moist, is bound by the common quality of heat to fire, by that of moisture to water. And again, water, being cold and moist, is bound by moisture to air and by cold to earth. Thus a single continuity of all is effected through the means and extremes (διὰ τῶν μέσων καὶ τῶν ἄκρων) being bound together with one another. ..." (I am indebted to Professor Harold Cherniss for reference to this passage.)

[9] Cf. 68 E, where the cosmos is called "the fairest and best of all things that become" and "the self-sufficing and most perfect god". See also the remarks of Proclus, II, 63.12-16, on the relationship between the well-being of the universe and the συμμετρία of its elements.

beset it from without and fall upon it immoderately (ἀκαίρως),[10] they undo it and cause it to waste away, bringing upon it illnesses and old age. For this reason, and so considering, he fashioned it as a single whole consisting of all these wholes, complete and free from age and sickness (33A, B).

Plato is obviously assuming here the hypothesis common in the medical writers, that there is an interplay of "powers" between the particles that constitute the body and those that surround it, so that it is possible for an excess of one or other element in the environment to disturb the proportion of elements *within* the body, causing disease and death.[11] This is developed at greater length when he discusses the human body (e.g., 81C-82B), where the effect of particles entering the body through diet and environment is made supremely important in accounting for deterioration and disease.

Since the universe, being without environment, is not subject to dissolution and decay, Plato nowhere describes the cosmic body as it would be in a state of disease or death. However, we can infer what the cosmos would become if its internal proportions were disrupted from Plato's description of the cosmic chaos *before* the imposition of these proportions by the Demiurge:

Now the nurse of Becoming, being made watery and fiery and receiving the characters of earth and air, and qualified by all the other affections that go with these, had every sort of diverse appearance to the sight; but because it was filled with powers that were neither alike nor evenly

[10] This is a slight departure from Cornford (p. 52), who translates: "... when hot things and cold ... beset that body and attack it from without, they bring it to *untimely dissolution.* ..." Thus he takes ἀκαίρως with λύει. It seems to me, however, that ἀκαίρως should go with the preceding participle, because: 1) the structure seems parallel with the coordinate phrase — περιιστάμενα ἔξωθεν καὶ προσπίπτοντα ἀκαίρως. Thus Diehl, in his edition of Proclus (II, 58.18), punctuates with a comma after ἀκαίρως. 2) In the medical writers we have seen that ἀκαίρως can be used for ἀμέτρως, signifying "without measure", and Plato's use of καιρός as equivalent to μέτρον at 85D (translated there by Cornford as "due mean") indicates that the terms may be synonymous in his mind also. Thus Proclus (II, 64.17): τὸ δὲ ἀκαίρως τὴν ἀσυμμετρίαν δηλοῖ. ... 3) Taken this way, the sentence becomes a much more pointed statement of the theory of disease and dissolution developed later in the dialogue (e.g., 82B), where only the *disordered* impact of particles from outside is thought to break down the ordered structure of the body.

[11] Thus Proclus, II, 62.31-63.9.

balanced, there was no equipoise in any region of it; but it was everywhere swayed unevenly and shaken by these things, and by its motion shook them in turn. And they, being thus moved, were perpetually being separated and carried in different directions ... by the Recipient, which itself was in motion like an instrument for shaking, and it separated the most unlike kinds farthest apart from one another, and thrust the most alike closest together; whereby the different kinds came to have different regions, even before the ordered whole consisting of them came to be. Before that, all these kinds were without proportion or measure (52D-53A).

Here the various powers, hot, cold, moist, dry, etc., exist in complete disorder, distinguished from their later condition by the fact that they are not similar (ὅμοιαι) or evenly balanced (ἰσόρροπαι) or in equilibrium (ἰσορροπεῖν), moving and being moved irregularly (ἀνωμάλως), without proportion or measure (ἀλόγως καὶ ἀμέτρως).[12] Such, we may presume, would be the condition to which the cosmos would return if its internal proportion and equilibrium were broken up by disease and old age, a condition which represents for Plato, as for the medical writers, the opposite of well being or health. This is confirmed by his description later of disease in the human body (cf. 81A ff.).

It should be noted that the "powers" or "qualities" from which earth, air, fire, and water are later formed by the Demiurge exist as *opposites* even in their most primitive state. Plato, in warning against conceiving the contents of the Receptacle as substances (50A), refers to these as "hot, or white, or *any of the opposites*" (ὁτιοῦν τῶν ἐναντίων). Taylor explains (p. 321): "... that is, any member of the pairs of 'opposites' — antithetical qualities, as we should call them in our more advanced terminology — which play such a prominent part in Ionian natural science, and consequently in Aristotle ..."[13] Moreover, through the movement of the Receptacle these opposing qualities tend to gravitate each towards

[12] Note that the description emphasizes the notion of *equilibrium*, or its absence: ἰσορρόπων ... ἰσορροπεῖν ... ταλαντουμένην. ... It is this especially that makes the difference between order and chaos.

[13] And cf. Cornford, pp. 183-184, where he assembles the pertinent passages from Proclus, showing that "the things that pass into and out of the Receptacle are simply the opposite qualities or groups of qualities characteristic of the four primary bodies".

its own "region", even before being formed into the primary bodies by the Demiurge (52D-53B; and cf. Cornford, pp. 198-199). Thus the opposite qualities and tendencies characteristic of earth, air, fire, and water (i.e., their ἴχνη; cf. 53B2) are native to the primal stuff from which these bodies are formed.

From this chaos of opposing factors moving at random the Demiurge forms the cosmos through the imposition of *proportion*. Just how all-embracing this principle of proportion, equality, and equilibrium is for Plato can be seen in his account of the biochemistry of the cosmic body (53C-68E). In forming the primary bodies from the Receptacle or Space (50A-C; 51D), in itself undifferentiated (50C-51B) but containing the ἴχνη of their nature (53C), the god gives earth, air, fire, and water distinct configuration "by means of shapes and numbers" (53C) with which the characteristic δυνάμεις of each are joined. Here Plato turns to mathematical analysis in attempting to penetrate to the ultimate explanation of the primary bodies postulated by Empedocles.

Assuming that 1) the work of the Demiurge is always most perfect, 2) that the primary bodies, as bodies, must be three dimensional, and 3) that they must be capable of changing into one another, Plato's problem is, as he says (53E): "What are the most perfect bodies that can be constructed, four in number, unlike one another, but such that some can be generated out of one another by resolution? If we can hit upon the answer to this, we have the truth concerning the generation of earth and fire and of the bodies which stand as proportionals between them" (τὰ ἀνὰ λόγον ἐν μέσῳ). On the assumption that the surfaces of the primary bodies are rectilinear and composed of triangles, Plato selects the right-angled isosceles, which is unique, and the "best" of the right-angled scalene, the half-equilateral, as the irreducible geometric units from which earth, air, fire, and water are formed (54A). Each is characterized by equality or proportion in that one has two equal sides while the other, when paired, forms the equilateral triangle. From the right-angled isosceles Plato constructs the *cube* (53C-55B), and from the half-equilateral are constructed the *pyramid*, the *octahedron*, and the *icosahedron*, so that these three can pass into one another by resolution into their common triangle.

The geometric properties of shape, size, and equality are then seen as determining the physical characteristics of each figure and its identification with one of the primary bodies (55D-56C). Of these, *earth* is most immobile and plastic. On the principle, basic in the medical writers, that stability arises from balance or equality of elements, earth is identified with the *cube*, since its base or unit is isosceles and so more stable than the unequal-sided triangle. Moreover, the square face of the cube, being four-sided and equilateral, "is necessarily a more stable base than the triangle, both in its parts and as a whole" (55E). The remaining three figures are then identified with their primary bodies on the basis of the proportion postulated earlier (32B), that fire is to air as air is to water, so that air forms a mean between fire and water. The characteristic δυνάμεις of each body are considered and a figure assigned to it accordingly (56A, B). Thus the particles of *fire* are identified with the *pyramid* because, of the three, that figure is most mobile (εὐκινητότατον), smallest, lightest (as composed of the smallest number of similar parts), and has the most acute angles so that sharpness and cutting power are the greatest. The particle of *water*, on the other hand, is shaped like the *icosahedron*, as being least mobile (δυσκινητότατον), largest, heaviest, and most blunt because of its obtuse angles. And the *air* particle is an *octahedron*, as constituting in all these respects a mean between the other two. The particles are invisible; what we see are the masses (ὄγκοι) of many joined together (56C). All these the Demiurge joins harmoniously according to proportion (συνηρμόσθαι ταῦτα ἀνὰ λόγον), fixing the proportions of their quantities (πλήθη) and motions (κινήσεις) and all other "powers" (αἱ ἄλλαι δυνάμεις) in every respect as accurately as their material nature will allow (56C).

That Plato regarded the four primary bodies as fundamentally opposed is clear from his assumption that conflict among them is natural and continual when a smaller quantity of one is surrounded by a larger quantity of another. This results in the dissolution of the weaker and, in the case of fire, air, and water, its transformation into another (56C-57C). Thus earth is broken up by fire, air, and water, but cannot combine with these, having a different base; water is broken up by air and fire, recombining as one particle of fire and two of air; air is broken up by fire, becoming two fire particles; and when a little fire is engulfed by a large quantity of any other, "fighting and being conquered", it is shattered and recombines into air. Dissolution continues as long as a "weaker contends with a stronger" (ἕως δ'ἂν ... ἧττον ὂν κρείττονι μάχηται). Unequal amounts of heterogeneous particles can, and must, modify each other until, by the transformation of the weaker into the form of the prevailing body (εἰς τὴν τοῦ κρατοῦντος ἰδέαν), it becomes homogeneous with it (ἓν ὅμοιον τῷ κρατήσαντι) and the conflict stops. "For no kind which is homogeneous (ὅμοιον) and identical can effect any change in, or suffer any changes from, that which is in the same condition as itself" (57A). Thus the primary bodies are in natural opposition which leads to conflict until their quantities be balanced or their quality reduced to homogeneity.

Similar principles govern motion and rest (57D-58C). Motion (κίνησις — change) does not exist in a state of *uniformity* (ἐν ὁμαλότητι — evenness, equilibrium), since the mover and moved can never be uniform (ὁμαλά). Rather, *stability* (στάσις — stationariness) prevails in what is uniform, while motion is attributed to lack of uniformity (εἰς ἀνωμαλότητα). The cause is the inequality (ἀνισότης) of the non-uniform structure (τῆς ἀνωμάλου φύσεως). These principles are later applied to the varieties of primary bodies and their compounds. Thus water is distinguished as: 1) the liquid, and 2) the fusible (58D, E). The liquid, whose particles are unequal (ἀνίσων ὄντων), is mobile because of its lack of uniformity (διὰ τὴν ἀνωμαλότητα), while the fusible, consisting of large and uniform particles, is more stable (στασιμώτερον) and solid because of its uniformity (ὑπὸ ὁμαλότητος). Likewise, earth forms the better type of rock when its parts are equal and uniform (τῶν ἴσων καὶ ὁμαλῶν ... μερῶν), while the opposite is inferior (60C).

For Plato, then, as for the other medical writers we have seen, equality, balance, uniformity produce peace and stability among opposing elements, while inequality is the basic cause of conflict and change. It becomes clear why *proportion*, which establishes equality among the cosmic masses and "powers" of every kind, is regarded as the unifying and stabilizing principle of the universe.

As was said at the outset, these things were in disorder and the god introduced into them all every kind of measure (συμμετρίαι) in every respect in which it was possible for each one to be proportioned and symmetrical (ἀνάλογα καὶ σύμμετρα) both with itself and with all the rest. For at first they were without any such proportion, save by mere chance, nor was there anything worthy to be called by the names we now use — fire, water, and the rest; but all these he first set in order, and then framed out of them this universe, a single living creature containing within itself all living creatures, mortal and immortal (69B-C).

This all-embracing principle is manifest in the very shape of the cosmic body, which is spherical. For the sphere is the perfect geometrical expression of proportion, uniformity, equilibrium:

And for shape he gave it that which is fitting and akin to its nature. For the living creature that was to embrace all living creatures within itself, the fitting shape would be the figure that comprehends in itself all the figures there are; accordingly, he turned its shape rounded and spherical, equidistant every way from center to extremity (ἐκ μέσου πάντῃ πρὸς τὰς τελευτὰς ἴσον) — a figure the most perfect and uniform (ὁμοιότατον) of all; for he judged uniformity to be immeasurably better than its opposite (33B).

The cosmos, being the closest possible copy of the ideal Living Being, and containing within itself all earth, air, fire, and water, has no need of external limbs or organs. For outside it there is nothing to see or hear, no air to breathe, no food to take in, but it is designed to "act and be acted upon entirely by itself and within itself; because its framer thought that it would be better self-sufficient (αὔταρκες) rather than dependent upon anything else" (33C, D). There is nothing outside to grasp, so it needs no hands, and nothing to stand upon, so it needs no feet. Its motion is proper to its bodily form, single, uniform, and circular (33D-34A). Within its sphere the heavenly bodies revolve in their orbits (38C, D) and at its center is the earth, a solid body poised in perfect equilibrium (τι στερεὸν κατὰ μέσον τοῦ παντὸς ἰσοπαλές).[14] Thus Plato conceives the body of the cosmos, the visible reproduction of the intelligible Living Being, as a compound of the four opposing pri-

[14] That the earth is situated at the center in the *Timaeus*, cf. Aristotle, *De Caelo* 293b; Proclus III, 133; and Cornford's discussion of the earth's position and motion, pp. 120-134, 265.

mary bodies, mixed in perfect proportion and characterized in every respect by the order of equilibrium, unchanging stability, and self-sufficiency.

All this, then, was the plan of the god who is forever for the god who was sometime to be. According to this plan he made it smooth and uniform, everywhere equidistant from its center, a body whole and complete, with complete bodies for its parts. And in the center he set a soul and caused it to extend throughout the whole and further wrapped its body round with soul on the outside; and so he established one world alone, round and revolving in a circle, solitary but able by reason of its excellence to bear itself company, needing no other acquaintance or friend but sufficient to itself. On all these accounts the world which he brought into being was a blessed god (34A, B).

B. THE SOUL OF THE COSMOS

It will be useful to look briefly at Plato's account of the soul which the Demiurge set in the center of the cosmic body "and caused it to extend throughout the whole". For the cosmic soul is, first, the model of the rational soul in man, and unless we know something of the structure and function of this most important element we cannot understand Plato's concept of the human organism. Secondly, in his account of the cosmic soul it becomes apparent that Plato employs the same basic principles in both physiology and psychology. For the cosmic soul, like the cosmic body, is a compound of opposing elements united by proportion, and its formation illustrates the integrating function of the *mean* (τὸ μέσον) in the *psychological*, as well as the physiological, sphere.[15]

In the passages which present Plato's account of the cosmic soul (35A-36D; 41D) the Demiurge is represented as compounding in

[15] In the *Timaeus* Plato treats the cosmic body before the soul, though he apologizes for doing so and declares the superiority of the latter γενέσει καὶ ἀρετῇ (31 B, C). This strikes the reader as curious. Cornford (p. 43) believes "the making of the body is taken first for convenience". I would suggest, from what we have seen of the medical writings, that it is a pedagogical device, that Plato follows this order of presentation so as to explain the less well known (psychological structure) through what was familiar to his readers (physiological structure), showing how the same basic principles govern both.

a κρατήρ a blend of Existence, Sameness, and Difference from which the soul of the cosmos is formed. The description is, of course, highly figurative, but the figure itself is significant, calling attention to the concept of κρᾶσις with its strong physiological connotations. As body is a mixture in which opposites are united, so rational soul is blended of diverse ingredients — Existence, Sameness, and Difference. Each of these is itself an intermediate blend of indivisible, eternal Existence, Sameness, and Difference and of the divisible, changing Existence, Sameness, and Difference characteristic of visible bodies.[16]

The things of which he composed soul and the manner of its composition were as follows: (1) Between the indivisible Existence that is ever in the same state and the divisible Existence that becomes in bodies, he compounded a third form of Existence composed of both (τρίτον ἐξ ἀμφοῖν ἐν μέσῳ συνεκεράσατο). (2) Again, in the case of Sameness and in that of Difference, he also on the same principle made a compound intermediate between (ἐν μέσῳ) that kind of them which is indivisible and the kind that is divisible in bodies. (3) Then, taking the three, he blended (συνεκεράσατο) them all into a unity, forcing the nature of Difference, hard as it was to mingle (δύσμεικτον), into union with Sameness, and mixing (μειγνύς) them together with Existence (35A).

Having compounded into a single, consistent whole (εἰς μίαν ἰδέαν) these intermediate blends of Existence, Sameness, and Difference, the Demiurge imposes *proportion* upon the mixture, first, by dividing it into lengths according to a geometric progression that advances through the first even and odd numbers (i.e., 2 and 3) to the cubes of each.

And having made a unity of the three, again he divided this whole into as many parts as was fitting, each part being a blend of Sameness, Difference, and Existence. And he began the division this way. First he took one portion (1) from the whole, and next a portion (2) double of this; the third (3) half as much again as the second, and three times the first; the fourth (4) double of the second; the fifth (9) three times the third; the sixth (8) eight times the first; and the seventh (27) twenty-seven times the first (35B, C).

The function of proportion here is the same as that of the proportions imposed upon the cosmic body (31B-32C), i.e., to link the

[16] This is the interpretation of Cornford (p. 59ff.), following Proclus.

parts into an intelligible whole.[17] And the geometric progressions advance to the cube numbers of two and three probably for the same reason given earlier in regard to the proportion determining the number of primary bodies in the cosmos (32A ff.), for the cosmic soul is to move the cosmic body, which is solid in form, "and solids are always conjoined not by one mean but by two."[18] There has been much speculation as to why the progression is double, advancing through both odd and even and involving seven terms instead of four,[19] but the only solution for which there seems to be direct support in the text is that the double progression provides a basis in the structure of the cosmic soul itself for the proportions which determine the orbits of the seven planets (cf. 36D; 38C).[20] But whatever be Plato's motive for choosing this progres-

[17] Cf. Proclus II, 198.29ff.

[18] This is the interpretation of Cornford (p. 66ff.). Taylor (p. 137) also remarks that the series stops with the third power because it "symbolizes the three dimensions of bodies", but his lengthy and learned treatment of this difficult passage (p. 106ff.) seeks rather to establish that Plato's description and the proportions involved are suggested by "the Pythagorean conception of a musical scale". Cornford, following Adrastus and Proclus, points out that the decision to end the series with cube numbers "has nothing whatever to do with the theory of musical harmony" (p. 67) and argues rather that the "geometrical framework of the whole *harmonia* is determined by arithmetical and physical preoccupations. ..." (p. 68). The symbolism of the mixing bowl and the division of the mixture by cutting (36A2, B5) seem to me much more suggestive of analogy with a physical compound such as the cosmic body, which is to be animated by the cosmic soul, though Taylor (p. 106) finds in this a reference to the blending of high and low tones according to definite mathematical ratios. However, the problems involved are too vast to consider here. Cf. Taylor's explanation, pp. 106-146; the reply of Cornford, pp. 59-72; and the caution of Proclus, II, 212.3-5.

[19] Cf. e.g., Proclus II, 200.21 ff.

[20] Cornford (p. 68) suggests this, with some misgiving, in answer to those who (like Taylor) seek an explanation in the science of harmonics. "No one, setting out to construct a musical scale, would start by arranging the terms of two geometrical progressions in the series: 1, 2, 3, 4, 8, 9, 27. The single series 1, 2, 4, 8 would yield a compass of three octaves. Plato is not content with this because Pythagorean arithmetical theory demanded that the odd numbers should be represented, *and also, perhaps, because he intends later to space the seven planets at distances corresponding to the terms,* and so needs seven numbers" (Italics mine; cf. also p. 79). The latter reason, offering an intelligible mathematical basis for the data of contemporary astronomy and suggested by the text (36D; 38C), seems much more compelling then the demands of "Pythagorean arithmetical theory".

sion, it is evident that *proportion* is the integrating principle of the cosmic soul, as it is of the cosmic body, and this is confirmed in the following passage, where the "material" of the cosmic soul, joined first in geometric proportion, is integrated further by the introduction of the other two types, arithmetic and harmonic proportion.

Next he went on to fill up both the double and the triple intervals, cutting off yet more parts from the original mixture and placing them between (εἰς τὸ μεταξύ) the terms, so that within each interval there were two means (μεσότητες), the one (harmonic) exceeding the one extreme and being exceeded by the other by the same fraction of the extremes, the other (arithmetic) exceeding the one extreme by the same number whereby it was exceeded by the other. These links (δεσμοί) gave rise to intervals of 3/2 and 4/3 and 9/8 within the original intervals. And he went on to fill up the intervals of 4/3 (i.e., fourths) with the interval 9/8 (the tone), leaving over in each a fraction. This remaining interval of the fraction had its terms in the numerical proportion of 256 to 243 (semitone). By this time the mixture from which he was cutting off these portions was all used up (35C-36B).

Here again we may leave detailed discussion of the passage to the commentators and be content to underscore the central fact that the harmonic and arithmetic μεσότητες constitute bonds (δεσμοί) uniting further the portions of the mixture already ordered according to geometric proportion. The terms recall that passage in which proportion is first explained as the bond of the universe, and the μεσότητες of air and water made to link the extreme constituents of the cosmic body (31C-32B). Now, to form the cosmic soul, the compound of Existence, Sameness, and Difference is united in exact proportion by the three μεσότητες — geometric, harmonic, and arithmetic.[21]

In accordance with its ingredients and proportions, the Demiurge next gives the cosmic soul the *structure* which enables it to function 1) as the source of all rational (i.e., ordered, measured, regular) change or motion in the universe, and 2) as the subject of knowledge and true opinion (36B-D). Splitting the proportioned strip

[21] Proclus (II, 198.29-200.21), in fact, finds a correlation between the three ingredients and the three types of proportion: Existence — geometric; Sameness — harmonic; Difference — arithmetic.

of material lengthwise, he joins the two halves at the center in the form of an X and bends them into concentric circles, one outside the other. To both he imparts uniform circular motion, the outer being called the circle of the Same, revolving to the right by way of the side, while the inner, called the movement of the Different, revolves to the left by way of the diagonal. He gives predominance (κρατός) to the uniform movement of the Same, which is left single and intact, but splits that of the Different into seven unequal circles according to the geometric progression through two and three. These are to move in opposite directions, three at uniform speed, four at different speeds from each other and from the three, but according to definite ratio (ἐν λόγῳ). The body of the cosmos is then fitted into this system of circles in the cosmic soul, center to center, and thus given life and intelligence (36D, E).

The circles of the cosmic soul are the source of all regular, uniform motion in the cosmos, and the seven planets, living gods set in the circuits of the Different, mark this regularity, constituting Time (38C, D). But the circles in the soul are also the mechanism of intelligence, and it is in this that we are chiefly interested.

> Seeing, then, that soul had been blended (συγκραθεῖσα) of Sameness, Difference, and Existence, these three portions, and had been in due proportion (ἀνὰ λόγον) divided and bound together, and moreover revolves upon herself, whenever she is in contact with anything that has dispersed existence or with anything whose existence is indivisible, she is set in motion all through herself and tells in what respect precisely and how, and in what sense, and when, it comes about that something is qualified as either the same or different with respect to any given thing, whatever it may be, with which it is the same or from which it differs, either in the sphere of things that become or with regard to things that are always changeless (37A, B).

Proclus gives the key to the theory of knowledge implied here, and to Plato's choice of ingredients of the soul, when he remarks: "Since the soul consists of three parts, Existence, Sameness, and Difference, in a form intermediate between the indivisible things and the divisible, by means of these she knows both orders of things; ... for all knowing is accomplished by means of likeness between the knower and the known."[22] The cosmic soul, being

[22] II, 198.2ff. and quoted by Cornford, p. 94. See Aristotle, *De Anima*

a μέσον between the eternal, indivisible reality of the Forms and the changing, divisible existence of bodies, is capable of responding to both upon contact, and in knowing both unites them in herself. Plato sketches the functions of the soul circles in producing correct distinction, affirmation, and negation regarding the objects of knowledge and opinion as follows:

Now whenever discourse (λόγος) that is alike true, whether it takes place concerning that which is different or that which is the same, being carried on without speech or sound within the thing that is self-moved, is about that which is sensible (αἰσθητόν), and the circle of the Different, moving aright, carries its message throughout all its soul — then there arise judgments and beliefs (δόξαι καὶ πίστεις) that are sure and true. But whenever discourse is concerned with the rational, and the circle of the Same, running smoothly, declares it, the result must be rational understanding and knowledge (νοῦς ἐπιστήμη τε). And if anyone calls that in which this pair come to exist by any name but "soul", his words will be anything rather than the truth (37B, C).

The cosmic soul, being "the best of things brought into being" (37A), functions in perfect equilibrium, so that the circles of the Same and Different constantly produce within it true opinion and knowledge, imparting to the cosmos "ceaseless and intelligent life for all time" (36E). These ordered functions of the cosmic soul can be traced in the movements of the fixed stars and the planets, living divinities set in the circles of the Same and the Different (38C, D; 40A, B). This is important for man since, as we shall see, his rational soul is but a miniature copy of the cosmic soul, so that by contemplation of the heavenly bodies he may restore the ordered functions of his own soul circles, producing in himself true opinion and knowledge and fulfilling his destiny in the imitation of the gods (90C, D).

Granted, then, that Plato's account is clothed in myth and symbol, still the picture of the cosmic soul as we have it is that of a psychological organism, compounded of immaterial ingredients, in themselves incompatible but bound together by proportions (μεσότητες) and functioning in equilibrium. Its composition and

---

404b16ff. and Taylor's long account summarizing the interpretations of this passage.

structure are conceived as closely parallel to those of the cosmic body which it animates. And the cosmos, body and soul, being the best imitation of the ideal Living Being, is the prototype and model of the human soul and body.

C. THE HUMAN SOUL AND BODY

Since for Plato all visible living beings are copies of the one intelligible Living Being they necessarily exhibit the same general structure and characteristics as the most perfect copy, the cosmos; and the higher each is in the scale of living beings the closer it approximates the body and soul of the cosmos. Thus the stars and planets are formed by the Demiurge as divinities, living beings intelligent and wholly immortal, "made for the most part of fire ... bright and fair to see; and after the likeness of the universe he gave them well-rounded shape" (40A) in keeping with their motions, which, like those of the universe, are uniform and circular. The symmetry and perfection of the universe demand, moreover, that there be living beings to dwell in the other primary bodies as the gods do in fire — birds for the air, fish for the waters, and animals to walk the earth (41C). And among these the highest is man (41E-42A).

1. *The Rational Soul and Its Destiny*

The Demiurge cannot himself form the whole of these creatures, for all he makes must of necessity be immortal and divine. Hence he undertakes only the immortal element, leaving the mortal part to the gods whom he has made (41C). This immortal element is the rational soul in man. The Demiurge blends it just as he did the cosmic soul.

> ... he turned once more to the same mixing bowl wherein he had mixed and blended the soul of the universe, and poured into it what was left of the former ingredients, blending them this time in somewhat the same way, only no longer so pure as before, but second or third in degree of purity (41D).

Thus the rational soul in man is a less pure blend of intermediate Existence, Sameness, and Difference. That it is bound together by the same proportions as the cosmic soul, and given the same functional structure, becomes clear in later passages which refer to "the revolution of the Same and uniform" (42C), "the revolution of the Different", the "intervals of the double and triple", and the "connecting means (μεσότητες) of the ratios 3/2, 4/3, and 9/8" (43D) in the soul of men. These passages make clear also that the revolutions of the Same and Different have identical functions in the human and the cosmic soul. For 1) the revolution of the Same is to bring order to the unruly components and motions of man's body (42C), and 2) it is through these revolutions that man's intelligence distinguishes the same and the different (43D).

Each rational soul, then, comes from the hand of the Demiurge, a miniature of the cosmic soul, properly proportioned and functioning in equilibrium. And the highest task of every man on earth is to master his body and rectify the circuits of his soul, disturbed by its mortality (41E ff.). For the mortal body in which the rational soul must be confined is by its nature a complex of changing elements and "passions" which tend to dislocate and hamper the revolutions of the Same and the Different.

Whensoever, therefore, they should of necessity have been implanted in bodies, and of their bodies some part should always be coming in and some part passing out, there must needs be innate in them, first, sensation (αἴσθησις), the same for all, arising from violent impressions; second, desire blended with pleasure and pain, and besides these fear and anger and all the feelings that accompany these and all that are of a contrary nature; and if they should master these passions, they would live in righteousness (δίκη); if they were mastered by them, in unrighteousness (ἀδικία; 42A, B).

Those who lead good lives on earth return to the star-divinity to which each soul was sent upon creation (41D, E), while one who continues in ἀδικία is reincarnated as a woman, or subsequently as that form of beast which typifies his peculiar vice. Nor is he released from the cycle of reincarnation until, "letting the revolution of the Same and uniform within himself draw into its train all that turmoil of fire and water and air and earth that had later grown about it, he should control its unregulated (ἄλογον) turbulence by rule (λόγῳ) and return once more to the form of his first and best condition" (42C, D).[23] The "first and best condition" (ἕξις) of man's rational soul, then, is that characterized by λόγος — rule, guiding principle, measure — provided by the "revolution of the Same and the uniform" functioning smoothly within him and regulating his bodily components. This condition is δίκη, moral excellence in man.

Thus when the rational soul is confined within a body its condition becomes closely linked with that of the bodily components. The degree of their mutual dependence, and the magnitude of man's task in subjecting his body to the control of reason, is indicated in the description of the rational soul newly implanted in its body by the created gods who, "having received the immortal principle of a mortal creature, imitating their own maker ... borrowed from the cosmos portions of fire and earth, water and air ... and cemented together what they took ... making each body a unity of all the portions. And they confined the circuits of the immortal soul within the flowing and ebbing tide of the body" (42E-43A). The effect of confining the rational soul in a changing body is, Plato says (43A ff.), like plunging its balanced circuits into the strong current of a river, where they neither control nor are controlled but move and are moved by violence (βίᾳ). The result is that the new creature, the compound of body and soul, moves at random in all directions, without rule and order (ἀτάκτως ... ἀλόγως). In this context Cornford remarks (p. 147): "The language hints at a certain analogy between the task of the human reason and the task of the Demiurge himself, who 'took over all

[23] Cornford (p. 144) renders ἄλογον and λόγῳ by "irrational" and "discourse of reason". However, cf. Taylor's remarks, p. 263, and Cornford's translation of ἀλόγως elsewhere, e.g., 43 E3 (p. 149): "unregulated"; 53A (p. 198): ἀλόγως καὶ ἀμέτρως — "without proportion or measure"; 43 B2 (p. 148): ἀτάκτως ... ἀλόγως — "without order or method". Obviously there is no single English equivalent for λόγος, but in these contexts it seems to mean an intelligible principle, rule, or measure — one applied, or capable of application, by an intelligence to produce order, proportion. Cf. e.g., the usage in Proclus II, 19.21-22. To be ἄλογος, therefore, is to be without intelligibility, intelligence, rule, measure, order, proportion, while to possess, or be characterized by, λόγος is to have all these qualities.

that was visible, not at rest but in discordant and unordered motion, and brought it from disorder into order' (30A)." Recall also the condition of the contents of the Receptacle "before the ordered whole consisting of them came to be" (53A), where they are said to be "without proportion or measure "(ἀλόγως καὶ ἀμέτρως). This is the opposite of the "first and best condition" of the soul, characterized by λόγος.

The source of the constant and disturbing changes in the body is twofold: 1) Within the body the tide of elements involved in the nutritive processes is ever "flooding and ebbing away", while 2) the "qualities" (παθήματα) of fire, earth, water, and air assail the organism from outside whenever it encounters them. Their impact — called "sensation" (αἴσθησις) — is carried along through the body and falls upon the soul, causing even greater turmoil. These combined disturbances disrupt the normal function of the newly-embodied soul completely.

And so at the moment we speak of, causing for the time being a strong and widespread commotion and joining with that perpetually streaming current in stirring and violently shaking the circuits of the soul, they completely hampered the revolution of the Same by flowing counter to it and stopped it from going on its way and governing; and they dislocated the revolution of the Different. Accordingly, the intervals of the double and the triple, three of each sort, and the connecting means of the ratios, 3/2 and 4/3 and 9/8, since they could not be completely dissolved save by him who bound them together, were twisted by them in all manner of ways, and all possible infractions and deformations of the circles were caused; so that they barely held together, and though they moved, their motion was unregulated (ἀλόγως), now reversed, now side-long, now inverted (43C-E).

This dislocation of the soul-circles through disturbances arising from nutrition and sensation breaks down and reverses the functions of the rational soul (43E-44A). Everything appears the opposite, as to a man standing on his head, and it pronounces things the same and different contrary to the truth, thus appearing mistaken and foolish. Then, since there is no soul-circuit functioning as ruler and guide, certain sensations, borne in from the outside, sweep along with them the whole composite, body and soul, and the soul is itself mastered when it should be master. So it is that the soul of an infant is foolish (ἄνους).

This confused condition of body and soul can, and should, be relieved in time. As man matures the processes of nutrition and growth die down and a balance is struck between waste and repair. When the bodily turbulence subsides the circuits of the soul tend to readjust to their normal orbits and functions. If the process is aided by proper regimen and education, the individual may become truly rational and virtuous, returning to the condition of δίκη which his soul possessed when it came from the hand of the Demiurge.

But when the current of growth and nutriment flows in less strongly, and the revolutions, taking advantage of the calm, once more go their own way and become yet more settled as time goes on, thenceforward the revolutions are corrected to the form that belongs to the several circles in their natural motion; and giving their right names to what is different and to what is the same, they set their possessor in the way to become rational (ἔμφρονα). And now if some right nurture (ὀρθὴ τροφή) lends help towards education, he becomes entirely whole and healthy (ὁλόκληρος ὑγιής τε), having escaped the worst of maladies (τὴν μεγίστην ... νόσον); whereas if he be neglectful, he journeys through a life halt and maimed (χωλὴν ... ζωήν) and comes back to Hades imperfect and without intelligence (ἀτελὴς καὶ ἀνόητος; 44B, C).[24]

Noteworthy in this passage is, first, the extreme dependence which the *rational* soul is thought to have upon the purely physical elements involved in bodily nutrition and growth, so that it cannot function properly and assert its control until the physical turbulence of childhood and youth dies down. Nor is freedom from

[24] This is a slight departure from Cornford (p. 150), who translates ὁλόκληρος ὑγιής τε as "whole and unblemished" and ἀτελής as "uninitiate", and explains that here "Plato uses terms borrowed from Mystery ritual" (note 2). This seems to me to obscure the primarily *physiological* meaning of the terms. Taylor's explanation (pp. 273-274) keeps the proper perspective, and his whole treatment of this passage is most helpful. That Plato intended the terms primarily in their physiological sense seems more consistent with later references to "diseases of the soul" (e.g., 86B, C). This is the interpretation of Proclus, who explains ἀτελής as ὥσπερ ὁ λελωβημένους ἔχων τοὺς πόδας (III, 352.16). Plato may very well have in mind the use of some of these terms in the mystery ritual, but is probably giving them his own interpretation, as he does at *Phaedo* 69C, D.

physical disturbances sufficient for the recovery of the soul, but its harmonious function must be positively promoted by ὀρθὴ τροφή — proper *regimen* — as well as education (παίδευσις). Now since for Plato the condition of the rational soul determines not only its intellectual efficiency but its moral status and destiny in the next life as well, it becomes clear how intimately the physical, intellectual, and moral aspects of man are linked in his thought.

Secondly, the closeness of this interdependence is emphasized by the *physiological* terminology used to describe the intellectual and moral conditions of man. For the turbulent condition which renders the soul ἄνους and causes a man to live in ἀδικία (cf. 42B, C) is called ἡ μεγίστη νόσος. Man recovers from this if, through the help of proper τροφή and παίδευσις, he brings the composite of body and soul to the state in which he can "control its unregulated turbulence by rule (λόγῳ) and return once more to the form of his first and best condition" (42C, D), i.e., the state of δίκη (42B), in which he is said to be "completely *whole* and *healthy*" (ὁλόκληρος ὑγιής τε παντελῶς). If he fails in this, his life is *maimed* (ζωὴ χωλή) and he returns to Hades imperfect (ἀτελής — incomplete, i.e., not ὁλόκληρος) and without intelligence (ἀνόητος).

Thus the rational soul in man, like the soul of the cosmos, is conceived as a mixture of incompatible immaterial ingredients linked by proportion and functioning in equilibrium. The circuits of the Same and the Different, when operating properly, not only produce knowledge and true opinion, but bring order and proportion to the rest of the soul and to the physical elements of the body in which they are confined. On the other hand, the interdependence of rational soul and body is so close that the bodily elements, when disturbed and disordered, interfere with the harmonious operation of the soul. This condition is called ἀδικία and ἡ μεγίστη νόσος; its opposite, the "first and best condition" of the soul, is δίκη and makes a man intellectually and morally ὁλόκληρος ὑγιής τε.

2. *The Physical Organ of the Soul*

For Plato the entire body is, in a broad but very real sense, the physical organ of the rational soul. Its ultimate purpose is to serve as a mortal container and vehicle for the immortal soul, and this τέλος determines its general structure, its individual organs and members, and the conditions of its health and disease. To understand fully what these last two terms imply for Plato, then, we must understand his concept of the *total* human organism.

We have already seen that the created divinities, at the command of the Demiurge, received the immortal element from him and, "imitating their own maker, they borrowed from the world portions of fire and earth, water and air", made "each body a unity of all the portions", and confined the circuits of the immortal soul within it (43A). Since the rational soul is a small copy of the cosmic soul whose uniform circular motion is contained within the perfect sphere of the cosmic body, the gods, "copying the round shape of the universe ... confined the two divine revolutions in a spherical body — the *head*, as we now call it — which is the divinest part of us and lord over all the rest" (44D). Thus the human rational soul embodied in the head imitates as closely as possible the soul and body of the cosmos.

However, since the human head, unlike the cosmic body, is not unique but exists within the cosmos and must cope with environment, it needs the help of other organs for local motion, sensation, respiration, nutrition, etc. This determines the structure and functions of the rest of the body.

To this the gods gave the whole body, when they had assembled it, for its service, perceiving that it possessed all the motions that were to be. Accordingly, that the head might not roll upon the ground with its heights and hollows of all sorts, and have no means to surmount the one or to climb out of the other, they gave it the body as a vehicle for ease of travel; that is why the body is elongated and grew four limbs that can be stretched out or bent, the god contriving thus for its travelling. Clinging and supporting itself with these limbs, it is able to make its way through every region, carrying at the top of us the habitation of the most divine and sacred part. Thus and for these reasons legs and arms grow upon us all (44D-45A).

Likewise the senses of sight and hearing, and the power of speech, with their particular organs and location, are ultimately explained by their relation to the rational soul as means to help it function properly when confined in the body. These are fixed in the face on the front of the head as "organs for all the forethought (προνοία) of the soul" (45A, B). The true purpose (αἴτια) of vision, lying beyond the mere mechanics (συναίτιαι) of its operation, is that "we might observe the circuits of intelligence in the heaven" to help the troubled revolutions of our own thought, which is akin to them, i.e., by studying the motions of the heavenly bodies to "reproduce the perfectly unerring revolutions of the god and reduce to settled order the wandering motions in ourselves" (47B, C). Sound and hearing are given for the same purpose, since not only is *speech* of the greatest help in learning (cf. 75D, E), but *harmony* and *rhythm*, motions also "akin to the revolutions of the soul within us", are intended to help "against the inward discord that has come into the revolutions of the soul, to bring it into order and consonance with itself" (47C, D). Thus the ultimate reason for these powers and their organs — eyes, ears, tongue and mouth — lies in the nature and condition of the rational soul.

Not only the head, then, but the body and its members are necessary for the rational soul in its mortal condition; and these, if they are to have the power of self-movement, must also contain soul. Hence the created gods, in forming the mortal body, animate it with soul, but of a mortal kind.

They, imitating him, when they had taken over an immortal principle of soul, went on to fashion for it a mortal body englobing it round about. For a vehicle they gave it the body as a whole, and therein they built on another form of soul, the mortal, having in itself dread and necessary affections. ... (69C).

For the moment we may postpone the discussion of these "dread and necessary affections" of the mortal part of the soul, and look at that physical substance in which, beyond all others, both immortal and mortal soul are thought to be rooted.

The seat of the soul, its organ in the restricted sense, is the substance located within the cranium and the bony structure, particularly the spine, called "marrow" (μυελός). Marrow is considered the vital center and core of the physical organism, "for the bonds of life, so long as the soul is bound up with the body, were made fast in it as the roots of the mortal creature; while the marrow itself is formed of other things" (73B). The constituents of the marrow are, as we should expect, the most excellent forms of earth, air, fire, and water, compounded in due proportion.

The god set apart from their several kinds those triangles which, being unwarped and smooth, were originally able to produce fire, water, air, and earth of the most exact form. Mixing these in due proportion to one another (ἀλλήλοις σύμμετρα), he made out of them the marrow, contriving thus a mixture of seeds of every sort for every mortal kind (73B, C).

The mixture is then animated with both the immortal and mortal parts of soul, being shaped accordingly:

Next he implanted and made fast therein the several kinds of souls; also from the first, in his original distribution, he divided the marrow into shapes corresponding in number and fashion to those which the several kinds were destined to wear. And he molded into spherical shape the ploughland, as it were, that was to contain the divine seed; and this part of the marrow he named "brain" (ἐγκέφαλος), signifying that, when each living creature was completed, the vessel containing this should be the head. That part, on the other hand, which was to retain the remaining, mortal kind of soul he divided into shapes at once round and elongated, naming them all "marrow". From these, as if from anchors, he put forth bonds to fasten all the soul; and now began to fashion our whole body round this thing, first framing round the whole of it a solid shield of bone (73C, D).

Thus the marrow, a mixture of the finest earth, air, fire, and water blended proportionately (ἀλλήλοις σύμμετρα), constitutes the vital center of the organism, containing the bonds between body and soul. It is the marrow also that forms the seed by which new life is produced (cf. 91A, B). And the whole structure of bone and flesh is designed to preserve and protect this central substance.

Bone is made of pure smooth earth, soaked in marrow, and plunged into fire and water twice to render it insoluble by either. From this is formed a sphere to enclose the brain and the vertebrae to enclose the marrow of the neck and back (73E-74A). Flesh is designed to serve a double purpose: 1) to protect the brittle bones in falls, and 2) to temper the extremes of heat and cold, which can cause bone to disintegrate (74B). In explaining how this last is accomplished, Plato introduces the medical theory of proportionate response of the organism to environment.

Flesh, it is said (74C), contains within itself a warm moisture (θερμὴ νοτίς) so that in summer by sweating it may cool the body through its own moisture spread over the outside, while in winter through its innate fire or heat it may ward off *proportionately* (μετρίως) the frost surrounding it. For this reason he who made man formed the flesh soft and moist (ἔγχυμος), mixing in proper adjustment (συμμείξας καὶ συναρμόσας) water, fire, and earth, along with a ferment of acid and saline. Thus flesh is formed to respond proportionately to external heat and cold and so to maintain the proper balance of these qualities *within* the organism.

Finally, to bind the bones together and yet allow movement, sinew is formed as a mean between bone and flesh (74D). For the god "blended (συνεκεράσατο) the nature of sinew from a mixture (ἐξ ... κράσεως) of bone and unfermented flesh, a single nature from the two of them, being midway between them in quality" (μέση δυνάμει).[25] Sinew, the product of this κρᾶσις, is a φύσις μέση δυνάμει in that its δύναμις is to be more tense and glutinous than flesh, more soft and moist than bone. Plato here speaks the language we have found in the medical writers, and offers an example of a μέσον δυνάμεων in the physical order.

With the compounding of sinews the materials for the gross structure of the human body are complete, so "with these the god enveloped the bones and marrow, binding the bones to one another with sinews, and he then buried them all under a covering of flesh" (74D). Thus the bonds of the soul, both mortal and immortal, are made fast in the marrow, which, provided with its protective covering of bone and flesh and sinew, constitutes the central structure of the human body and the organ of the rational soul.

[25] The sentence is here literally translated, departing from Cornford, p. 297.

3. *The Mortal Soul and Its Organs*

We have been told that, when the created gods formed the body of man as a vehicle for his rational soul, "therein they built on another form of soul, the mortal, having in itself dread and necessary affections" (69C). In this passage Plato goes on to identify these δεινὰ καὶ ἀναγκαῖα παθήματα, representing them as ingredients from which the mortal soul is blended:

... first, *pleasure* (ἡδονή), the strongest lure of evil; next, *pains* (λῦπαι) that take flight from good; *temerity* (θάρρος) moreover and *fear* (φόβος), a pair of unwise counsellors; *anger* hard to manage (θυμὸς δυσπαραμύθητος) and *expectancy* too easily led astray (ἔλπις εὐπαράγωγος). These they blended (συγκερασάμενοι) with *irrational perception* (αἴσθησις ἄλογος) and *desire* that shrinks from no venture (ἐπιχειρητὴς παντὸς ἔρως), and so of necessity compounded (συνέθεσαν) the mortal kind (69D).[26]

These παθήματα are said to be ἀναγκαῖα because they are regarded as the necessary concomitants of mortal existence. That they are δεινά seems to be explained by their potency for evil, suggested by the epithets "lure of evil", "fleeing from good", "unwise counsellors", etc., for all are represented as undesirable extremes. Further, pleasure and pain, temerity and fear, anger and expectancy are conceived as *pairs of opposites*. This suggests another passage we have seen (42A), where Plato, describing the necessary consequences of rational souls being confined in changing bodies, says that "there must needs be innate in them, first, *sensation* (αἴσθησις) the same for all, arising from violent impressions (ἐκ βιαίων παθημάτων); second, *desire* blended with *pleasure* and *pain* (ἡδονῇ καὶ λύπῃ μεμειγμένον ἔρωτα), and besides these *fear* (φόβος) and *anger* (θυμός) and all the feelings that accompany these and *all that are of a contrary nature*" (ὁπόσα ἐναντίως πέφυκε διεστηκότα).

These two statements, condensing much of Plato's psychology of sensation and emotion, can only be interpreted adequately in the light of other dialogues such as the *Philebus* and *Theaetetus*.

[26] Cornford's translation, p. 281, has been altered slightly here in the direction of greater literalness.

For the present discussion it must suffice to sketch briefly the general theory of sensation, pleasure, and pain as they appear in the *Timaeus*, and to indicate their connection with desire and the emotions. For all these are regarded as the proper activities of the mortal soul, and of the parts of the body animated by this soul.

The theory of sensation in the *Timaeus* follows from the basic physiological concept of the body as a compound of primary particles in equilibrium. Their union is dynamic, involving constant movement and transformation according to the principles we have seen in connection with the cosmic body. The mobility of the particles in most of the body, particularly in the sense organs and vascular system, allows a disturbance created by the sensible object at the periphery to transmit its motion through the body to the soul, producing sensation. This can take place, however, only if the disturbance is strong enough; hence αἴσθησις ἐκ βιαίων παθημάτων.[27] We have already seen the effect of this upon newly embodied rational souls (43B, C) where "a yet greater tumult was caused by the qualities of the things that assailed them" (τὰ τῶν προσπιπτόντων παθήματα), i.e., when their bodies encountered "alien fire from outside" as well as earth, water, and air, and "the motions caused by all these passed through the body to the soul and assailed it" (προσπίπτοιεν). And Plato adds: "For this reason these motions were later called by the name they still bear — 'sensations'." The process is more clearly described in a later passage (64A-C):

Now in seeking the explanation of any affection (παθήματος), whether perceptible or imperceptible (αἰσθητοῦ καὶ ἀναισθήτου), we must begin by recalling the distinction drawn earlier between what is mobile in structure and what is immobile. ... When something that is naturally mobile is invaded by even a slight affection, it spreads it all round, one

[27] This is Proclus' interpretation of the phrase (III, 286.7 ff.), which seems to differ somewhat from that of Taylor (p. 286), who associates it with "sensation attended by violent organic shock". However, Plato appears to regard all sensation as a kind of shock or disturbance passed on through the body to the soul, for if the impulse (πάθημα) is not forceful enough (βίαιον) the soul is not affected and no sensation is produced. Hence the soul is thought to be disturbed (especially when newly embodied, as in 43 B, C) by *any* sensation, not merely by those arising from what we ordinarily consider violent stimuli, burns, bruises, etc. Cf. *Philebus* 33 D; Proclus III, 329-332.

particle passing on the same effect to another, until they reach the consciousness and report the quality (δύναμιν) of the agent. The immobile, on the other hand, being too stable to spread the motion round, merely suffers the affection without setting any of its neighbors in motion; accordingly, since the particles do not pass it on to one another, the original affection remains in them incapable of transmission to the living creature as a whole and leaves the subject without sensation.

As we have seen, particles of earth are most stable and difficult to move, so that structures like bone and hair, being composed mainly of earth, are without sensation, while sight and hearing are most sensitive because the δύναμις of fire and air, the most mobile particles, is greatest in them (64C).

Thus fire issuing from the eye forms with the sunlight a single homogeneous medium which transmits the impulses from whatever it falls upon through the body to the soul, causing the sensation of *sight* (45B-D). Color is explained (67C ff.) as a flame streaming off each object, consisting of particles proportioned to the fiery medium of sight (ὄψει σύμμετρα μόρια) so as to cause sensation. Some of these are larger than the particles of the visual medium, others equal, others smaller, and when they encounter the medium (ἐμπίπτοντα εἰς τὴν ὄψιν), the equal are unperceived or transparent, the larger *compress* it (συγκρίνοντα), causing a sensation of black, and the smaller *expand* it (διακρίνοντα), causing that of white. Thus the particles forming the organ of sight in their normal state constitute a sort of balance or equilibrium which is upset by the visible object in the direction of expansion or contraction and produces in the soul the corresponding sensation of white or black. Other colors are explained analogously by the intrusion of different varieties and proportions of fire mingled with the moisture of the eyeball.

Plato is careful to indicate an analogous process in the senses of touch and taste (67D, E). There, too, motion is imparted to the skin and the tongue by the impact of particles from outside, and variation of sensible quality (cold — hot; astringent — pungent) in both is accounted for by σύγκρισις and διάκρισις, i.e., by the action of larger particles which compress, or smaller particles which enter and expand, the sense organ in its normal condition (61D-62B; 65C-66C). In fact, Plato generalizes to the extent of declaring that most παθήματα come about διὰ συγκρίσεών τινων καὶ διακρίσεων (65C). The organs of these sensations, therefore, are considered to be in a middle or *mean* condition, poised to respond proportionately to external stimuli either by expansion or contraction.

Some sensations, however, do not seem to operate according to this principle. The sensation of hard and soft, for example, is determined by whether the object yields to the flesh or the flesh to the object (62B, C), though here, too, the normal condition of the flesh constitutes a sort of mean between them. The important sense of hearing seems to function as follows (67A-C): Sound (φωνή) is conceived as a blow (πληγή) transmitted through the ears by the air of both brain and blood to the soul, and the motion caused by it, beginning from the head and ending in the region of the liver, is (the sensation) hearing.[28] Pitch, smoothness, and volume are determined by the rapidity, regularity, and strength of this motion. Even here the basic concept of sensation is the same, the disturbance of equilibrium by the impact of external particles which evokes a proportionate response in the organism.

Plato relates his theory of pleasure and pain specifically to his physiology and psychology of sense perception. His explanation of the general nature of sensation (64A, B) is introduced as fol-

[28] Plato's psychology of hearing presents a number of thorny problems which cannot be treated here. Cf. Taylor, pp. 476-479; Cornford, 275-276; 320-326. The interpretation given here seems most consistent with Aetius' *Placita* IV, 16.4 (*Doxogr. Graec.*, p. 406) and Theophrastus *De Sensu* 6 (*ibid.*, p. 500) and 85 (*ibid.*, p. 525), both cited by Taylor, pp. 476-477. Analogy with the sense of sight in which daylight and fire from the eye form a continuous medium of sensation leads one to suspect that Plato conceived a similar medium for hearing formed by the air outside the body and that of the ear.

The present interpretation also suggests that for Plato the seat of hearing, and all sensation, is not the head, as Taylor and Cornford assume, but the central region of the body (and soul) between head and liver. This is supported by the passage (65 C) in which the sensoria of taste are traced to the *heart*, and is consistent with the fact that sensation, and related pain and pleasure, are regarded as proper to the *mortal* soul, situated in the body (61 C; 65A; 69 D), whereas the brain is the seat of the immortal, rational soul whose proper activities are νοῦς and ἐπιστήμη. It may be significant, too, that while the blood and blood vessels are thought to have both nutritive and sensory functions (77 C, E), the heart is discussed only in connection with, and as the organ of, sensory functions.

lows: "Concerning the affections (παθήματα) common to the whole body, the most important point left to be considered is the cause of the pleasant and painful ... and what affections, attaining sensation through the particles of the body, also have pain and pleasures following along in them" (64A).[29] Then, after describing the nature of sensation as a πάθος passed on through the body from one moving particle to another until it reaches consciousness and reports the δύναμις of the agent, he links this to pleasure and pain in the following principles (64C-65B): 1) An affection (πάθος) is painful which is unnatural (παρὰ φύσιν) and violent in that it is produced in us suddenly (γιγνόμενον ἀθρόον παρ' ἡμῖν), while that affection which returns again suddenly to a natural condition (εἰς φύσιν) is pleasant. 2) On the other hand, one that occurs gently and little by little causes no sensation at all; it is only those with the opposite qualities that produce sensation, pleasure, and pain. 3) Again, that which comes about easily and readily (μετ' εὐπετείας) produces the keenest sensation, but brings neither pleasure or pain. The visual medium, for example, can be cut or burnt without pain and return to its normal state without pleasure, yet gives the clearest sensation insofar as it is affected by what it encounters; for there is no violence at all in its expanding and compressing. On the other hand, bodies composed of larger parts, which yield only with difficulty to what acts upon them, experience pleasure and pain in passing on the motions to the whole — pain in being changed from their normal condition and pleasure in returning to it. 4) Finally, a body which suffers a gradual privation (κένωσις), and then suddenly has the privation filled or satisfied in large measure, is not sensibly aware of the privation but does become aware of its fulfillment (πλήρωσις) and so experiences no pain but the greatest pleasure. Conversely, those that suffer sudden changes from the normal and return only gradually and with difficulty experience the opposite, as in the case of burns and cuts. Other illustrations of these principles appear elsewhere in the dialogue (cf. 60A; 66C; 67A).

With regard to sensation and its pleasure and pain, then, there

[29] This is a departure from Cornford's translation, pp. 266-267, in the direction of greater literalness.

are three types or degrees of παθήματα, three ways that the organism is affected by the environment: 1) If the stimulus is weak, or falls upon an unresponsive part of the body, its motion is not passed on to the soul and no sensation occurs. 2) If the stimulus is stronger and the bodily organ responds, the motion is passed on to the soul and reports the δύναμις of the agent in sensation. 3) If the stimulus is strong enough to produce sensation, and at the same time occurs so suddenly, or with such difficulty, that it is not σύμμετρον with the condition of the organ but disturbs its normal state unduly, then the experience is painful. If, on the other hand, it produces a rapid return of the organ to a normal state (as honey does with the pores of the mouth; cf. 60A) or brings the quick fulfillment of a privation, the experience is pleasant. Basic in all this is the concept of *proportionate* response of an organism in equilibrium to the impact of environment.

It must be noted, however, that the theory outlined here is meant to explain the pleasurable and painful aspects of *sensation*, and should not be thought to represent Plato's complete explanation of pleasure and pain on all levels.[30] For a much broader view can be found expressed in connection with his discussion of death (81D, E) where, surprisingly enough, he remarks that a normal death, coming from the natural disintegration of the body after its proper span of years, may even be accompanied by a certain amount of pleasure:

And at last, when the conjoined bonds of the triangles in the marrow no longer hold out under the stress, but part asunder, they let go in their turn the bonds of the soul; and she, when thus set free in the course of nature (κατὰ φύσιν), finds pleasure in taking wing to fly away. For *whereas all that is against nature is painful, what takes place in the natural way is pleasant* (πᾶν γὰρ τὸ μὲν παρὰ φύσιν ἀλγεινόν, τὸ δ' ᾗ πέφυκεν γιγνόμενον ἡδύ).

This last is a generalization which goes beyond the explanation

[30] This, it seems, was the mistake made by the κομψοί of the *Philebus* (53 C-55 C), who conceived *all* pleasure as a process of return from an abnormal state, and so rejected it as an element of the good life, since such pleasure implies pain and so is always "mixed". Taylor, in his long account of Plato's theory of pleasure and pain, tends to overemphasize the aspect of κένωσις-ἀναπλήρωσις.

of pain and pleasure through κένωσις and ἀναπλήρωσις (since these are not involved in the process of dying) and associates pain and pleasure with *any* (πᾶν) process or change (τό γιγνόμενον — "becoming") that is παρὰ φύσιν or κατὰ φύσιν, even though it be the process of dissolution. "So death itself, on this principle, is painful and contrary to nature when it results from disease or wounds, but when it comes to close the natural course of old age, it is, of all deaths, the least distressing and is accompanied rather by pleasure than by pain." This principle formulates the general assumption of the medical writers.

Further discussion of the "dread and necessary affections" proper to the mortal soul is possible only in the light of other dialogues, since the *Timaeus* offers very little more than what has been sketched here. In the *Philebus*, for example, we can pick up the discussion of pleasure and pain, finding the physiology and sense psychology of the *Timaeus* assumed there, while the theory of pleasure and pain is developed further. There we learn that not only do pleasure and pain accompany sense perception, but that *any* distortion of the equilibrium of the body's constituents brings pain, and any restoration to normal, pleasure (31D-32B), though this principle may be implicit in the account of disease in the *Timaeus*. Further, pleasures and pains occurring in the body alone give rise to another group of pleasures and pains on the level of the soul alone, through anticipation and memory of bodily pleasures and pains (32B, C; 33C). The mixture of pleasure and pain on this level produces other emotions (47D ff.), each of which, like hot and cold, constitutes an ἄπειρον (52C, D), being capable of indefinite progression in either direction. Such are the emotions mentioned in the *Timaeus* — fear and temerity, anger and hope, "and all the feelings that accompany these and all that are of a contrary nature".

It is in the *Philebus* too that the phrase ἡδονῇ καὶ λύπῃ μεμειγμένον ἔρωτα finds its explanation, for there desire is analyzed as a mixture of the painful consciousness of a present want with the pleasant anticipation of its fulfillment (34D ff.). The *Timaeus* makes it clear that man has appetites (ἐπιθυμίαι) on various levels (88B) directed toward food and drink for the body and wisdom (φρόνησις) for the rational soul, though generally ἐπιθυμία is associated with bodily cravings. But a full appreciation of Plato's concept of ἔρως can be formed only with the help of the *Symposium*.

These, then, are the παθήματα from which the created gods blended the mortal soul — sensation and desire, pleasure and pain, and the whole range of conflicting emotions that spring from them. It should be recalled that these are of prime importance to men in the fulfillment of their destiny (42B), for "if they should master (κρατήσοιεν) these, they would live in righteousness (δίκη); if they were mastered by them, in unrighteousness" (ἀδικία).

The created gods confine this mortal soul, with its "dread and necessary affections", in the body, forming the neck to separate it from the head, lest it defile the divine element in man and disturb it more than necessary (69D ff.). Moreover, the mortal soul itself consists of a nobler and baser part, and so the first is housed in the chest, closer to the head, and separated from the second by the diaphragm.

The nobler part, located in the chest, is the center of man's sensitive[31] and affective life and constitutes an intermediary or mean between the rational soul and the baser part of the mortal soul, whose function is nutrition. For though itself without the power of reason, the affective soul is responsive to the rule of reason (τοῦ λόγου ὑπήκοον) and can cooperate with it in controlling the lower appetites. Hence courage (ἀνδρεία), spirit (θυμός), the urge to conquer (φιλόνικον ὄν) are regarded as its specific characteristics, implanted that it may restrain by force the nutritive soul when appetite prompts it to disobey the rule and command of reason.

The chief organ of the affective soul is the heart, a knot of blood vessels whence the blood springs forth to flow through all the body. It should be noted that the vascular system is regarded as the means by which sensation is transmitted through the body (cf. 77E), and the heart as the center which mediates the commands of reason to the other members. It is established, as it were, "in the guard-room, in order that, when the spirit (θυμός) should boil with anger

[31] Cf. note 28 above.

at a message from reason that some act of wrong is taking place in the members, whether coming from outside or, it may be, from the desires within, then every sentient part of the body should quickly, through all the narrow channels, be made aware of the commands and threats and hearken with entire obedience, and so suffer the noblest part to be leader among them all" (70B).

The throbbing of the heart in rising anger or fear is thought to come by means of fire; and so, to cool its heat and cushion its pulsations the lungs are devised, soft, bloodless, spongy substances fed with breath and drink through the windpipe, "so that, when the spirit therein was at the height of passion, the heart might leap against a yielding substance and be cooled down, and so being in less distress be the better able to help the spirited element in the service of reason" (70D). Thus the lungs are to temper by the coolness of air and water the heat of fire in the heart, and so maintain the balance of these qualities within the body. This function of the heart as organ of the affective soul is the only function assigned it in the *Timaeus*.

Finally, in the lower part of the trunk, between diaphragm and navel, is situated the third part of soul, which functions in the nutritive processes (70D ff.). This is the same kind of soul that is found in plants (77B), and has no share in belief (δόξα), reasoning, or understanding, but only in pleasant and painful sensation along with the appetites for food and drink necessary to physical survival. It must be maintained, like a necessary but untamed beast, but is consigned to the position furthest from the head "with the intent that, always feeding at its stall and dwelling as far as possible from the seat of counsel, it might cause the least possible tumult and clamor and allow the highest part to take thought in peace for the common profit of each and all" (70E).

The organs housed in the lower part of the body — liver, spleen, stomach, and intestines — all serve some need of this part of the soul. The liver is the organ by which it is brought, to some extent, under the control of reason. For though it cannot understand the discourse of reason and would not heed it if it could, this kind of soul does respond to images and phantasms. Hence the liver is made smooth, bright, and firm, with a capacity for sweetness and bitterness, so that the influence (δύναμις) of the thoughts (διανοήματα) issuing from the intellect can be reflected in it and evoke a proportionate response in this part of the soul. Thus the influence of the intellect can frighten the nutritive soul by using the liver's bitter aspect to threaten it, suffusing bitterness through it, causing a bilious color and contracting it in wrinkles and roughness, shrinking the lobe and obstructing the vessels, to cause pain and nausea. Or again, a gentle thought will reflect its sweetness in that of the liver, setting it straight, smooth, and free, which makes the lower soul thrive in gentleness. Thus the liver is the organ which translates thought into pleasant oι painful sensation in this part of the soul. And the position and structure of the spleen is determined by its function in this process, which is to serve as a sort of napkin for the liver, absorbing impurities and keeping it bright and clean.

The stomach and intestines, finally, are meant to serve as a receptacle and reservoir of food (72E-73A). The gods foresaw that man's appetites would prompt him to eat more than the μέτριον καὶ ἀναγκαῖον and so to prevent disease and destruction, made the stomach and coiled the intestines below it so that food would not pass through too quickly and lead the body to call constantly for more. For that would "render all mankind incapable, through gluttony, of all cultivation and philosophy, deaf to the divinest part of our nature" (73A).

Such then are the forms and functions of the mortal kind of soul which the created gods "built on" (προσῳκοδόμουν) to the immortal to animate the body (69C). We have seen that, in Plato's view, the immortal element, confined in the head, would need neither the body nor the mortal soul if it were *unique* like the cosmos. Its need for both arises from the condition of its mortal existence as part of a larger whole in which the primary bodies in various forms surround the head, constituting its environment. This makes local motion, sensation, nutrition, and respiration both possible and necessary, functions which demand the addition of the body and the mortal soul. We have seen how Plato conceives sensation as the response of the organism to the impact of environment, by which the sensible qualities of external objects are transmitted to the soul. We must now indicate the importance of environment for the life of the body as a whole.

4. *Environment and the Nutritive Process*

To understand the full effect of environment upon the human body, we must recall that Plato conceives the cosmos and all it contains as a compound in which there is no void, so that the particles of the four primary bodies composing it are contiguous and constantly interacting according to the laws of motion and transformation (58A; 80C; and cf. pp. 86-88 *supra*). This results in perpetual heterogeneity and change.

Man's body is but one small compound within this compact universe and, being composed of the same elements, is subject to its laws of constant change. He "must spend his life surrounded by fire and air and be consequently dissolved and depleted by them and so waste away" (77A) unless his body can be replenished. "For the elements besetting us outside are always dissolving and distributing our substance, sending each kind of body on its way to join its fellows ... " (81A). If it is to survive, therefore, man's body must be renewed as quickly and constantly as it is depleted by his environment.

For this reason the gods formed plants to be man's food (76E), and designed his *respiratory* system so that the food contained within the stomach and intestines would be broken up by the regular oscillation of air and fire particles passing through the trunk, forming blood and carrying it through the vascular system to nourish the entire body, i.e., to supply the particles depleted by environment (77C-81B). There is no need to enter into the complications of the respiratory system, which is thought to involve not the lungs alone but the entire trunk. It is enough to note that the same principles are operative in the respiratory and nutritive processes within man's body — the microcosm — as account for the constant alterations of the macrocosm. Since compounds of smaller particles are impervious to larger, while those of the larger particles can be penetrated by the smaller, fire and air can pass through any composed of the others. Both of these, therefore, can pass through the trunk of the body. Again, on the principle that what is heated has a natural drift upward, while what is cooled tends to go down, the breathing process is explained as the exhalation of warm air which displaces the surrounding cool air, thrusting it around where it penetrates the trunk and is itself warmed so that the respiration cycle ensues. The fire particles within the trunk, following the alternate motion of the breath, pass through the stomach, come upon the food, break it down into its constituents to form blood, and force it through the outlets into the blood vessels and through them to all members. On the cosmic principle that like seeks like, the various particles which constitute the blood then seek and are assimilated to their several kinds within the body. At the same time respiration serves as a cooling system to temper the heat of the fire within through the cool air introduced from outside, keeping the balance between them in the body as it is in the universe. Thus the principles which act in the macrocosm to break down the microcosm also operate within the microcosm to sustain it and replenish its losses.

> The manner of this replenishment and wasting is like that movement of all things in the universe which carries each thing towards its own kind. For the elements besetting us outside are always dissolving and distributing our substance, sending each kind of body on its way to join its fellows; while on the other hand the substances in the blood, when they are broken up small within us and find themselves comprehended by the individual living creature, framed like a heaven to include them, are constrained to reproduce the movement of the universe. Thus each substance within us that is reduced to fragments replenishes at once the part that has just been depleted, by moving towards its own kind (81A, B).

The growth of childhood and the disintegration of old age are both explained through the ability of the bodily constituents to master and assimilate the constituents of its food (81B-D). In the perpetual process of depletion through environment and repletion through nourishment, when more flows in than out, the body increases, while when the process is reversed, it disintegrates. The particles of the body must break down its nourishment into *like* particles in order to assimilate them, for like must be joined to like to fill up what is depleted. When the body is new the triangles of its constituents are fresh, their links strong, the whole supple and close knit, so that it easily breaks up and masters (ἐπικρατεῖ) the triangles of food and drink, since these are older and weaker, and grows large ἐκ πολλῶν ὁμοίων. But when the roots of the body's triangles are loosened by many conflicts so that they cannot cut up food and drink into similar triangles but rather are themselves divided by those coming from outside, then the body is overcome and wastes away, and this is old age. Finally, when the bonds of the triangles in the marrow part under this stress, the soul slips her moorings and departs (81E). Thus the whole vital process is conceived as a struggle between the organism and its environment, and to survive in well-being the organism must counter each fresh assault of environment with a proportionate response from within, so that the element depleted by environment is restored by the organism and its internal equilibrium kept intact.

Finally, it must be noted that in outlining the general principles governing the impact of environment upon the body, Plato does not descend to details of climate and topography as do other medical writers. However, he does allude to the general theory that equable climate favors health, intelligence, and a well balanced character when, early in the *Timaeus*, the Egyptian priest is made to speak of the founding of Athens as follows:

> All this order and system the goddess had bestowed upon you earlier when she founded your society, choosing the place in which you were born because she saw that the well-tempered climate (τὴν εὐκρασίαν τῶν ὡρῶν) would bear a crop of men of high intelligence. Being a lover of war and wisdom, the goddess chose out the region that would bear men most closely resembling herself and there made her first settlement (24C, D).

This same theory is referred to a number of times throughout the dialogues, and in a way which shows that Plato accepts it as his own.[32] It is summarized in the *Symposium* by the physician Eryximachus, who discourses on the two loves that rule the uni-

[32] Cf. e.g., *Critias* 111E4; *Epinomis* 976A1; *Philebus* 26A6, B1.

verse, one Heavenly, the god of union and harmony, the other Popular, disordered and insolent:

Note how even the system of the yearly seasons is full of these two forces; how the qualities I mentioned just now, heat and cold, drought and moisture, when brought together by the orderly Love, and taking on a temperate harmony as they mingle, become bearers of ripe fertility and health to men and animals and plants, and are guilty of no wrong (οὐδὲν ἠδίκησεν). But when the wanton-spirited Love gains the ascendant in the seasons of the year, great destruction and wrong does he wreak. For at these junctures are wont to arise pestilences and many other varieties of disease in beasts and herbs; likewise hoarfrosts, hails, and mildews, which spring from mutual encroachments and disturbances in such love-connexions as are studied in relation to the motions of the stars and yearly seasons by what we term astronomy (188A, B).[33]

Thus the climate in which hot, cold, moist, and dry take on a *harmonious blend* (ἁρμονίαν καὶ κρᾶσιν λάβῃ) is considered the cause of fertility and health and characterized by δίκη, while the opposite, in which there is mutual "encroachment and disturbance" (πλεονεξία καὶ ἀκοσμία) among these qualities, i.e., ἀδικία, produces "pestilence and many other varieties of disease". The connection between medicine and astronomy is likewise pointed up, as we have seen in *Airs, Waters, Places.* Hence it is clear that Plato was well acquainted with the details of the theory as we have seen it in the medical writers, and so it is reasonable to assume that he is drawing upon this background when he formulates his general principles concerning the effect of environment upon the body. And if he considers a balanced and well-blended climate healthiest for man, it must be because he believes it best preserves a similar balance and blend within the human organism.

D. DISEASE AND HEALTH IN MAN

The *Timaeus* depicts, as we have seen, the various powers or qualities from which the cosmic body was formed as existing originally in the Receptacle without order or similarity, equilibrium or regular movement, proportion or measure (52D-53A); but then "the

[33] Translated by W. R. M. Lamb, Loeb Classical Library, *Plato* V (London, 1946), pp. 128-131.

god introduced into them all every kind of measure in every respect in which it was possible for each one to be proportioned and symmetrical (ἀνάλογα καὶ σύμμετρα) both with itself and with all the rest" (69B), making the cosmos "supreme in greatness and excellence, in beauty and perfection ... " (92C). Similarly, the microcosm, the composite of body and soul which is man, is thought to be in excellent condition when measure, proportion, symmetry, equilibrium, and regularity prevail in each part and in the composite as a whole. This ordered condition is health, and all that disrupts it, disease.

1. *Disease of the Body*

In stating the fundamental cause of disease in the body, Plato's manner seems to indicate that he is proposing nothing new, but simply recalling a theory quite familiar to his audience:

> The origin of diseases is no doubt evident to all. Since there are four kinds which compose the body, earth, fire, water, and air, disorders and diseases (στάσεις καὶ νόσους) arise from the unnatural prevalence or deficiency of these (ἡ παρὰ φύσιν πλεονεξία καὶ ἔνδεια); or from their migration from their own proper place to an alien one; or again, since there are several varieties of fire and the rest, from any bodily part's taking in an unsuitable variety, and from all other causes of this kind (82A).

This general statement of the fundamental causes of disease is obviously based on the same theory we have seen in the medical writings, that disease results from a disturbance of the balance of opposing elements within the body. This happens, according to the *Timaeus*, first, when one of the primary bodies grows excessive or deficient, or, secondly, when they are not properly distributed in the body, but one or other shifts to a strange location, setting up conflict according to the laws of transformation. Finally, disease is said to arise from the attempt of some part of the body to assimilate a like element, but in the wrong variety. For Plato postulates several grades of the same primary body (cf. 57C, D; 58C; 67C). There are, for example, various types of fire — the type that forms flame, another that forms daylight and the visual medium, a third found in embers, and the varieties that cause different colors (58C; 67C ff.). It seems that these cannot form a homogeneous mass unless they are of the same type, as in the case of daylight and the fire in the eye (45B, C). Hence, when the wrong type is introduced the result must be heterogeneity, change, and strife (57A, B, E; and cf. 58D), traceable ultimately to the *inequality* of the particles (58A).

It must be recalled, moreover, that the primary bodies are not to be conceived merely in terms of quantity, but that each has distinctive *qualities* or powers associated with its size and shape, so that balance and distribution of the primary bodies involves balance and distribution of these fundamental, and opposing, qualities: "For when any one of the kinds is formed or shifts its place contrary to nature, parts that were formerly cold are heated, the dry become moist, and so also with the light and the heavy, and they undergo changes of every kind" (82A, B). Thus, health demands that proper proportion and distribution, both quantitative and qualitative, be maintained amid the constant depletion and restoration of the elements in the body.

> The only way, as we hold, in which any part can be left unchanged and sound and healthy is that the same thing should be coming to it and departing from it with constant observance of uniformity and due proportion (ἀνὰ λόγον); any element that trespasses beyond these limits in its incoming or passing out will give rise to a great variety of alterations and to diseases and corruptions without number (82B).

Specific diseases are associated with the excess of each primary body according to its proper qualities. Thus (86A), excess (ὑπερβολή) of fire produces continuous fever; excess of air, quotidian; excess of water, tertian, since water is more sluggish than fire and air; and excess of earth, quartan, because earth is the most sluggish and its excess most difficult to purge away.

However, Plato considers the cause of disease not only in terms of the relationship of the primary bodies in the organism as a whole, but also from the point of view of the secondary structures — blood, flesh, sinew, bone, and marrow (82B ff.). These are composed of earth, air, fire, and water of various types and proportions, and most diseases arise in them from a disproportion or displacement of the elements as described above. The most serious, however, occur when the process of their γένεσις — their growth and nourishment — is reversed, so that the structures themselves disintegrate and the products of this disintegration back up into the circulatory system. Normally flesh and sinew are produced from blood, and in turn themselves produce a viscous, oily material that binds flesh to bones and nourishes bone. The purest of this substance, consisting of the finest and smoothest triangles, slips through the bone and nourishes the marrow. The reversal of this order brings serious disease. When flesh decomposes, the products back into the blood vessels, forming black and yellow bile, white and acid phlegm, and the serum purged as sweat and tears. These destroy the blood itself, and are carried through the vascular system, affording no nourishment to the body but feuding among themselves and making war on all that remains sound in the body. If the secretion from flesh and sinew fails, the bones are starved, the binding material loses its smooth and viscous quality, falls back into flesh and sinew and thence to the blood stream, where it aggravates existing diseases. Bone that is overheated decays and crumbles into the nutritive system where it makes diseases even worse. And finally, when an excess or deficiency develops in the marrow the case is most desperate, since the whole normal process is then completely reversed. *Corruptio optimi pessima.* Not only must the normal proportion of the elements be preserved, therefore, but the normal order of their distribution through the body must be kept as well, if the organism is to remain in health.

Finally, Plato considers the causes of disease from the aspect of breath, phlegm, and bile (84C-85E), and several features of this discussion should be mentioned. First, we have seen how breath and "the condition of the body" were accounted a source of disease by Philistion of Locri. Plato holds that if the lungs or pores are blocked, breathing is irregular and breath becomes deficient in some parts, excessive in others — "the breath fails to reach some parts and causes them to putrefy for lack of refreshment; while too much of it passes into other quarters, where it forces its way through the veins and contorts them, dissolves the body ... " (84D). Breath is the cooling element, and its regulaı passage is needed to counterbalance the innate heat of the body. When this equilibrium is broken, disease ensues.

Secondly, bile is said to be the cause of all inflammations in the body (85B), and most dangerous (85C, D) when it mixes with the blood and dislodges the fibrin from its appointed function (τάξις). For fibrin is distributed through the blood to keep it in a condition of due proportion (συμμέτρως) between thickness and thinness, so that it will not be too thick to move easily through the blood vessels, nor yet so thin as to flow through the pores of the body. "The fibrin, by the composition of its substance, preserves the due mean" between these extremes (καιρὸν δὴ τούτων ... φυλάττουσιν). Bile, collecting in quantity, may dominate the fibrin and destroy this ordered function by its boiling heat.

The terminology of this passage is significant, showing clearly that Plato conceived the condition of a particular physical compound like blood to be proportionate (συμμέτρως) when it holds to a mean (here καιρός) between excess (thickness) and deficiency (thinness), and that as long as this condition is preserved the compound is healthy.

Finally, the discussion of disease caused by phlegm (85A, B), the source of all catarrhs, brings home the close interdependence of the physical and the psychic, and introduces our next consideration. For phlegm is made responsible for the mental derangement of epilepsy "when it is mixed with black bile and is diffused over the divine circuits in the head so as to throw them into confusion" (85A).

2. *Disease of the Soul*

We have already seen something of the close interdependence of body and soul in the passage where Plato describes the condition of the newly embodied rational soul (42E-44D) and the disturbing effect upon it of nutrition and sensation, which dislocate the circuits of the Same and Different and distort the μεσότητες that bind them in harmony, causing them to operate ἀλόγως and to become ψευδεῖς καὶ ἀνόητοι. This, Plato tells us, is why the soul of every infant is ἄνους. However, as time goes on and the tumult of the growing body dies down, the circuits readjust themselves, and man is put on the way to becoming ἔμφρων. Then if he is helped by ὀρθὴ τροφή and παίδευσις, he becomes ὁλόκληρος ὑγιής τε παντελῶς, τὴν μεγίστην ἀποφυγὼν νόσον; otherwise he goes through life maimed, and dies ἀτελὴς καὶ ἀνόητος. Thus the body is made the source of ἄνοια, called ἡ μεγίστη νόσος because it affects the immortal element in man, and his happiness in this life and the next.

It appears now that Plato applies the terms "health" and "disease" literally to the soul, and that he is quite serious about this close interdependence of soul and body, for after the discussion of bodily diseases he proposes to explain (86B ff.) how diseases (νοσήματα) of the *soul* are brought on by the condition of the body (διὰ σώματος ἕξιν). Assuming, no doubt, the teaching on the nature of man and his destiny outlined early in the dialogue, he begins with the proposition that disease of the soul is ἄνοια. This is divided into two types, madness (μανία), which suggests a state of excessive or frenzied activity of the mind, and stupid ignorance (ἀμαθία), in which mental activity is defective and slow. Both are states of the *rational* soul. However, since these are induced by disturbances in the mortal parts of the soul and the body, "*any* affection that brings on either of these must be called a disease" (86B).

Accordingly, states of excessive pleasure or pain must be accounted first among diseases of the soul. For "when a man is carried away by enjoyment or distracted by pain, in his immoderate (ἀκαίρως) haste to grasp the one or to escape the other he can neither see nor hear aright; he is in a frenzy and his capacity for reasoning is then at its lowest" (86B, C).

As an example of a morbid state of soul in which the dominant feeling-tone is one of pleasure, though mixed with severe pain as well, Plato cites the condition of sexual hypersensitivity (46C, D), in which the victim is maddened through most of his life by intense pleasures and pains, "his soul rendered sick and senseless (νοσοῦσαν καὶ ἄφρονα) by the body." The influence of the body is thought to be so strong, in fact, that Plato registers a protest against condemning such a man as deliberately bad.

But the truth is that sexual intemperance is a disorder of the soul arising, to a great extent, from the condition of a single substance which, owing to the porousness of the bones, floods the body with its moisture. We might almost say, indeed, of all that is called incontinence in pleasure that it is not justly made a reproach, as if men were willingly bad; the bad man becomes so because of some faulty habit of body and unenlightened upbringing (διὰ πονηρὰν ἕξιν τινὰ τοῦ σώματος καὶ ἀπαίδευτον τροφήν), and these are unwelcome afflictions that come to any man against his will (86D, E).

Just as a morbid appetite for bodily pleasure brings sickness on the soul, so also, Plato goes on to explain, "where pains are concerned, the soul likewise derives much badness from the body" (86E). Bile and phlegm had been named previously as the cause of epilepsy, when "diffused over the divine circuits in the head" (85A). Now Plato indicates some further painful effects of these upon the soul, when they concentrate in various regions of the body. And in doing so he offers a significant insight into his concept of the soul.

When acid and salt phlegms or bitter bilious humors roam about the body and, finding no outlet, are pent up within, mixing the vapor that arises from them with the motion of the soul, they are blended with it and induce all manner of disorders (νοσήματα) of the soul, of greater or less intensity and extent (μᾶλλον καὶ ἧττον καὶ ἐλάττω καὶ πλείω). Making their way to the three regions of the soul, according to the region each of them invades, they beget many divers types of *ill-temper* (δυσκολία) and *despondency* (δυσθυμία), many types of *rashness* (θρασύτης) and *cowardice* (δειλία), as well as of *forgetfulness* (λήθη) and *stupidity* (δυσμαθία).[34]

Here an excessive concentration of bile or phlegm is said to induce various diseases of the soul, depending upon the part of the body in which they settle. Apparently these mingle their vapor with the motion of the portion of soul animating that part of the body in which they settle, and interfere with its characteristic activity. Plato then cites six specific disorders of the soul as being brought

[34] 86E-87A. The translation is a departure from Cornford, p. 345, in the direction of greater literalness.

on in this way, clearly dividing them into three pairs.[35] The problem then arises: Are we to consider these as cited at random, or are they significant as representing a deliberate classification of predominantly painful disorders of the soul according to the general principle given? The latter explanation seems more consistent with the careful selection of details manifest throughout the *Timaeus*.

Taylor (p. 617) suggests that "the 'vapours' of φλέγμα and χολή which find no vent are supposed to spoil the temper and make a man moody if they collect in the region of the ἐπιθυμητικόν, to make him passionate, arrogant, or unduly timorous if they affect that of the θυμοειδές, dull and stupid if they invade the brain." The disorders of λήθη and δυσμαθία clearly pertain to the *knowing* soul, whose highest activity on the rational level is ἐπιστήμη. Likewise, θρασύτης and δειλία are evidently disorders of the *affective* soul whose characteristic activity is θυμός. But how are δυσκολία and δυσθυμία related to the *nutritive* soul, whose activity concerns τροφή? This is not so clear.

In the *Republic* the term δυσκολία (δύσκολος) is associated with disorders of the θυμοειδές, though not so definitely as to exclude the possibility that Plato may regard it as originating in the ἐπιθυμητικόν when he wishes to use the term properly and exactly.[36] It may be that in the present passage, which is directly concerned with the distinction of various disorders and their point of origin, Plato is being more precise and suggesting that δυσκολία

[35] The three pairs are set off stylistically thus: ... ποικίλλει μὲν ... ποικίλλει δὲ ... ἔτι δὲ. ...

[36] Cf. the use of δυσκολία especially in *Republic* 411 C and 586 C, where it is associated with disorders of the *affective* soul. However, in *Republic* 590A αὐθάδεια and δυσκολία are attributed to τὸ λεοντῶδες (the affective soul) and τὸ ὀφεῶδες (the *nutritive* soul), and the parallel order may indicate that Plato conceives αὐθάδεια as a disorder of the affective soul and δυσκολία as a disorder of the nutritive soul *respectively*, drawing a distinction here which is not called for in the other passages cited above. Hence, as to the source of δυσκολία, a definite conclusion on the basis of usage elsewhere in Plato is difficult. An examination of various passages makes it very clear, however, that δυσκολία is regarded as a disorder of the soul *by way of excessive activity*. Thus *Republic* 590A: ... ὅταν τὸ λεοντῶδές τε καὶ ὀφεῶδες αὔξηται καὶ συντείνηται ἀναρμόστως. ... For usage cf., in addition to the above, *Laws* 649 E, 757 E, 777 B, 791 C, D, 792A, 935A; *Theaetetus* 161A, 174 D.

has its source in the nutritive soul. There is an etymological basis for this which, to judge from the *Cratylus* and the instance at *Timaeus* 43C 5-7, might well have appealed to him. According to Athenaeus[37] the stem — κολ — is equivalent to ἡ τροφή, apparent in the etymology of κόλον, βουκόλος, κοιλία, and κόλαξ, so that δύσκολος applied to persons means, properly, "hard to satisfy *with food*" (σικχός), hence "hard to please, discontented, fretful, peevish, ill-tempered". This might have led Plato to regard δυσκολία primarily as a disorder of the nutritive soul, the effect of overactivity and oversensitivity of the appetite for food. Consistent with this would be the teaching of *Laws* 791D that τρυφή — luxury and daintiness in satisfying the lower appetites — renders the character (ἤθη) of the very young *peevish* (δύσκολα) and choleric (ἀκράχολα) and oversensitive (σφόδρα ἀπὸ σμικρῶν κινούμενα).

The same passage in the *Laws* teaches that the opposite treatment, when imposed upon children, has an opposite effect, making them dejected (ταπεινούς), mean-spirited (ἀνελευθέρους), and misanthropic (μισανθρώπους), qualities which are summed up fairly well by the term δυσθυμία, melancholy or despondency. Δυσθυμία is apparently a rather rare term in Plato. It is used in the *Laws* (666B) to designate the depressed and unresponsive disposition of later life (as opposed to the ἐμμανὴς ἕξις of youth), for which Plato prescribes as a remedy the copious use of wine. In the treatise *On Ancient Medicine* (ch. 10) some of the effects of missing one's regular lunch aie listed as a feeling of depression in the pit of the stomach (τὰ σπλάγχνα δοκεῖ οἱ κρέμασθαι), dizziness (σκοτοδινίη), listlessness (δυσεργείη), and *despondency* (δυσθυμίη). There seems little doubt, then, that δυσθυμία denotes a disordered condition *opposed* to δυσκολία as implying a *deficiency* of activity in the soul. And there is some basis for tracing its origin to the *nutritive* soul, so that it is probable that Taylor's suggestion is correct.

[37] 6.262A (ed. G. Kaibel, Leipzig, 1887-1890, 3 vols.), cited by Liddell-Scott (9th ed.; s.v. κόλον, τό), who point out that the term is "applied to some form of preserved food" in a papyrus of the third century B.C. (*Papiri greci e latini* [Firenze, 1912-], 5.535.39, 46). Cf. Liddell-Scott, s.v. δυσκολία, ἡ; δύσκολος, ον.

We can feel more confident, however, that Plato is employing another, or further, principle of classification in citing these six disorders of the soul. He says that phlegm and bile, when pent up in the body, mix their vapor with the motion of the soul "and induce all manner of disorders of the soul μᾶλλον καὶ ἧττον καὶ ἐλάττω καὶ πλείω." This phrase, which Cornford translates (p. 345) "of greater or less intensity and extent", strikes one as rather oddly elaborate in the context. I suggest that Plato is here deliberately speaking the language of that part of the *Philebus* in which Socrates discourses on the classes of πέρας, ἄπειρον, and their combination (23C ff.), assigning to the ἄπειρον (25C) the hotter and colder, the "drier and wetter, *more and less* (πλέον καὶ ἔλαττον), quicker and slower, greater and smaller, and all that we assigned before to the class which unites *more and less*" (μᾶλλον καὶ ἧττον).[38] The phrase μᾶλλον καὶ ἧττον καὶ ἐλάττω καὶ πλείω, understood in the technical sense of the *Philebus*, then, implies that the disorders of soul mentioned are to be conceived as *opposites* like hotter and colder, drier and wetter, each pair constituting an ἄπειρον in which the terms are related by way of *excess* and *deficiency*.

It becomes immediately evident that θρασύτης and δειλία fall easily into this classification, being related as an excess and deficiency with regard to θυμός, the characteristic activity of the affective soul. For θρασύτης implies frenzied overactivity of the "spirited" element, while δειλία implies a lack of proper spirit. We have seen also how δυσκολία and δυσθυμία are opposite disorders, distinguished by excessive and deficient activity of the soul, probably of the appetites and functions related to τροφή. But can this relationship exist between λήθη and δυσμαθία, the disorders of the *knowing* soul?

Plato's understanding of the matter may be indicated by his own use of the terms elsewhere in the dialogues. In the *Republic* (618D) δυσμαθία is contrasted with εὐμαθία as "stupidity" with "quickness of perception", and the discussion of the two in the *Charmides* (159E) makes it clear that δυσμαθία implies *deficient*

[38] Translated by H. N. Fowler, Loeb Classical Library, *Plato*, III (London, 1942), pp. 249-251.

mental activity, an undesirable indolence and slowness in learning (ἡσυχῇ καὶ βραδέως μανθάνειν). Hence δυσμαθία is a disorder of the knowing mind by way of *deficiency*. But how is λήθη related to this by way of excess? In the *Philebus* (33C ff.) Socrates insists that λήθη means the loss of something *once acquired* by the mind, and so is not to be confused with insensibility. Hence λήθη may be contrasted with δυσμαθία as a state in which the mind cannot *recall* certain knowledge it *once acquired*, while δυσμαθία implies dullness *in acquiring* that knowledge in the first place.

But perhaps the best insight into Plato's mind is provided in the *Theaetetus* by the famous comparison of the knowing soul to a wax tablet (191C ff.), which is in some cases too hard or too soft, and in others properly proportioned (μετρίως ἔχων). There we are told (194C) that when the wax is plentiful, deep, smooth, and kneaded or tempered proportionately (μετρίως ὠργασμένος) impressions are deep and lasting, the possessors learn quickly and retain their knowledge (εὐμαθεῖς ... μνήμονες). However, when the wax of the soul is too soft (ὑγρὸς σφόδρα) they *learn quickly but are forgetful* (εὐμαθεῖς μέν, ἐπιλήσμονες δὲ), and when the wax is too hard (σκληρός) the *opposite* is true, i.e., they are δυσμαθεῖς. For the impressions on the hard wax are shallow, while those in the soft wax quickly become blurred by melting together. However inadequate the figure may be found to illustrate all knowledge and error, it suggests at least that Plato conceived λήθη and δυσμαθία as opposite extremes arising from abnormal conditions of overactivity or deficient activity in the knowing soul.[39]

Hence we may say that the three pairs of psychic disorders mentioned are probably intended to represent diseases characteristic of each part of the tripartite soul, and each pair is quite clearly proposed as opposite extremes constituting an ἄπειρον and related by way of excess and deficiency. But all these are attributed to an

[39] This same pattern appears in *Regimen I*, ch. 35, where a proportionate blend of the finest fire (hot-dry; active) and water (cold-wet; passive) makes the soul most intelligent and retentive (φρονιμωτάτη καὶ μνημονικωτάτη), while an excess of fire causes mental instability, oversensitivity, flightiness, and an excess of water produces dullness, stupidity. And cf. Aristotle, *On Memory and Recollection*, 450 b.

accumulation of bile or phlegm in the body, and this suggests a final correlation.

Bile and phlegm were said to mingle their vapors with the motion of the soul and so to bring on the disorders mentioned. Now there can be no doubt that Plato regarded *bile* as an accumulation of excessive *heat* (fire), while it is almost as sure that for him *phlegm* represented an excess of *cold* (air).[40] The effect of heat upon an

[40] For *bile* as excess of *heat* (fire) and the cause of all inflammation in the body, cf. 85 B5 ff. where we are told that "inflammations" in any part of the body, so called because it is being burned or "inflamed", are all due to bile, which is the product of disintegrating flesh (water-fire-earth; cf. 74 C), "hot and liquid" (85 D). In quantities it masters the fibrin in the blood "with its own hotness and by boiling up" (85 E). Clearly, then, bile is conceived as a fiery excess of heat.

Phlegm is generally regarded as *cold* (and moist) by the medical writers (cf. pp. 46, 47, 49 *supra*) and Galen is most emphatic on the point (Kühn II, 130): ... καὶ οὐδεὶς οὐδὲ μαινόμενος ἂν ἄλλο τι ἢ ψυχρὸν καὶ ὑγρὸν εἴποι ἂν αὐτον. This is the opinion of Diocles of Carystos (cf. Wellmann, M., *Die Fragmente der sikilischen Ärzte* [Berlin, 1901], pp. 141-142, fr. 55, 57) who, according to Wellmann ("Diokles", *RE*, V, Pt. I, col. 803) considers blood as the mean between bile (hot) and phlegm (cold), making bile the cause of inflammations and phlegm that of catarrhs as Plato does at 83 C and 85 B. Cornford (pp. 337-338) believes that Plato is following Diocles and Philistion in this as he does in other matters, e.g., the role of *breath* in health and disease (cf. Cornford, p. 341). Aristotle likewise teaches that phlegm is cold and the cause of catarrhs: *Problems* 862b 27, *Parts of Animals* 653a 2, *On Sleeping and Waking* 458a 3. Therefore, since Plato regards disease as caused by excess and deficiency in the bodily components, and bile and phlegm as morbid humors, the cause of inflammation and catarrh respectively, of which bile represents an excess of heat, we should normally conclude on the basis of the general theory and tradition that Plato considered phlegm an excess of *cold*, though this is not stated explicitly in the *Timaeus*.

However, Taylor, in detecting another link between Timaeus and Pythagoreanism, makes some difficulty about this. The Menon papyrus reports of Philolaus (XVIII, 41-46): "Again, while most say phlegm is cold (!), he lays it down that it is by nature hot, 'phlegm' being derived from φλέγειν (to burn). Wherefore too it is by participation in phlegm that inflamed parts are inflamed" (τὰ φλεγμαίνοντα μετοχῇ τοῦ φλέγματος φλεγμαίνει). In the passage cited above (85B5 ff.) Plato flatly contradicts this conclusion, and in a way which shows he is aware of the similarity of stem pointed out by Philolaus (ὅσα δὲ φλεγμαίνειν λέγεται τοῦ σώματος, ἀπὸ τοῦ κάεσθαί τε καὶ φλέγεσθαι, διὰ χολὴν γέγονε πάντα). Nevertheless, in keeping with his central thesis, Taylor maintains (p. 595) that: "Since Timaeus says at 83 D7 that ἱδρώς and δάκρυον (both of which have already been declared to have 'fire' in them) are the 'lymph' of φλέγμα and never alludes to the view that this humour is cold, he probably is taking Philolaus's view of it, though he does not accept the theory that 'inflammations' are caused by it." To explain this discrepancy Taylor postulates (pp. 599-600) that, after using Philolaus to explain the formation of phlegm at 83 C, D, Plato switches to another authority for his account of specific diseases caused by bile and phlegm (84 D-86A), an authority who, he concedes, "may have held the more common view that φλέγμα is ψυχρόν" (p. 600). Thus Taylor does not deny that Plato is thinking of phlegm as excessive cold in the part of the dialogue with which we are concerned.

Taylor's chief reasons for maintaining that Plato follows Philolaus earlier seem to be the two indicated (p. 595, and repeated, pp. 600 and 604), i.e.: 1) Timaeus (or Plato) nowhere says that phlegm is cold. 2) Sweat and tears are said to be a lymph of phlegm (83 D) and Taylor regards both of these as containing fire, implying that they come from a fiery source. The first, an argument from silence, is inconclusive. In fact we might conclude more readily from Plato's silence that he assumes the common opinion of medical men and feels no need to labor the point. As to the second reason, it does not seem at all clear that the lymph from which sweat and tears are formed need itself contain fire. For while the flesh contains a warm moisture, being composed of fire, water, and earth, the moisture which appears on the surface (ἱδρώς *properly*) is designed as a *cooling* device against the summer heat, just as the fire in flesh is to warm the body in winter (74 B, C). As to tears, if it is true that, as Taylor says (p. 482, ad 68A1), they are "fire and water commingled", the fire, as he also points out (*ibid.*) comes from the *eye*, and we need not assume that the moisture which is the lymph of phlegm contains any fire of its own. The moisture, in fact, is said to quench the fire that enters the eye (68A).

Furthermore, it is difficult to see how white phlegm, at least, could be considered warm, since it is said to be the product of decomposing flesh *in combination with air*, "inflated by air and enveloped in moisture so as to form bubbles" invisible and innumerable (83 C, D). And for Plato, as for Philistion, the δύναμις of air is to *cool*. Moreover, it is this white phlegm which, with black bile, is given as the cause of epilepsy (85A) in an account which follows closely the theory of the Hippocratic work on *The Sacred Disease*, as Taylor himself points out (p. 600), and in this work it is the *coldness* of phlegm which, by chilling and freezing the blood, brings loss of consciousness and motion (cf. chs. 9, 10).

Finally, in the eighth book of the *Republic* the dangerous proletariat that appears in the diseased forms of government, oligarchy and democracy, and is the cause of their change to a worse, is compared to bile and phlegm in the diseased body (564 B), being divided into two groups, the *active* trouble-makers, the criminals or "drones with stings", and the great, *passive* mob, the beggars or "stingless drones". This confirms the notion that Plato regarded phlegm and bile as morbid humors with opposite properties, bile being compared to the active, exciting, "stinging" element, while phlegm is like the great, inert mass that swells the body politic, sapping its resources and serving no useful function. This leads Jowett and Campbell to comment (*Plato's Republic* [Oxford, 1894], III, 400): "According to Aristotle, *Prob.* I, 29, χολή is hot and φλέγμα cold. The hot humour answers to the stinging, the cold to the stingless drones." Adam agrees with this judgment (*The Republic of Plato* [Cambridge, 1902], II, 251) and quotes the words of Aristotle: τῶν κατὰ τὸν ἄνθρωπον ἡ χολὴ μέν ἐστι θερμόν, τὸ δὲ φλέγμα ψυχρόν.

object, according to the *Timaeus*, is to quicken, stir up, agitate its motion, while the effect of cold is to constrict and slow it down.[41] When Plato says, then, that the ἀτμίς of bile and phlegm is somehow capable of affecting the motion (φορά) of the soul, it is reasonable to assume that he regards *bile*, which is excessively *hot*, as agitating the soul's normal motion to *excess*, and *phlegm*, which is *cold*, as slowing it down, rendering it *deficient*. Hence bile may be thought to induce the disorders characterized by excessive activity of the soul (λήθη, θρασύτης, δυσκολία), while phlegm is responsible for those in which activity is deficient (δυσμαθία, δειλία, δυσθυμία).[42]

The pattern emerging from this passage, then, may be represented diagrammatically thus:

Part affected:		*Excess:*	*Deficiency:*
Body:	head	bile (hot)	phlegm (cold)
Soul:	knowing (ἐπιστήμη)	λήθη	δυσμαθία
Body:	below head to diaphragm	bile	phlegm
Soul:	affective (θυμός)	θρασύτης	δειλία
Body:	below diaphragm	bile	phlegm
Soul:	nutritive (τροφή)	δυσκολία	δυσθυμία

We need not suppose that the disorders mentioned are any more than examples typical of each level of soul, though Plato says there are various forms (παντοδαπὰ εἴδη) of each. Nor should we assume that these particular diseases of the soul are induced *only*

[41] Cf. the effect of fire (hot) and air (cold) upon "fusible" water, 58E-59A; the sensation of "hot" (through rending of flesh by small, active fire particles) and "cold" (through compression caused by larger particles of air, etc., bringing *immobility*), 61E-62B; hot is said to dilate, cold to compress, 67D; the heart in anger throbs violently and swells through fire, is cooled and quieted by air and liquid in the lungs, 70C, D; passage of air through body cools it, 78E; bone, when not properly cooled by air, becomes overheated and disintegrates, 84B.

[42] This general pattern explains why Plato (*Laws* 666B) recommends the copious use of wine as a φάρμακον against the δυσθυμία characteristic of *old age*, which the medical writers consider cold-moist or phlegmatic. On the other hand, he forbids it to youth, an age characterized as hot-moist, because it would add "fire to fire in body and soul" and aggravate the excited condition (ἐμμανὴς ἕξις) of the young.

by an accumulation of bile and phlegm.[43] But their arrangement as terms in an ἄπειρον, related by way of excess and deficiency, and their close correlation with bodily disorders similarly related, provide an insight of capital importance into Plato's conception, not only of the interdependence of body and soul, but especially of the *close parallel of basic structure in the bodily and psychic organism*.

For just as the body is conceived as a complex of opposing qualities, tendencies, capacities (δυνάμεις), hot-cold, moist-dry, heavy-light, quick-slow, etc. — which bring disorder or disease when one becomes excessive or deficient, so the soul on each level seems to be conceived as an organism constituted of opposing tendencies or capacities, each of which, becoming unbalanced through excess or deficiency, brings its own specific *psychic* disease. Thus the body has a capacity for heat in varying degrees (μᾶλλον καὶ ἧττον) and for its opposite, cold, in the same way. When the capacity for heat is activated to excess, it results in inflammation; when it is deficiently activated, so that its opposite prevails in excess, catarrh or "cold" ensues. So too the affective soul, for example, has the capacity to face danger, θυμός or "spirit", and a capacity for the opposite, φόβος, fear. When the capacity for θυμός is activated in excess, the resulting disorder is θρασύτης; when it is deficient, so that its opposite prevails in excess, the disordered condition is δειλία. The present passage tells us that an excess and deficiency in the one can bring on a corresponding excess and deficiency in the other. And a disordered condition in either is properly called "disease".

When we bear this pattern in mind it becomes clearer why Plato speaks of extreme pleasure and pain as diseases of the soul (86B). For they form an ἄπειρον, and if one or other becomes excessive the equilibrium of the psychic organism is broken and disorder or disease ensues just as truly as when heat or cold become excessive in the bodily organism. Moreover, it will be recalled that Plato earlier (69C, D) associated pleasure and pain with other "dread and necessary" affections of the soul. These also were

[43] Cf., for example, the deleterious effect of the wrong kind of music in the *Republic*, 411Aff.

grouped in pairs, and in the light of the general pattern or internal structure of the soul it appears that they too are intended to represent opposing terms in an ἄπειρον, existing in an extreme state and hence to be accounted as diseases of the soul. In expressing this relationship Plato is hampered by lack of proper terminology,[44] but tries to make it clear by qualifying his terms. Thus (69D), grouped with pleasure and pain, as opposing and undesirable extremes (... "lure of evil" ... "fleeing from good" ...), are: 1) *Temerity* (θάρρος) and *fear* (φόβος), considered as excessive and deficient (i.e., "unwise counsellors") and apparently equivalent to θρασύτης and δειλία in the present passage, representing disordered states of soul in the face of impending danger. 2) *Unmanageable anger* (θυμὸς δυσπαραμύθητος) and *expectancy* or *hope* that is too easily led, yielding, acquiescent (ἔλπις εὐπαράγωγος), both apparently disordered attitudes toward present opposition or difficulty. These two pairs are obviously disorders of opposing capacities within the *affective* soul. And it is probably this tendency to conceive the soul's capacities in terms of opposites constituting an ἄπειρον which lies behind Plato's remark in the earlier passage (42A) that mortal souls must contain "fear (φόβος) and anger (θυμός) and all the feelings that accompany these and *all that are of a contrary nature.*"

With the internal structure of the soul conceived thus analogously to that of the body, the implication becomes immediately evident from all we have seen of the physical aspect of man and the cosmos that disorders of the soul, like those of the body, must be brought under control and the soul restored to well-being by the imposition of order through *measure* and *proportion.* Just as the opposite extremes of hot and cold, moist and dry, are united in the cosmic body by the Demiurge, who balances one against the other, setting up all possible proportions, tempering each extreme by its opposite and uniting them in a mean, so the same

[44] E.g. θυμός by itself can imply either proper "spirit" or excessive wrath, as φόβος can be a legitimate and reasonable fear or excessive fear, panic. Aristotle remarks on the lack of proper terminology for expressing the various degrees of order and disorder in the habitual reactions of the human soul; cf. *N.E.* 1107b30ff.

opposing elements in man's body must be ordered if he is to imitate the perfection of the cosmos and enjoy physical health. But since man's soul, like his body, is a complex of opposing tendencies or capacities, these same principles must order it on each level if he is to enjoy health of soul. Like the hot and cold, moist and dry in his body, man's capacities for anger and fear, for quick perception and abiding memory, must be controlled and integrated so that one tempers the other and neither grows excessive to dominate his psychic organism. This means that, like the δυνάμεις of the body, they must be kept *proportionate* (σύμμετρα) by the imposition of *measure* (μέτρον) or *rule* (λόγος) so that each balances the other and the soul maintains its equilibrium ἐν μέσῳ. The soul, or any level of the soul, whose opposing capacities are so disposed possesses health, or perhaps more properly, ἀρετή.

It is true that Plato does not develop this implication in the *Timaeus*, where he is more concerned with physics and physiology than with psychology and ethics. We must turn to the *Philebus* to find the principles formulated in universal terms. But there are also indications in other later dialogues that Plato assumes a structure of the soul such as we have described, and that he applies the principle of proportion or equilibrium as governing its health or ἀρετή. We have seen in the *Theaetetus* how opposite disorders are attributed to excess and defect in the powers of the knowing soul, while its well-being or proper functioning is represented as dependent upon proportionate tempering (κηρός ... μετρίως ὠργασμένος), though the figure of the wax tablet must be taken with reservations. But in the *Laws* (792C, D) the Athenian Stranger declares that the man who lives rightly should neither pursue pleasure nor shun pain entirely, but "*cleave to the mean*" (ἀσπάζεσθαι τὸ μέσον). And the sequel to the *Timaeus*, the *Critias*, presents the ideal citizen of ancient Athens (112C) as "pursuing the *mean* between arrogant pride and pettiness" (τὸ μέσον ὑπερηφανίας καὶ ἀνελευθερίας).

It will be recalled that the Hippocratic work *Airs, Waters, Places*, after treating the effect of environment upon the body and soul of man, concluded by indicating how the φύσις of man thus conceived might be modified by his νόμοι for better or worse. It is not surprising, then, to find Plato concluding his account of the diseases of body and soul by alluding to the effect of good and bad government upon the human organism, and the need for both proper τροφή and παίδευσις to correct the disordered tendencies in body and soul.

Besides all this, when men of so bad a composition dwell in cities with evil forms of government, where no less evil discourse is held both in public and private, and where, moreover, no course of study that might counteract this poison is pursued from youth upward, that is how all of us who are bad become so, through two causes that are altogether against the will. For these the blame must fall upon the parents rather than the offspring, and upon those who give, rather than those who receive, nurture; nevertheless, a man must use his utmost endeavor by means of nurture (διὰ τροφῆς) and by his pursuits and studies (δἰ ἐπιτηδευμάτων μαθημάτων τε) to escape from badness (κακίαν) and lay hold upon its contrary. This subject, however, belongs to another kind of discourse (87A, B).

That "other kind of discourse" will be found in such dialogues as the *Republic*, *Statesman*, and *Laws*, for which the *Timaeus* provides the physical, or physiological, basis.

3. *The Total Organism and Its Care*

Plato's analysis of man, part by part, body and soul, is complete. He has described the body, its elements, gross structure, organs, and sensitive, nutritive, vascular, and respiratory systems, the whole designed as container and vehicle of the immortal soul. He has discussed the structure and function of this "divine" element, the rational soul, by which man is to attain the truth and reproduce the order of the cosmos in himself, in the affective and nutritive parts of the mortal soul and in the body which they animate. He has shown the close interdependence of body and soul, and pointed to lack of proportion and equilibrium as the basic cause of disease or malfunctioning in the body and in each part of the soul. Plato now turns, in the final, and culminating, portion of the *Timaeus*, to the compound as a whole, to the total human organism.

Proportion and symmetry, he points out, must be maintained not only among the various factors which make up each part of man's body and soul, but also, and especially, between body and soul themselves. For the relationship of the whole soul to the whole body is of supreme importance, and only if these two be proportionate can man be said to be truly healthy, completely καλὸς κἀγαθός.

Now the good is always beautiful, and the beautiful is never disproportionate (πᾶν δὴ τὸ ἀγαθὸν καλόν, τὸ δὲ καλὸν οὐκ ἄμετρον); accordingly a living creature that is to possess these qualities must be well-proportioned (σύμμετρον). Proportions of a trivial kind we readily perceive and compute; but the most important and decisive escape our reckoning. For health or sickness, goodness or badness, the proportion or disproportion between soul and body themselves is more important than any other (πρὸς γὰρ ὑγιείας καὶ νόσους ἀρετάς τε καὶ κακίας οὐδεμία συμμετρία καὶ ἀμετρία μείζων ἢ ψυχῆς αὐτῆς πρὸς σῶμα αυτό); yet we pay no heed to this and do not observe that when a great and powerful soul has for its vehicle a frame too small and feeble, or again when the two are ill-matched in the contrary way, the creature as a whole is not beautiful, since it is deficient in the most important proportions (ἀσύμμετρον γὰρ ταῖς μεγίσταις συμμετρίαις); while the opposite condition is to him who can discern it the fairest and loveliest (κάλλιστον καὶ ἐρασμιώτατον) object of contemplation (87C, D).

Just as when the legs or other members are disproportionately (ἄμετρον) large the whole body is ugly (αἰσχρόν) and its functioning defective and painful, so it is with the total human organism when body and soul are disproportionate (87E-88B). A soul that is too ardent and strong, through intensive intellectual pursuits, rivalry, and ambition, racks the body and burns it out, filling it with disease that most physicians mistakenly blame on the body; while in a body that is too large and strong the appetite for nourishment prevails over that for wisdom (φρόνησις) and the body grows at the expense of the soul, resulting in the greatest disease, stupid ignorance (ἀμαθία). So body and soul must be kept proportionate. How is this to be done?

Plato states his solution in the following principle (88B): "Now against both these dangers there is one safeguard: not to exercise (κινεῖν) the soul without the body, nor yet the body without the soul, in order that both may hold their own and become equally balanced and healthy" (ἵνα ἀμυνομένω γίγνησθον ἰσορρόπω καὶ ὑγιῆ). The statement is important for a number of reasons. First, since Plato says that body and soul must each *defend* itself, hold its own against the other, it is clear that he conceives them as *opposed* in the human compound, each tending to usurp the rights of the other. Secondly, when they are exercised and developed proportionately so as to hold their own against the other, the resulting state is conceived as an *equilibrium* (ἰσορρόπω), and this is identified with *health* (ὑγιῆ). We have here Plato's concept of health in its most basic form, as the condition which prevails when the opposing factors in a compound are so proportioned as to constitute an equilibrium. This is clearly the concept of Alcmaeon and the other medical writers we have seen, which Plato applies in this passage not to the body alone, but to the compound of body and soul.

Finally, Plato states that each part of the compound must be developed proportionately to the other through exercise of its proper activity. He will develop this at greater length in the subsequent paragraphs. Now he merely clarifies by citing the case of the mathematician or any other engaged in intellectual pursuits, who must compensate by giving the body its proper exercise, just as the athlete, engaged in developing his body, must exercise his soul in the liberal arts and philosophy. "Thus", Plato concludes (88C), "one may deserve to be called in the true sense both fair and good" (ἅμα μὲν καλός, ἅμα δὲ ἀγαθός).

By the last phrase Plato indicates that in the foregoing discussion he has offered the true explanation of the epithet καλὸς κ'αγαθός. Returning to these terms at the beginning of the discussion (87C), and noting the terms or conditions associated with them and their opposites, we can detect the pattern of values underlying this passage, and the whole of the *Timaeus*. Plato began with the principle that all *good* (πᾶν τὸ ἀγαθόν) is fair or beautiful (καλὸν), and that the *beautiful* (τὸ καλόν) is not disproportionate (οὐκ ἄμετρον) but *proportionate* (σύμμετρον). Then we are told that *proportion* (συμμετρία) and *disproportion* (ἀμετρία) are implied in *health* (ὑγιεία) and *disease* (νόσος), (moral) *excellence* (ἀρετή) and *bad-*

ness (κακία). Again, that which is *proportionate* (σύμμετρον) in the *most important proportions* (ταῖς μεγίσταις συμμετρίαις) is called *best* or *most beautiful* (κάλλιστον) and *most desirable* (ἐρασμιώτατον), while the *disproportionate* (ἄμετρον) is said to be *ugly* (αἰσχρόν) and the cause of evils (κακῶν αἴτιον), bringing diseases of the body, and the greatest disease of the soul, stupid *ignorance* (ἀμαθία), which is opposed to *wisdom* (φρόνησις). These evils are avoided when body and soul are *in equilibrium* or *balance* (ἰσορρόπω) and *healthy* (ὑγιῆ). When this condition prevails, man is truly *fair* (καλός) and *good* (ἀγαθός).

The pattern of values implied here may be arranged as follows:

The good	- (evil)	τὸ ἀγαθόν	- (κακόν)
beautiful	- ugly	καλόν	- αἰσχρόν
proportionate	- disproportionate	σύμμετρον	- ἄμετρον
balanced	- (unbalanced)	ἰσόρροπον	- (ἀνισόρροπον)
health	- disease	ὑγιεία	- νόσος
moral excellence	- moral evil	ἀρετή	- κακία
wisdom	- stupid ignorance	φρόνησις	- ἀμαθία

The order represents a progression from the more general (the good) to the more specific (wisdom). The *good* is said to be *beautiful*; the beautiful is that which is *proportionate*; the proportionate in the order of concrete realities is *balanced*; the balanced in man, body and soul, is *health*; the health of man's soul, especially the rational and affective, is ἀρετή, i.e., *excellence* of habitual reaction; and the ἀρετή of the rational soul is *wisdom*. Hence the human organism is pronounced καλὸς κ'αγαθός when proportion or balance prevails both in the parts (συμμετριῶν τὰ μὲν σμικρά) and in the whole (τὰ δὲ κυριώτατα καὶ μέγιστα), i.e., 1) among the opposing elements of the body, 2) on each level of the soul, 3) among the three levels or kinds of soul, and 4) between body as a whole and soul as a whole in the total compound. Man, the microcosm, then approximates most closely the perfection of the cosmos, among whose opposing factors or constituents "the god introduced ... every kind of proportion (συμμετρία) in every respect in which it was possible for each one to be proportioned and symmetrical (ἀνάλογα καὶ σύμμετρα) both with itself and with all the rest" (69B). And the cosmos, it will be remembered, is the closest possible reproduction of the ideal Living Being, the intelligible reality which is "best and perfect in every respect" (κάλλιστον καὶ κατὰ πάντα τέλεον: 30D).

Exercise, Plato has implied, is the means by which the organism as a whole and each part of body and soul are to be developed proportionately. Before closing his discussion of man, Plato elaborates this notion, which might be considered the basis of the educational methods recommended in the *Republic* and *Laws.* The parts, he says (88C), must be cared for individually. The body and the three kinds of soul have their own proper motions (κινήσεις) and each grows strong through the exercise of these:

Accordingly, we may say now as briefly as possible that whichever of these lives in idleness and inactivity with respect to its proper motions must needs become the weakest, while any that is in constant exercise (ἐν γυμνασίοις) will be strongest; hence we must take care that their motions be kept in due proportion (συμμέτρους) one to another (89E-90A).

Later, in connection with the *rational soul,* Plato generalizes (90C): "Now there is but one way of caring for anything, namely to give it the nourishment and motions proper to it" (τὰς οἰκείας ἑκάστῳ τροφὰς καὶ κινήσεις ἀποδιδόναι). Thus the whole educational process of body and soul is conceived as a kind of dietetics and gymnastics.

In the case of the body, first of all, constant exercise is necessary for the same reason that nourishment is constantly needed. For, Plato says (88D, E), the body is always being heated and cooled, dried and moistened by what enters it and what assails it from outside, and if it is allowed to remain passive and to suffer the consequences of this double attack it will be overcome and destroyed. So man must imitate the innate motion of the Receptacle or Nurse of the cosmos and never allow the body to rest idle. He must keep it moving, constantly imposing a kind of shock (σεισμοὺς ἀεί τινας ἐμποιῶν) so as to counter (ἀμύνηται) the motions within and without in a natural way, holding them in balance. By agitating the body moderately (μετρίως σείων) he brings into order according to their affinities the "qualities" (παθήματα) and particles (μέρη) wandering about the body (as happens in the Receptacle through its innate motion), not leaving them as "foe against foe" to engender wars and disease, but gathering them together as friend to friend to produce health.

In this the best motion is *self-motion*, being most like that of thought and the cosmos. Inferior is the motion that originates in another, and the worst is to allow the body to be moved inertly by another agency. Hence to purify and brace the body, gymnastics are best, the swaying of ship or carriage inferior, and the use of drugs least desirable (89A).

In these principles we may perhaps see Plato's solution to the difficulty found in the medical writers,[45] namely, that while a perfectly blended climate like that of Asia naturally produces health, growth, and intelligence in men, its very moderation and freedom from the shock (ἔκπληξις) which accompanies sudden climatic change make the inhabitants soft, physically and morally. According to Plato's principles, the solution would not be to locate the state in a more rugged climate where the citizens might develop a hardier, more fierce type of physique and character through the influence of environment. For shock or motion induced by environment is less desirable, being imposed from outside, and may be dangerous in its violence. Rather, ruggedness of body, and soul, should be developed by self-movement, by exercise that provides sufficient shock to body and soul, yet can be moderated at will. Thus νόμοι imposing such discipline might compensate for a less hardy φύσις, a possibility which the medical writers themselves recognize.

These, then, are the principles which, according to Plato, must guide a man if he is to live according to rule (κατὰ λόγον). But if the body is to be exercised, much more is that part which should govern man, his soul (89D ff.).[46] Of the three types of soul each

[45] Cf. above, p. 60 and note 60 (Chapter II, *supra*).

[46] In the *Theaetetus* (152D-153C) the practical precept of constant exercise for body and soul is connected with the concept of the sensible universe as a blend of opposites which is always "becoming". "Becoming" is produced by *motion* and *change*, while by *rest* it perishes. In the *Laws* (790C-791C) provision is made that children, especially the youngest, be rocked constantly by their nurses to quiet the disordered motions of fear in body and soul and thus develop ἀρετή in the soul.

has its own proper motion and each must be exercised in such a way as to keep their motions in due proportion to one another. The proper function or activity, and the relative importance, of each have been discussed already, and the subject provides the central theme of the *Republic.* Here (90A ff.) Plato merely warns that the man who is completely taken up with the appetites (ἐπιθυμίαι) of the nutritive soul and the urge to conquer (φιλονικία) characteristic of the affective soul will increase the mortal parts of the soul at the expense of the immortal, and so become mortal as far as possible. But the man who has earnestly applied himself (ἐσπουδακώς) to learning and true wisdom and exercised himself (γεγυμνασμένος) in them will think divine thoughts; the divine element (δαίμων) dwelling in him will be well ordered (εὖ κεκοσμημένος), and he will be blessed (εὐδαίμων).

Finally, the means by which the divine element must be cultivated is that which Plato has recommended before, contemplation of the universe.

Now there is one way of caring for anything, namely to give it the nourishment and motions proper to it. The motions akin to the divine part in us are the thoughts and revolutions of the universe; these, therefore, every man should follow, and correcting those circuits in the head that were deranged at birth, by learning to know the harmonies and revolutions of the universe, he should bring the intelligent part, according to its pristine nature, into the likeness of that which intelligence discerns, and thereby win the fulfillment of the best life set by the gods before mankind, both for this present time and for the time to come (90C, D).

E. CONCLUSION: THE TIMAEUS AND THE MEDICAL WRITERS

We are now in a position to appreciate why the Menon papyrus treats Plato first, and at greatest length, among those who "hold that our bodies are composed of a combination of the elements" (XIV. 9-11) and that health and disease are "due to the elements" (IV. 28). The common features of this theory, as we have found them indicated in Menon's short accounts and elaborated in the works of the Hippocratic corpus, are clearly discernible in the

Timaeus, but worked out with a depth, consistency, and vastness of scope that surpass any other treatment we have examined and make it clear why the work evoked the admiration of a medical scientist and philosopher like Galen.

In *common* with the other medical writers we have seen, Plato conceives the body in health as a complex of opposing elements, mixed in proportion (σύμμετρα) with respect to both quantity and quality, and given a relative stability[47] by being blended (κεκραμένα) to form an equilibrium (ἰσόρροπα) which is dynamic because of the constant interaction of the elements even in combination. The basic cause of disease is the loss of this equilibrium when the elements become disproportionate through excess or deficiency, or when the blend is disturbed by improper distribution of the elements. Environment, nourishment, and exercise are of first importance in maintaining, or disturbing, the equilibrium of health. The elements of the macrocosm, which constitute environment, continually assail those of man's body and deplete it, so that it must be proportionately renewed by new elements from food and drink. This calls for constant adjustment or proportionate response of the organism, by which it masters and assimilates these new elements, with the help of exercise, and holds its own against the environment. Likewise, the body must respond to *changes* in environment, maintaining the internal balance of hot and cold, for example, by the coolness of perspiration in summer and the warmth of the skin in winter. Diet and climate are both considered healthy when they best contribute to the maintenance of the inner equilibrium, and when well-balanced, are thought to have a beneficent influence not only upon the body but upon the intelligence and character as well. These basic features, in one form or another, are common to Plato and the medical writers we have seen.

The notable differences in treatment of this common theory

[47] That stability results from proportionate blend is clear in the cosmic body, which is the most stable of compounds because unified most perfectly by proportion of every kind. Compare the instability of the contents of the Receptacle, where proportion and equilibrium are lacking (52D-53A). Cf. also *Republic*, 556E, where an unhealthy person and an unhealthy state are said to be easily upset, so that they fall into sickness and dissension at the slightest provocation.

as it appears in the medical writings and in the *Timaeus* arise, of course, largely from the different interests of their authors. Plato, as a philosopher, employs this physiological theory in explaining not only man's body but the entire universe, so that the scope of his work is cosmic.[48] He is interested in the theory as manifesting the ultimate principles operative in *all* visible things, and sees in its practical application the means of bringing men's souls as well as their bodies into harmony with the order of the cosmos. For him the body is the less important part of the human compound and its health and disease significant mainly for their effect on the soul. But the medical writers, as we should expect, are interested primarily in the body, and the theory is assumed or explained for the most part as a help to understanding specific diseases and the treatment recommended for their cure.

Plato, then, conceives not only man's body in health, but *all* visible or "created" things in their natural or normal condition as compounds of opposing or incompatible factors united by the application of measure or rule (μέτρον, λόγος) to form a proportion, balance, or equilibrium (συμμετρία, μετριότης, ἰσορροπία) in which the opposing extremes are blended and temper each other

48 Comparison between the microcosm and macrocosm is implicit, of course, in the common assumption of the medical writers that both are composed of the same elements and governed by the same principles. And the comparison is drawn explicitly in the Hippocratic works *On Sevens* (ch. 6) and *Regimen I* (ch. 9, 10). Cf. A. Olerud, *L'Idée de Macrocosmos et de Microcosmos dans le Timée de Platon* (Upsala, 1951), esp. pp. 13-98. The author remarks (p. 220): "Nous trouvons en Grèce un point d'origine de la doctrine macro-microcosmiques — dans sa forme pseudo-scientifique — dans certains des écrits d'Hippocrate, surtout le π. ἑβδομάδων et les variantes macro-microcosmiques du διαίτης." He believes, however, that such theories were transmitted to Plato through the Pythagoreans especially.

Aside from the problem of the origin of the notion and its possible transmission to Plato, we may ask whether in his use of the comparison Plato tends to reason from the nature of the microcosm to that of the macrocosm, or vice versa. The fact that he conceives the cosmos as a living being and is at pains to account for its freedom from sickness and old age, its lack of limbs or senses, of respiratory and nutritive organs, etc. seems to indicate that he is reasoning from the nature which is better known to that which is less well-known. A specific example of this manner of reasoning from microcosm to macrocosm may be found in *Philebus* 29A-30C. Likewise, in the *Symposium* (186A-188D), Eryximachus the physician is represented as tracing the operation of the two-fold Love from the human body to the cosmos.

(κρᾶσις), thus constituting a mean (μέσον, μεσότης). The human body and its perfect prototype, the cosmic body, are compounds of the opposites fire and air, water and earth, unified by proportion to a blend in which no one of them dominates the whole. The cosmic soul and its miniature, the rational soul in man, are represented as blended of incompatibles, Being, Sameness, and Difference of both the temporal and eternal order, bound together by proportion, and moving in equilibrium through the circuits of the Same and Different. The created gods are said (69D) to synthesize the mortal soul of man by blending (συγκερασάμενοι) opposing "dread and necessary affections" like pleasure and pain, temerity and fear, with sensation and desire. The concept of body as a compound of particles in equilibrium lies behind the theory of sensation, which is explained basically as a disturbance of the equilibrium of the sense organs transmitted through the body to the soul; and sense pleasure and pain are seen as a violent distortion of this same equilibrium, or its sudden restoration. Man's whole tripartite soul is a compound in which opposites, the rational and the nutritive soul, are joined by an intermediate, the affective soul, capable of obeying the rational and of commanding the nutritive. Finally, in the total human compound the whole body and whole soul are regarded as opposites which, when joined in proportion, make the whole man well-balanced and healthy.

Thus the concept of the living organism as a proportionate blend of opposites is applied by Plato not only to the human body but analogously to the soul on all levels and to the human compound as a whole. It becomes the basis of his psychology of sensation, pleasure, and pain. The microcosm is seen as an approximation of this basic structure manifest in its most perfect visible form in the macrocosm, which in turn copies most faithfully the ideal Living Being. The cosmology of the *Timaeus*, then, is rather a cosmic physiology and psychology. And this is grounded metaphysically in the concept of visible or "created" Being as Becoming.[49] The scope and consistency of Plato's treatment, then, marks the difference between the philosopher and the medical writer.

In Plato's discussion of microcosm and macrocosm, moreover,

[49] Cf. *Theaetetus*, 152D-153C; and cp. *Timaeus*, 49A-50A.

there is a *depth* which distinguishes it from the medical writings. Galen remarks[50] that Plato, as a philosopher, "was not content simply with the qualities (δυνάμεσιν) which appear in the primary bodies, but inquired into the cause of their generation, a speculation useless for the medical man. For to know why water moistens, fire burns, water flows, etc. ... is of no help in the cure of disease. It is enough for one who is to preserve health and cure disease to know that health comes about through a good blend (εὐκρασίαν) of hot, cold, dry, and moist, and disease arises through improper blending" (δυσκρασίαν). Plato apparently accepted the postulates of contemporary science (e.g. that the body of man and the cosmos are proportionate blends of earth, air, fire, and water, each with distinctive qualities or "powers") and sought for an ultimate explanation beyond these (i.e. why *four* primary bodies, why they differ in characteristic qualities, etc.).[51] Since experimental methods and micro-techniques were unknown the only remaining method of probing into the nature of material things was through the application of mathematics, and this Plato used, as we have seen. In the *Timaeus* the universal concept of proportion, with its power

50 Kühn, V, pp. 667-668. The translation is mine.

51 That these postulates were current in medical thought in the time of Plato and the generation before seems clear from what has been said (in the notes to the preceding chapter) about Empedocles and Philistion in particular, and about the date of composition of the Hippocratic works examined. The golden age of Hippocratic medicine is thought to fall between 430 and 380 B.C. (L. Bourgey, *Observation et expérience chez les médicins de la collection hippocratique* [Paris, 1953], p. 99.), which coincides with Plato's formative period, and it is estimated that most of the Hippocratic corpus was composed between 440 and 350, the period of greatest production being 430 to 390 (*ibid.*, p. 41).

In searching beyond the postulates of the science of his day, Plato seems to follow the method recommended in the *Phaedrus* (270C, E), where Hippocrates is commended for his position that "one cannot understand the nature of the body without knowing the nature of the whole".

"Then see what Hippocrates and true reason say about nature. In considering the nature of anything, must we not consider first, whether that in respect to which we wish to be learned ourselves and to make others learned is simple or multiform, and then, if it is simple, enquire what power (δύναμιν) of acting it possesses, or of being acted upon, and by what, and if it has many forms, number them, and then see in the case of each form, as we did in the case of the simple nature, what its action is and how it is acted upon and by what?" (Trans. H. N. Fowler, Loeb Library [London, 1953], *Plato*, I, p. 549).

For a discusion of this passage, and bibliography, cf. Bourgey, pp. 88-96.

to unify opposites or extremes, is analyzed in mathematical terms (τὰ ἔσχατα, μεσότης). Geometric proportion and the theory of "plane" and "solid" numbers provide a rational explanation for the existence of four primary bodies, no more or less. The characteristic "powers", motion, stability, weight, interaction and transformation, and the sensible qualities of the primary bodies and their compounds are ultimately traced to the geometric properties of shape and size. Other inscrutable aspects of the macrocosm and microcosm, — the regular motion of the heavenly bodies, the ordered workings of the rational soul — are explained in terms of circular motion based on specific mathematical ratios. Through this exploitation of the mathematical aspect of material things, then, Plato is able to offer a rational account which searches beyond the postulates of the medical writers.

Finally, Plato as a philosopher is not satisfied to explain simply *how* things are, as do the medical writers, but thinks it far more important to understand *why* they are as they are, to discover the purpose (τέλος) behind the universe and to trace its determining influence in the formation of all things. This he calls the *true* cause (αἰτία) as distinguished from the accessory causes (συναίτια, συμμεταιτίαι) which explain the mere mechanics of operation (46B ff.). The *Timaeus* describes not only the ultimate natuie of the elements and the manner in which the Demiurge and created gods form the cosmos and man's body, with its various members, organs, and senses; it also explains *why* the cosmos is "created" in the first place, why man's head is round and attached to a body, why the body is equipped with arms, legs, and the five senses, why its internal organs are so arranged, etc. This type of cause, which Aristotle calls "final", originates in intelligence and is for Plato the first and most important. To explain man and the universe merely in terms of the interaction of physical elements, as the medical writers generally do, is insufficient. For "a lover of intelligence and knowledge must necessarily seek first for the causation that belongs to intelligent nature, and only in the second place for that which belongs to things that are moved by others and of necessity set yet others in motion" (46D, E).

While the medical writers assume a regularity and proportion in the normal operation of nature, Plato sees behind these the working of an Intelligence which originally *designs* both microcosm and macrosm and deliberately fixes the proportions which the art of medicine, and every other art, discover and seek to maintain or restore. In fact, from the existence of medicine and the other arts in man Plato (*Philebus* 30A-C) argues to the existence of an Intelligence in the universe: There exists in man a cause which, "producing in him the art of physical training (σωμασκίαν) and of medicine (ἰατρικήν) for the ailing body, and generally synthesizing and healing in other matters also, is called, in a universal and comprehensive sense, wisdom" (σοφία); therefore, Plato infers, there exists in the universe an analogous cause which "orders and arranges the years and seasons and months, and can most justly be called Wisdom (σοφία) and Intelligence" (νοῦς).

The fact that in making this important point in the *Philebus* Plato selects physical training and medicine as typifying art, and characterizes wisdom generally as a synthesizing (συντιθέν) and healing (ἀκούμενον) power, suggests that this dialogue is worked out against the same background of medical and physiological theory that we find in the *Timaeus*. It may not be out of place, then, to introduce in this connection some comparisons between the two dialogues, since the *Philebus* seems in many ways to complement the *Timaeus* and to formulate its basic principles.[52]

The function of the cosmic Intelligence discovered in the *Philebus* is to form a third class or kind (εἶδος, γένος, φύσις) of beings by mixing those of the first (τὸ ἄπειρον) with those of the second (τὸ πέρας; cf. 26B, C and 30C). This has with good reason been compared to the function of the Demiurge in the *Timaeus*,[53] who mixes opposing ingredients according to measure (μέτρον) or rule (λόγος) to form a new kind of being which is proportionate (σύμ-

[52] Some links between the two dialogues have already been noted in regard to: the theory of sensation (note 27); pleasure and pain (p. 112); the opposites conceived as an ἄπειρον (p. 128).

[53] Cf. W. J. Oates, "The Doctrine of the Mean", *Philosophical Review*, XLV (1936), p. 388: "... in the *Philebus* ... the mean is distinctly a cosmological principle ... Involved as it is in the *Philebus* with the idea of mingling and mixture, we cannot fail to be reminded of the *Timaeus*, where the Creator mingles various elements in his construction of the universe."

μετρον, ἀνάλογον) in every respect, — the visible universe and all it contains. For in the *Philebus* the beings which Socrates (24A ff.) assigns to the class of the unlimited (ἄπειρον)[54] are, as we have indicated earlier, the same kind of opposing qualities or "powers" which are elemental in the macrocosm and microcosm of the *Timaeus* and the medical writings, — the hotter and colder, moister and drier, quicker and slower, forceful and gentle, greater and smaller, more and less. And in the class of limit (πέρας) Socrates includes (25B, C) "the equal or *equality* (τὸ ἴσον καὶ ἰσότητα) first of all, and then after the equal the double and all things whatever that are related as number to number (πρὸς ἀριθμὸν ἀριθμός) or measure to measure" (μέτρον πρὸς μέτρον).[55] Now equality of number or measure is nothing more than proportion,[56] and just as in the *Timaeus* measure is imposed upon the opposites to unify them in proportion and form the universe, so in the *Philebus* (25D, E) the second class (φύσις), which is "the offspring of limit", is combined with the first, that of the unlimited, to form a third, that of "the equal and the double, and whatever kind *settles the differences among the opposites* (παύει πρὸς ἄλληλα τ'αναντία διαφόρως ἔχοντα) and by the imposition of number renders them *proportionate* (σύμμετρα) and harmonious" (σύμφωνα).

Socrates' first illustration of the concept (25E) is the remark that "in the case of sickness, it is the right combination of these two (i.e. limit and the unlimited) that produces the nature of *health*" (τὴν ὑγιείας φύσιν ἐγέννησεν). This again seems to bear out the suggestion that in the *Philebus* Plato is working against a back-

[54] The terms ἄπειρον and πέρας may very well be Pythagorean in origin, but are adapted by Plato to the meanings explained in the *Philebus* (24A ff.). Cf. A. Grant, *The Ethics of Aristotle*, 3 ed. (London, 1874), I, pp. 251-262, cited by W. J. Oates, *loc. cit.*, p. 384: "... as Grant has stated, in the *Philebus* there comes the transition from regarding the limit as absolutely fixed law to an idea of the limit as a proportion or mean, μεσότης, that is to say, law or form as relative". With this adaptation Plato can use the terms to formulate the "creative" process as described in the *Timaeus*.

[55] This and all subsequent quotations, except those from the *Timaeus*, will be presented in my own translation.

[56] Cf. *Gorgias* 508A, where Plato uses the expression ἰσότης γεωμετρική for geometric proportion. See also Proclus, II, 18. 29-19.3; 22. 23-24.

ground of medical theory. Next, Socrates observes that the whole art of music is established by the application of this principle to the ἄπειρα formed by high and low tonal qualities and by fast and slow beat. Then he remaıks that, with regard to wintry cold and stifling heat, the proper combination of limit and the unlimited does away with excess and irregularity, and produces what is measured (ἔμμετρον) and proportionate (σύμμετρον). Hence, "the seasons and all fair things we know "come into being when the unlimited is mingled with that which has limit. Finally, there are, Socrates says, countless other examples that could be mentioned, — "beauty and strength of body, as well as health, and besides these all the many noble qualities of soul" (ἐν ψυχαῖς αὖ πάμπολλα ... καὶ πάγκαλα).

Thus Plato in the *Philebus* conceives health, strength, and beauty of body, the harmony of sounds, the regularity of seasons and the beauties of the world, and finally, all the virtues of the soul as arising from the proper combination of limit and unlimited, producing that which is measured and proportionate. This is effected by a universal cause which Socrates refers to (26B) as a goddess who, "looking down upon the insolence and utter wickedness of all ... set up law and order based on limit". This cause, as we have seen, is later identified with Wisdom or Intelligence (30C), and corresponds to the Demiurge in the *Timaeus*. We might remark here that the universal agent of both these dialogues corresponds closely to the "orderly Love" responsible for all the good of the universe sketched in the *Symposium* by the physician Eryximachus.[57]

[57] It is noteworthy that the *Symposium* represents Eryximachus as speaking from the medical point of view (186B) and of matters which he learns from his art, as dramatic propriety, so carefully observed in the dialogue, demands. His discourse exhibits the same basic principles, and a striking similarity of topic, as those treated in the present passage of the *Philebus* and found incorporated in the *Timaeus*. He begins with the proportionate blend of opposites in the healthy body, finds the same principle applied in harmonics, in moral life and training, and in the regular changes and movements of the macrocosm. Taylor (*Plato, the Man and His Work* [New York, 1929], p. 217) holds that his "science" represents "the views of the Sicilian medical school, the very type of biology from which both Plato and Aristotle draw the biological analogies which play so large a part in their ethics".

Eryximachus and his father, Acumenos ("The Healer"), are mentioned in

For this god too governs the arts of medicine, gymnastics, agriculture, and music, and effects health of body, virtue of soul, harmony and rhythm of sound, temperate climate, and the regular movement of the heavenly bodies by the proportionate blending of opposites.

In comparing the *Timaeus* with the medical writings seen, we may conclude, then, that Plato agrees with the medical writers in conceiving the basic structure of the human body and the physical universe as a proportionate blend of opposites, but applies this concept with far greater scope and consistency to all visible beings, body and soul, and to "created" Being itself. Thus in Plato this postulate of contemporary science is established as a *universal philosophic principle.* Moreover, he goes beyond the medical writers in probing more deeply into the roots of body and soul by the application of mathematics. And, unlike them, he traces out in all things, from the shape of the smallest particle to the motion of the stars, the purpose and design of a transcendent Intelligence.

Again, both Plato and the medical writers, as we have seen, conceive the physical life of man as a constant effort to maintain the internal equilibrium of opposing elements against the onslaught of elements constituting the environment. To survive in health, the human body must respond *proportionately* at all times to the environment, assimilating new particles to replace those dissipated by the environment, compensating by its native heat or coolness for the cold or heat of the environment, adjusting proportionately to changes in environment. The body in health makes these adjustments easily because its internal equilibrium is properly disposed to respond proportionately, and so remains in health. When the internal balance is broken, however, the response be-

the *Protagoras* (315C) and again in the *Phaedrus* (268Aff.), where they are cited as authorities in the art of medicine, and compared to Sophocles and Euripides in tragedy and "Adrastus" and Pericles in statesmanship. Taylor (*ibid.*) suggests that they are the same Acumenos and Eryximachus said by Andocides (I, 18 and 35) to have been implicated with Alcibiades and Phaedrus in the mutilation of the Hermae. Cf. also, Xenophon, *Memorabilia,* III, 13; and L. Edelstein, "The Role of Eryximachus in Plato's *Symposium*", *Transactions of the American Philological Assn.* LXXVI (1945), 85-103.

comes correspondingly disproportionate, and there is malfunction, pain, disease.

This second basic postulate of the medical writers, the proportionate response of the healthy organism to environment, is also given a much broader application by Plato as governing not only man's physical life but his sensitive, affective, rational, in fact, the *total* life of the human organism. This is consistent with the extension of the first postulate (the conception of the organism as a proportionate blend of opposites) to body and each level of soul, and to the human organism as a whole. Plato sees the total environment in which man lives as affecting more than the mere vegetable life of his body. The particles surrounding his body are also capable of stimulating *sensation* when they assail (ἐμπίπτουσι) the sense organs with force sufficient to disturb their equilibrium and communicate the motion through the body to the soul. This response to the environment, involving both body and soul, makes man aware of the sensible qualities (δυνάμεις) of the objects around him. And when particles assail the senses with a suddenness or violence contrary to their normal state or operation (παρὰ φύσιν), the response set up in body and soul is *pain*; while particles that perceptibly restore the normal state or function (κατὰ φύσιν) evoke a *pleasurable* response. Moreover, the environment in a comprehensive sense includes objects, persons, situations, and actions perceived as causing, or capable of causing, pleasure or pain in present, past, or future. These evoke their proper responses in the affective soul particularly, giving rise to the various *emotions*. Thus an object perceived as threatening pain arouses courage and fear; one perceived as inflicting present pain, anger and hopeful expectancy. Again, the perceptible and intelligible aspects of the total environment in which man finds himself, — the visible world of Becoming and the invisible, eternal realm of Ideas — evoke in his knowing soul the responses of δόξα and ἐπιστήμη. For the rational soul is said to *make contact* (ἐφάπτηται; 37A6) with both "created" and eternal Being, while the circuits of the Same and Different *encounter* (περιτύχωσιν; 44A2) external objects belonging to the categories of Sameness and Difference. The rational soul is then "set in motion all through herself" (37A-C) and reports on sameness and difference in the realms of both Being and Becoming. The report of the circle of the Different concerning visible things gives rise to δόξα, and the declarations of the circle of the Same concerning eternal Being produce ἐπιστήμη. In Plato's comprehensive view the human organism, body and soul, responds to its proper object or stimulus on *every* level, and this total response of the whole organism to the total environment constitutes the life of man.

There is, of course, a direct connection between the internal equilibrium of elements and the proper response of the organism. If the opposing constituents are proportionate, balanced, in equilibrium, the organism will *respond proportionately* to the impact of its proper object in the environment. For example, the healthy and virtuous man will, according to the *Philebus* (45C-E), experience moderate pleasure and pain, whereas the sick and the profligate will feel the extremes of each. On the other hand, exercise in proportionate response on the various levels will establish and maintain the internal proportion or equilibrium of health, virtue, intelligence, etc. This is the basis of Plato's educational method, as we have seen. As Socrates remarks in the *Republic* (444C): "Healthy activities produce health, and unhealthy, sickness ... just action produces justice, and unjust, injustice."

In Plato's comprehensive view, then, this internal blend of opposites and proportionate response to environment produce on all levels what is *best* for man: in his body, health; in body and soul, true perception and moderate pleasures and pains; in the affective soul particularily, moral excellence (ἀρετή); in the knowing soul, wisdom, true δόξα and ἐπιστήμη; and in the total compound of body and soul, the good life, making man in the true sense καλὸς κ'αγαθός. For man, the microcosm, then reproduces most closely the perfection of the macrocosm, which is proportionate and symmetrical in every way possible.

We have noted the pattern of values underlying the view of man and the cosmos presented by the *Timaeus*: the *good* is somehow identified with the *beautiful*, and the beautiful with the *proportionate*, which among concrete realities means the *balanced*, and balance produces in man wisdom, moral excellence, and health, —

the good life. The *Philebus*, attacking the problem from the standpoint of the good life, with particular reference to the relative value of wisdom and pleasure, arrives at the same pattern. Socrates there discusses the various ingredients of the good life, seeking to learn from "the best and most stable (ἀστασιαστοτάτην) mixture and blend" (κρᾶσιν) of these "what is by nature good in man and the universe, and what we should divine its peculiar nature (ἰδέαν) to be" (63E-64A). The "most valuable (τιμιώτατον) factor and the cause of such a disposition being attractive to all" is, he concludes, measure and proportionate structure (μέτρον καὶ ἡ σύμμετρος φύσις), without which there is no κρᾶσις (64C-E).[58] "And now the power of the good has run off into the nature of the beautiful. For *proportion* or *symmetry* (μετριότης καὶ συμμετρία) happen to be everywhere identical with *beauty* and *excellence*" (κάλλος καὶ ἀρετή). Accordingly, at the end of the dialogue (66A ff.) Socrates declares that "the eternal nature chooses first" *measure* (μέτρον) and *moderation* (τὸ μέτριον) or *timeliness* (καίριον) and the like, while second place is held by the proportionate or beautiful (τὸ σύμμετρον καὶ καλόν) and the complete or sufficient (τὸ τέλεον καὶ ἱκανόν). Wisdom and the perfections of the knowing soul (ἐπιστήμη, τέχνη, δόξα ὀρθή) are valued next as most closely related to μέτρον, and last are the pure and true pleasures that accompany them and sensation. The *Philebus* develops the hierarchy of values no further.

It must be clear that the less consciously elaborated pattern of values assumed by the medical writers coincides with this in essentials. For them also, as we have seen, measure marks all that is good in man's body, soul, food, and environment because it brings *proportion* (μετριότης), equilibrium, a disposition of the elements in which no extreme dominates but opposites moderate each other by blending (κρᾶσις), and blending gives to the organism its greatest possible unity, permanence or stability, and self-sufficiency. These values are concretely manifest in the perfectly healthy body, which is for them, as medical men, the ultimate objective and norm.

Plato, of course, provides a new ontological basis for these

[58] Cp. here *Politicus*, 283E-284C.

values by establishing as his ultimate constitutive norm the ideal Living Being, the "intelligible thing which is best and in every way complete" (*Timaeus* 30D). This, the model for macrocosm and microcosm, is supreme in unity, permanence, completeness, and self-sufficiency, since its Being is indivisible (37A) and eternal (37D), and it contains within itself all intelligible living beings (30D). All visible or "created" things can only approximate these attributes of the intelligible Model. The best of them, the cosmos, is unified as perfectly as possible by means of *proportion*, the most perfect bond of all "created" beings, and is rendered most stable, permanent, and self-sufficient by the perfection of its proportions and the fact that it contains the sum of all elements so that nothing remains to disrupt these proportions. In the human composite, likewise, it is proportion on all levels that orders and unifies, bringing man the greatest stability of body and soul, the greatest autonomy and self-control, and the longest, most noble life. Through proportion man shares the unity, eternity, and self-sufficiency of the ideal Living Being as far as his mortality will allow.

Consequently, while the ideal Living Being provides the ultimate constitutive norm, those visible beings which imitate and manifest its perfections best, — the cosmos, the just or virtuous man, the healthy body, the just form of government — each in its own order provides a *guiding* norm for the physician, the moral philosopher, the educator, the statesman. Thus Socrates can say in the *Republic* (373E) that the society in which each citizen performs his function according to natural aptitude "seems to me to be *the state conformed to the truth* (ἀληθινὴ πόλις), like the *healthy individual*" (ὥσπερ ὑγιής τις). And later, after drawing an exact parallel between health in the body, and justice in the soul and in the state (443A-444E), he sets up the just man and the just constitution as standards by which other states and dispositions of soul are to be judged bad and defective (449A).

The *Timaeus*, then, is not a practical document in the same sense that most medical writings are, descending as they do to minute details of disease, therapeutics, and the peculiar needs of particular types of patient. On the other hand, Plato undoubtedly considered the cosmic view and standards of value presented in the *Timaeus* as supremely important not only for the practicing physician, but for the educator, the moral guide, the orator and statesman as well. This was the knowledge of the whole which, he believed, distinguished the true artist and scientist from the mere practitioner in each field. And in the treatment of his vast subject, it is surprising that Plato should trace his principles through as far as he does in discussing the basic sources of disease and outlining the principles of correct regimen. Even in this limited discussion he displays a knowledge and control of medical lore which, to judge by the impression made upon Menon and Galen, far surpassed that of the ordinary cultivated layman of his time. This fact alone attests to his profound interest in medicine and medical theory, even if we disregard the frequent allusions and comparisons to health and disease, the physician and trainer, physiological theory and the art of medicine occurring throughout the dialogues.

This interest especially, we may assume, was shared by his pupil, Aristotle, and when Aristotle discusses moral excellence in his own ethical writings, comparing it to bodily health, we may likewise assume that he has in mind not merely current medical theory, but that theory as formulated and applied to the total human organism in the writings of his master.

SOCRATES: To produce health is to bring it about that the elements of the body, according to nature, hold firm and be held firm (κρατεῖν ... κρατεῖσθαι) by each other; whereas to cause disease is to bring it about that, contrary to nature, they dominate and are dominated (ἄρχειν ... ἄρχεσθαι) one by the other.
GLAUCON: It is.
SOCRATES: To produce justice, then, is to bring it about that the constituents of the soul, according to nature, hold firm and be held firm by one another; while to cause injustice is to bring it about that, contrary to nature, they dominate and are dominated one by the other. Is this not so?
GLAUCON: Certainly.
SOCRATES: Moral excellence (ἀρετή), then, it seems, must be a kind of health and beauty and well-being of the soul, and badness (κακία) a kind of disease and ugliness and weakness.
GLAUCON: So it is.

— *The Republic*, 444D, E.

IV

PHYSIOLOGY AND THE DOCTRINE OF THE MEAN IN ARISTOTLE

Upon studying in broad outline the physiology of some of the medical writers earlier or contemporary with Aristotle we have found their thinking dominated by the principle of proportion, equilibrium, and the mean, a principle which even they apply, in a limited way, to the ethical and political sphere. Further, we have seen that for Plato a middle state or equilibrium of opposing factors explains not only health of the body but the best condition of the soul and of the whole man, reflecting the balance and harmony of the cosmos and, ultimately, the perfection of the Intelligible Living Being. Now the writings of Aristotle remain to be considered from this point of view, to determine whether, and to what extent, the same basic notions may be found in his thought, particularly his physiological and ethico-political theory.

Our procedure, then, with Aristotle as with the medical writers and Plato, must be to examine his total concept of the human organism in health and disease, in the context of his physics, physiology, and psychology, and then to study the leading passages of the *Ethics* and *Politics* where physiological analogy is used in explaining the moral life of the individual and the corporate life of the state. It is hoped that a deeper understanding of the physiological analogues may thus cast light upon Aristotle's ethical and political thought, and particularily upon his doctrine of the mean.

A. ELEMENTAL STRUCTURE OF THE ORGANISM IN HEALTH AND DISEASE

In the collection of short studies that comprise the *Parva Naturalia* Aristotle refers at least three times to a projected discussion of the nature of health and disease.[1] Had this treatise been written, or if written, had it survived, our task would be greatly simplified. As it is, no comprehensive discussion of health and disease is to be found among the extant works, so that Aristotle's understanding of the matter must be pieced together from remarks introduced in the course of his treatment of other topics.

1. *The Nature of Health and Other Bodily Excellence*

An interesting passage in the seventh book of the *Physics* (246a10-248a9) may provide a starting point. Aristotle is here defending the proposition that *change of quality* (ἀλλοίωσις), as well as of quantity and place, is brought about by direct contact with a perceptible agent, and can occur only in those things capable of being affected directly by such an agent. Habit (ἕξις) offers an apparent exception to this proposition,[2] and so he is obliged to present a brief analysis of habits, good and bad. These are, specifically, the excellences or virtues (ἀρεταί) and defects or vices (κακίαι) of body and soul.

Aristotle does not here define habit (ἕξις) in general, but we may assume that he has in mind what he explains elsewhere (*Cat.* 8b27-9a4), namely, that habit is a stable and abiding disposition (διάθεσις), and that disposition is "the order or arrangement (τάξις) of that which has parts, with respect either to place or 'power' (δύναμις) or kind" (*Met.* 1022b1-3).[3] Now since the parts may be habitually disposed either well or badly,[4] some habits will be

[1] Cf. 436a17-19; 464b32-34; 480b21ff. See also E. Heitz, *Die verlorenen Schriften des Aristoteles* (Leipzig, 1865), p. 56ff.

[2] For habits are said to qualify a man (*Cat.* 8b27), and their acquisition or loss seems to be a qualitative change (*Physics*, 245b8), yet not all habits (e.g. the intellectual virtues) are found in things capable of being affected directly and in themselves by a sensible agent, nor do all changes of habit seem to be brought about by direct contact with a sensible agent.

[3] This is an adaptation of the Oxford translation of Aristotle, which will be used throughout unless otherwise indicated.

[4] This aspect of habit is presented in the definition at *Met.* 1022b10-13: "'Having' or 'habit' means a disposition according to which that which is disposed *is either well or ill disposed*, and either in itself or with reference to something else, e.g., health is a 'habit', for it is such a disposition."

excellences (ἀρεταί) and others defects (κακίαι), as he does mention here (246a11-12). It should be noted, then, that for Aristotle, as for Plato and the medical writers, excellence and defect in general seem to imply ultimately the notion of *order* and *disorder*. And the various types of excellence or defect will be specified by the nature of the parts or factors involved on each level of body and soul.

The present passage provides some further insights into the nature of ἀρετή and κακία in general. First, Aristotle says (246a11-246b2), neither they nor their acquisition and loss are qualitative modifications (ἀλλοιώσεις) because excellence is rather a *perfecting* (completion, τελείωσις) or fulfillment of the nature of a thing,[5] while defect is a deviation from (ἔκστασις), or loss of (φθορά), that state. These may involve or depend on qualitative changes in the subject, effected by direct contact with a perceptible agent, but are not themselves ἀλλοιώσεις.

Furthermore, the states or habits that constitute excellences and defects cannot themselves be considered subjects of any change whatsoever. For they partake of the nature of *relation* (πρός τι),[6] and relations are not the subject of change. To illustrate this Aristotle presents his brief analysis of excellence and defect on each level of body and soul, including health, moral excellence, and intellectual virtue.

The excellences of the body are treated first, as follows (246b3-10):

Thus the bodily excellences such as health (ὑγίεια) and a good state of body (εὐεξία) we regard as constituted in a *blending of hot and cold elements within the body in due proportion* (ἐν κράσει καὶ συμμετρίᾳ θερμῶν καὶ ψυχρῶν), *in relation either to one another or to the surrounding atmosphere* (ἢ αὐτῶν πρὸς αὐτὰ τῶν ἐντὸς ἢ πρὸς τὸ περιέχον); and in like manner we regard beauty (τὸ κάλλος), strength (ἡ ἰσχύς), and all the other bodily excellences and defects. Each of

[5] Cf. *Met.* 1021b20-23: "And excellence is a completion; for each thing is complete and every substance is complete when in respect of the form of its proper excellence it lacks no part of its natural magnitude".

[6] Cf. *Cat.* 11a20ff., where Aristotle explains that dispositions and habits pertain to the category of *relation* as well as of quality (ποιότης). That relations are not subject to change, cf. *Met.* 1068a11-13.

them exists in virtue of a particular relation and puts that which possesses it in a good or bad disposition[7] with regard to its proper affections (περὶ τὰ οἰκεῖα πάθη εὖ ἢ κακῶς διατίθησι τὸ ἔχον), where by "proper" affections I mean those ... that from the natural constitution of a thing tend to promote or destroy its existence.

Aristotle then argues to the conclusion in which he is primarily interested, and in doing so further eludicates the nature of the bodily states or habits, and of excellence and defect in general (*ibid.*, 10-20):

Since, then, relatives are neither themselves alterations (ἀλλοιώσεις) nor the subjects of alteration or of becoming or in fact of any change whatever, it is evident that neither states (ἕξεις) nor the processes of losing and acquiring states are alterations, though it may be true that their becoming or perishing is necessarily ... the result of the alteration of certain other things, e.g., *hot and cold* or *dry and wet* ... or whatever they may be on which the states primarily depend. For each ... defect or excellence involves a relation with those things from which the possessor of the defect or excellence is naturally subject to alteration; thus excellence disposes its possessor to be unaffected (ἀπαθές), or to be affected in such and such a way,[8] while defect disposes its possessor to be affected or unaffected in the opposite way.

Aristotle proceeds immediately to explain how the same analysis applies to habits of the soul (246b20 ff.), but this comparison must be deferred until we have further explored his notion of health and disease. The present passage, then, explains that:

(1). Health and well-being, as also strength and beauty, are abiding states or habits of body in which the elements or parts are properly disposed or ordered.

(2). The factors involved in health and well-being are *opposites*, the hot and cold especially, but also the moist and dry. These are the "proper affections" (τὰ οἰκεῖα πάθη) through which the body,

[7] This is a departure from the Oxford translation. It seems better to say that health is "constituted in" a proportionate blend of hot and cold, rather than that it "consists in" such a blend, since this latter identifies health with the blend, against the warning of *Topics* 145b 7 ff. (cf. below). Moreover, the Oxford translator's "condition" is a little vague for διάθεσις; "disposition" better suggests the notion of *order* or *arrangement*.

[8] The Oxford translation is altered here to follow Ross' text (*Aristotle's Physics* [Oxford, 1936; corrected reprint, 1955]) which reads παθητικὸν ὡδί in place of ὡς δεῖ, although there is no great difference in the sense.

according to its material nature, suffers alteration, and which, therefore, "tend to promote or destroy its existence". Health exists when the body is properly disposed with regard to these.

(3). The body is properly disposed when the hot, cold, moist, and dry are blended (ἐν κράσει) in proportion (ἐν συμμετρίᾳ).[9] And this proper proportion involves not only the relationship of the opposing factors *within* the body, but also their relationship to those of the *environment*.

(4). When the hot, cold, moist, and dry are properly disposed in health, the body remains unaffected (ἀπαθές) by certain influences, or responds in a certain way; while the diseased body, in which the opposites are improperly disposed (i.e., unblended and disproportionate), is too easily affected and its responses quite contrary to those of the organism in health.

These principles are found illustrated and elaborated in other parts of the Aristotelian corpus. First, the *Physics* passage declares that health, strength, beauty, "and the other bodily excellences and defects" are dependent upon a relation of proportion or symmetry among the parts involved. This generalization appears again in the *De Anima* (407b33-408a2), where harmony is defined as "a certain proportion (λόγος) or composition of constituents in mixture" and "harmony of the body" is said to be properly associated with "health and the excellence of the body in general" rather than with the soul. Philoponus, in commenting on this,[10] cites a parallel from Aristotle's lost *Eudemus*. Here Aristotle says that lack of harmony in the living body is disease, weakness, and ugliness, of which *disproportion* (ἀσυμμετρία) in the *elements* is disease, in the *homoeomerous parts* (blood, bone, sinew, etc.), weakness, and in the *organs* or *members*, ugliness (*Eudemus*, fr. 45 Rose). The *Topics* (116b17 ff.) clarifies the meaning of "elements" here by explaining that health is better than strength or beauty because it is "inherent in the moist and dry, the hot and the cold, in fact in all the primary

[9] The alternative text of this part of the *Physics* (246b 21) states simply that "health is a certain proportion of hot and cold factors" (ἡ ὑγίεια θερμῶν καὶ ψυχρῶν συμμετρία τις). The fact that Aristotle objects to this definition at *Topics* 145b 7 ff. may speak against the authenticity of this version.

[10] *Commentaria in Aristotelem Graeca* XV, 144-145.

constituents of an animal",[11] while the other excellences arise in what is secondary, strength in the bones and sinews, beauty in a symmetry of the limbs.

Later the same work warns (*Topics* 145b7 ff.), however, that the bodily excellences are not to be defined *as* proportion or symmetry of parts, but as *dependent upon* such proportion or symmetry,[12] and the *Posterior Analytics* (78b18-20) cites disproportionate (ἀσύμμετρα) hot and cold elements as the *cause* (αἰτία) of disease, proportionate (σύμμετρα) hot and cold as the cause of health. These developments and qualifications of the basic notion presented in the *Physics* may be summed up in the words of Simplicius (on *De Anima* 407b32):[13]

For neither is health a harmony, nor is strength, nor beauty, nor those virtues which in common Aristotle calls "bodily" (since they are bodily perfections), but rather these *come into being in dependence upon* a harmony or proportion (ἐγγινόμεναι ἐπὶ ἁρμονίᾳ καὶ συμμετρίᾳ), either of the qualities, or of the homoeomerous parts, or of the non-homoeomerous parts; not the proportions (συμμετρίαι) but the *supervening perfections* (αἱ ἐπιγινόμεναι τελειότητες) are virtues.

The bodily excellences, then, are regarded as perfections consequent upon the presence of συμμετρία,[14] and each is specified by

[11] Cf. *Physics* 210a20, where health is also said to be "in" the hot and cold. The term στοιχεῖον is used by Aristotle of both the primary bodies, earth, air, fire, and water, and their fundamental qualities, hot, cold, moist, and dry. Cf. *De Gen. et Cor.* 330a30 and 331a15 ff.

[12] "Every disposition ... is formed naturally in that of which it is an affection or disposition ... Sometimes people make mistakes in matters of this sort ... Similar ... is the definition of 'health' as a 'balance (συμμετρία) of hot and cold elements'; for then health will be necessarily exhibited by the hot and cold elements ... People who define in this way put effect for cause, or cause for effect ..." This definition of health is also referred to in *Topics* 139b21.

[13] *Commentaria in Aristotelem Graeca* XI, 53. The translation is mine.

[14] This notion of the bodily excellences, — health, strength, beauty — based upon proportion or *symmetry* is, according to Werner Jaeger (*Aristotle*, translated by R. Robinson, 2nd. ed. [Oxford, 1948], p. 43), a "Platonic doctrine" which Aristotle adopts in the *Eudemus* but which is "*wholly foreign to the treatises*" (italics mine). The implication is that in his mature work, embodied in the treatises, Aristotle *rejects* this notion, along with the rest of the "Platonic doctrine" that "the correct ethical state of soul, just like the normal and natural state of the body, is only a special case of the universal cosmic law of symmetry ..." However, in a footnote to this statement (p. 44, note 1), Jaeger

the nature of the bodily constituents or parts involved. In fact, the notion of συμμετρία, — proportion, symmetry, equilibrium — and the related notions of measure (μέτρον), ratio or rational principle (λόγος), mean (μέσον), middle state (μεσότης), moderation (μέτριον), equality (τὸ ἴσον), etc. tend to dominate Aristotle's thinking on the living organism, as they do Plato's. This becomes clear from even a brief examination of Aristotle's biochemistry, anatomy, physiology, psychology, and etiology of health and disease.

2. *Elemental Structure of the Organism*

Following the tradition of Plato and the medical writers we have seen, Aristotle regards the body, the material part of the animal organism, as a unity arising from opposing constituents. These

qualifies his position somewhat by acknowledging that the doctrine of the bodily virtues does recur in the *Topics* (116b17, 139b21, 145b8) and at *Physics* 246b4ff.; but he adds that these two particular portions of the treatises are "early" and "the *only other places* (italics mine) in which the bodily virtues appear".

With all due respect for Professor Jaeger's great authority, I believe the evidence will show that this is not the case. Ross (*Aristotle's Physics*, p. 675 *ad* 246b4-6) has already questioned the accuracy of Jaeger's statement, citing the appearance of the bodily virtues in *N.E.* 1102a16 and *Rhet.* 1361a3, b3, 1362b14-18. Even more pertinent examples might be *N.E.* 1104a11-19 (moral virtues, like health and strength of body, are produced, increased, and preserved by τὰ σύμμετρα); 1173a24-27 (the συμμετρία of health need not be the same for all); *Politics* 1302b35-37 (the body is spoiled if its parts do not grow proportionately to maintain its symmetry; and cf. 1284b9, 1309b23-30, 1326a-38); *De Gen. An.* 767a14-24 (συμμετρία of male and female element needed for generation, since everything formed by art or nature exists through a mean proportional); *De Part. An.* 652b17-20 (in the animal body cold needed as a counterbalance for heat, to achieve moderation and a mean); etc.

It is interesting to note that in one instance at least Jaeger himself has apparently modified his view on the disappearance of the Platonic "bodily virtues" and their parallel with the "virtues of the soul" in the "later works" of Aristotle, for in his article on "Aristotle's Use of Medicine as Model of Method in His Ethics" (*Journal of Hellenic Studies*, LXXVII [1957], p. 57) he affirms that in *N.E.* 1104a12-14 Aristotle "illustrates what he said about the bad effect of excess and defect by the example of health and strength ... Strength and health are the 'virtues of the body'; they had already been paralleled with the 'virtues of the soul' by Plato in the *Gorgias* ... Aristotle follows Plato's fine observation ... etc.".

are the elemental substances earth, air, fire, and water, which compose not only animal bodies but the whole of the sublunary world.[15]

The four elemental bodies are *opposed*, first of all, in that they tend by nature to move in opposite directions, fire moving naturally upwards toward the periphery of the universe, earth downwards toward its center, making these two respectively "light" and "heavy" in an absolute sense; while the "intermediate" bodies, air and water, tend upward and downward, but only in a relative sense (e.g., air tends to rise above earth and water, but to sink below fire), and so have relative lightness and heaviness (*De Caelo*, 269a18-19; 269b30ff.; 276a33ff.; 308a14-33; 311a15ff.). These opposite tendencies follow from a fundamental opposition among the four primary bodies in that each is distinguished by a particular combination of essentially opposite qualities or "affections" (πάθη), — hot, cold, moist (or fluid), and dry (or solid).[16] Thus *fire* is hot

[15] Cf., for example, *Meteor.* 339a19-21. While the subsequent discussion will be largely occupied with these material elements and the process by which they combine in the animal body and its parts, we must never forget that for Aristotle the nature, essence, or form was by far the more important factor in the analysis of the living organism, determining its formation and structure. In the biological works themselves he warns against the one-sided view which takes into account only the material aspect. Cf. the discourse on method and causes at the beginning of the *De Partibus Animalium* (639b4ff.), and the brief recapitulation in *De Generatione Animalium* (778b2-11), as well as the passage on the role of the final-formal cause in generation in the same treatise (734b-28ff.). It is the final-formal-efficient cause, — the soul — which determines the matter at every stage of the animal's development (739b34-745b23). Aristotle sums up the determining influence of essence thus (*De Part. An.* 640a34-36, as translated by A. L. Peck in the Loeb edition): "So the best way of putting the matter would be to say that *because* the essence of man is what it is, therefore a man has such and such parts, since there cannot be a man without them." And again (640b1-4): "*Because* man is such and such, *therefore* the process of his formation must of necessity be such and such and take place in such a manner ... And thus similarly with all the things that are constructed by Nature." There is no question here, therefore, of an analysis of substance as such, but the discussion will be limited largely to the material aspect of compound substances of the sublunary sphere. For a masterful presentation of the subject in historical perspective see F. Solmsen, *Aristotle's System of the Physical World* (Ithaca, N.Y., 1960), Part Four: The Nature and Formation of Physical Objects, pp. 319-389.

[16] The nature of the four opposing qualities is described in *De Gen. et Cor.* 329b27ff.; and cf. *De Part. An.* 649b9ff. On the rendering "fluid" for "moist"

and dry, *air* (i.e. vapor) hot and moist, *water* cold and moist, and *earth* cold and dry (*De Gen. et Cor.* 330b1-6). However, each of the four elemental bodies is characterized more by one of its qualities than by the other. Aristotle summarizes the matter as follows (330b31-331a6):[17]

The "simple" bodies, since they are four, fall into two pairs which belong to the two regions, each to each; for Fire and Air are forms of the body moving towards the "limit", while Earth and Water are forms of the body which moves toward the "center". Fire and Earth, moreover, are extremes and purest; Water and Air, on the contrary, are intermediates and more like blends. And, further, the members of either pair are contrary to those of the other, Water being contrary to Fire and Earth to Air; for the qualities constituting Water and Earth are contrary to those that constitute Fire and Air. Nevertheless, since they are four, each of them is characterized *par excellence* by a single quality; Earth by dry rather than by cold, Water by cold rather than by moist, Air by moist rather than by hot, and Fire by hot rather than by dry.

Earth, air, fire, and water are capable not only of local motion, but subject to generation and destruction by changing into one another (304b23-305a33). This is possible because each is a combination of certain qualities and of matter which is capable of receiving the opposite qualities (*Physics* 189b6; *De Gen. et Cor.* 319a17-320a6; 323b30-324a14; 324b5-10; 329a24-b3). One elemental body then acts upon another in virtue of their *contrary qualities*, "for agent and patient are contrary to one another, and coming-to-be a process into the contrary" (324a12-13). However, the contraries could not affect one another unless they were embodied in matter capable of receiving each (324b5-23), "for hot and cold do not change reciprocally into one another, but what changes ... is the substratum" (322b18-19). Moreover, it is only when they have been brought together in contact that the primary

---

and "solid" for "dry", see A. L. Peck in his introduction to the Loeb edition of the *De Partibus Animalium*, p. 32.

[17] In this section on the elemental structure of the organism (IV, A, 2) all passages cited by Bekker number alone, as here, are from *De Generatione et Corruptione*. Aristotle is not consistent in assigning one quality to each elemental body; cf. 334b35-335a3, where *water* is characterized by *moist*, as also in *Meteor.* 382a3-4 and *passim*.

bodies can interact, "for action and passion ... can only occur between things which are such as to touch one another" (322b23-25; and cf. sqq.). But when different primary bodies do come into contact, their contrary qualities necessarily act and are acted upon by one another (cf. 324b7-10). And if one contrary quality prevails over the other, replacing it in the matter, the result is true generation of a new element and destruction of the old, since no perceptible substratum persists in its identity through the change.[18]

This transmutation takes place when the primary bodies coming into contact are such that the total "power" and quantity of one are much greater than those of the other;[19] for then the stronger body will prevail in the conflict of opposites, driving out the contrary qualities and assimilating the matter of the weaker to itself (328a23-27):

> Amongst those things ... which are reciprocally active and passive, some are easily divisible. Now if a great quantity (or a large bulk) of one of these easily-divisible "reciprocating" materials be brought together with a little (or a small piece) of another, the effect produced is ... increase of the dominant; for the other material is transformed into the dominant.

This situation Aristotle compares to the falling of a drop of wine

[18] Both *generation-destruction* (γένεσις-φθορά) and *alteration* (ἀλλοίωσις) or qualitative change in the primary bodies involve action and passion between opposite qualities, the difference being that in alteration a perceptible substratum or matter persists and keeps its identity through the change, while in generation-destruction "nothing perceptible persists in its identity as a substratum" (*De Gen. et Cor.* 319b 5-15; and cf. the whole of chapters 3 and 4 of Book I on this distinction). The change of one primary body into another is true generation and destruction because their matter has no separate existence (329a 24-26) and is in itself only *potentially* perceptible (319b 33-34; cf. also 332a 28 and 332a 36 - b1).

It should be noted that Aristotle, unlike Plato, holds that *all four* primary bodies can change into one another (331a13 ff.). However, the ease and speed of their interchange will differ, since those that have one of the contrary qualities in common — e.g., air, (hot-*moist*) and water (cold-*moist*) — will change into each other more easily and quickly than those without a common quality (331a 25-b2). Aristotle agrees with Plato that the cosmic mass of elemental bodies is *finite*, and that in this cosmic mass the four types of elemental body *balance* each other, maintaining this proportion through all their interchanges (cf. *Physics* 204b12-22; *Met.* 1066b 27-31; *De Caelo* 304b 23-24).

[19] Cf. 333a 21 ff., where *quantity* (τὸ ποσόν) and *power of action* (δύναμις) are distinguished; and 334b 22-25, where hot, cold, moist, and dry are said to change into one another "unless they are equally balanced" (ἐὰν μὴ ἰσάζῃ).

into ten thousand gallons of water, in which case the wine loses its form and is changed into the total mass of water (328a27-28). In other words, when two or more primary bodies come into contact, an excess of power and bulk in one and a corresponding deficiency in another will result in the destruction of the deficient body and the further generation or increase of the excessive body, which dominates the weaker and transforms it into itself.

However, when opposing elemental bodies are brought into contact in quantities of *more or less equal power* (i.e. when proportionate or balanced), the result is not destruction of one and generation of another, but a combination (μίξις) or blend (κρᾶσις) of the two.[20] And this process results in the production of a *new substance which combines the qualities of the constituents* (328a28-32):

> On the other hand when there is a certain equilibrium (ἰσάζῃ πως) between their "powers of action" (ταῖς δυνάμεσι), then each of them changes out of its own nature toward the dominant; yet neither becomes the other, but both become an intermediate with properties common to both (μεταξὺ καὶ κοινόν).

There is no question here of a mere mechanical aggregation of the elemental bodies in which they are mixed together in minute particles while preserving their identity intact. For each of the constituents of a combination or blend must "*change out of its own nature* toward the dominant",[21] a combination being a union of mutually interactive constituents which modify each other in uniting (328b20-23; ἡ δὲ μίξις τῶν μικτῶν ἀλλοιωθέντων ἕνωσις).

[20] This is to accept the conclusion of H. H. Joachim, "Aristotle's Conception of Chemical Combination", *Journal of Philology*, XXIX (1904), 74, that "in *strict* Aristotelian terminology we must recognize (1) σύνθεσις as meaning mechanical mixture, and (2) μίξις as meaning chemical combination. And we must regard (3) κρᾶσις as the name for the common and conspicuous type of μίξις, viz. the chemical combination of liquids". Joachim's excellent discussion should be consulted for a detailed analysis of the notion and process of κρᾶσις; see also his commentary on the pertinent passages of *De Generatione et Corruptione* in *Aristotle: On Coming-to-Be and Passing-Away* (Oxford, 1922).

[21] In a note on this phrase (328a30-31) the Oxford translator (Joachim) remarks: "Each of the constituents, *qua* acting on the other is *relatively* 'dominant'. Neither of them is *absolutely* 'dominant', for each 'suffers action' from the other. Hence each meets the other half-way, and the resultant is a compromise between them".

Aristotle remarks (328a8-9) that "so long as the constituents are preserved in small particles, we must not speak of them as 'combined' (μεμῖχθαι); for this will be a 'composition' (σύνθεσις) instead of a 'blending' (κρᾶσις) or 'combination' (μίξις)." Whereas when the constituents enter into true combination or blend, the resultant is such that each smallest portion will "exhibit the same ratio (λόγος) between its constituents as the whole" (328a9-10), i.e. the compound will be "*uniform ... throughout* (ὁμοιομερές), any part of such a compound being the same as the whole, just as any part of water is water" (328a11-12).

It is evident that, since each must "change out of its own nature" in blending, the primary bodies are not preserved in actuality in the compound, but rather it is the *compound* that now exists actually (327b24-26). On the other hand neither are they destroyed, neither all nor any one of them, but they exist δυνάμει in that (1) just as they coalesced to form the blend, so they can be separated out again to exist independently, and (2) even in the blend their "power of action" (δύναμις) is preserved (327b26-33),[22] though moderated by its opposite.

The "power of action" of each primary body is identified with its characteristic quality or combination of qualities, — hot, cold, moist, and dry (cf. *De. Part. An.* 646a13-18) — which Aristotle analyzes in terms of their peculiar functions in the compound, hot and cold being active "powers", moist and dry passive (329b24-34; *Meteor.* 378b10-388a9; *De Part. An.* 648a37ff.).[23] The qualitative characteristic or "power of action" of each *compound*, then, will be the resultant of the qualities or "powers of action" of its primary constituents, as combined and modified by each other in the

[22] For δύναμις in the sense of "power of action" cf. also 328a 28-31; 328b18-23; 333a 23-27; 334b 8-30; *De Part. An.* 646a15 ff.; 646b15 ff,; *De Caelo* 281a 2-28; 307b 6; etc. It is precisely because their "power of action" persists in the compound that the elemental bodies can be said to *combine* in the formation of a new substance rather than being *destroyed* in the *generation* of a new substance. For unless something of them besides their base matter were preserved in the compound, the elemental bodies would simply *pass away*, and so could not be said to *combine*. Cf. 327a 33-b 33; 328a 28-31; 334b 8-30; and Joachim's commentary on these passages, especially that on 327b 22-31, pp. 179-181.

[23] For a detailed account of the interaction of the opposites in the compound, cf. Joachim in *Journal of Philology*, XXIX, 81-86.

blend (cf. 334b14-17, quoted below; *Meteor.* 381b24-382a11).

Now while there must be a "certain equilibrium between their 'powers of action'" if the opposing elemental bodies are to blend (328a28-32, quoted above), still they need not be absolutely equal. Instead, their ratio or proportion (λόγος) may vary within a certain range. And Aristotle explains the qualitative differences between one compound and another (e.g. between flesh and bone) according to the particular proportion of the four primary bodies entering into each (*Met.* 1092b17ff.; *De Anima* 408a14ff.; *De Gen. An.* 734b33).[24] For example, one compound will be harder or warmer than another according as it contains more earth or fire than the other. The proportion determines qualitative structure.

Consequently, the λόγος or proportion of the primary constituents in each compound substance is linked closely to its particular *nature*, which demands a distinctive qualitative structure by which the compound will be enabled to perform its function and fulfill its purpose. Thus, the λόγος of earth to water in each determines the softness of flesh and the hardness of bone; and bone can fulfill its peculiar purpose in the body only because it is hard, flesh because it is soft (*De Gen. An.* 743a36-b6).[25] In fact, Aristotle says that everything produced by art or nature exists in virtue of some λόγος or proportion of opposites (*De Gen. An.* 767a17-18; 740b33; and cf. *Meteor.* 378b32-379a1), and he speaks as though this λόγος belongs to, and is identified with, the substance (οὐσία), essence (τὸ τί ἦν εἶναι), form (εἶδος), or nature (φύσις) as the object of definition of the thing (*Met.* 993a17-22; 1092b17ff.; *De Gen. et Cor.* 333b8-20; 335b6-7; *De Part. An.* 642a18-23; *De Anima* 416a15-18; *De Gen. An.* 734b31-735a5). Hence each perishable compound retains its nature as long as the proper proportion is maintained

[24] The variety of homoeomerous bodies produced by different proportions of the primary bodies or elements is discussed in great detail in the fourth book of the *Meteorology* (388a10ff.). On the authenticity of this book, cf. I. Düring, *Aristotle's Chemical Treatise* (Göteborg, 1944), p. 18ff.; H. D. P. Lee, in his introduction to the Loeb edition of the *Meteorology* (London, 1952), pp. xii-xxi.

[25] On the other hand, when the λόγος of ingredients changes, bone becomes too soft, flesh too hard to function properly, so that disease and disintegration result; cf. *Parva Nat.* 479a8-14. On the relation of the λόγος to the function or purpose of each thing, cf. *De Gen. An.* 778b11-13; *Meteor.* 389b28ff.

among its opposing constituents (*Meteor.* 380a1), so that in this λόγος lies its well-being (τὸ εὖ) and its good (τὸ ἀγαθόν: 333b17-20; 335b6-7; and cf. *De Part. An.* 639b12-21).[26]

These principles may be illustrated from a passage in the *De Generatione et Corruptione* (334a22 ff.) in which Aristotle explains briefly the process by which the "homoeomerous" parts of the animal body, — flesh, bone, marrow, etc. — come to be out of earth, air, fire, and water. Granted that the primary bodies have a common substratum of matter and can change into each other (334a23-25), the problem is how will any third thing arise from two or more of them taken together, — flesh, for example, from cold and hot, or from fire and earth (334b5-8):

For if flesh consists of both and is neither of them, nor again is a "composition" of them in which they are preserved unaltered, what alternative is left except to identify the resultant of the two "elements" with their matter? For the passing-away of either "element" produces *either* the other *or* the matter.

Aristotle offers the following solution (334b8-14):

There are differences of degree in hot and cold. Although, therefore, when either is fully real without qualification, the other will exist potentially; yet, when *neither exists in the full-completeness of its being*, but *both by combining destroy one another's excesses* so that there exist instead a hot which (for a "hot") is cold and a cold which (for a "cold") is hot; then what results from these two contraries will be neither *their matter* nor *either of them existing in its full reality* without qualification. There will result instead an "*intermediate*" (μεταξύ).

And the qualitative characteristic or "power of action" of this

[26] Joachim (*Aristotle*: *On Coming-to-Be* and *Passing-Away*, p. 235 *ad* 333b16-20) offers the following summary: "The 'formula expressing the essential nature' of a ὁμοιομερές (like bone) is the λόγος τῆς μίξεως of its constituents, i.e. the scheme of proportions constituting the plan of the combination. This 'combining formula' (a) adequately expresses the 'form' (and is therefore the scientific definition) of the ὁμοιομερές; and (b) states the normal or perfect development of the ὁμοιομερές, its φύσις in the sense of τὸ τέλος τῆς γενέσεως (cf. e.g. *Met.* 1015a10-11), i.e. its 'good'". It must always be remembered, however, that for Aristotle the nature, form, or essence of the animal *as a whole* is primary and it is this which *determines* the proportion of elements in the formation of its individual parts; cf. *De Part. An.* 639b4 ff., esp. 640a10-b4; *Meteor.* 389b28 ff.

intermediate or compound depends upon the *proportion of the ingredients* and their "powers of action" (334b14-17):

And this "intermediate", according as it is potentially more hot than cold or vice versa, will possess *in the same proportion* (κατὰ τοῦτον τὸν λόγον)[27] a power-of-heating that is double or triple its power-of-cooling, or otherwise related thereto in some similar ratio.

Aristotle then adds that all material compounds come to be through this proportionate blending or "combining", asserting that "thus *all the other bodies* will result from the contraries, or rather from the 'elements', in so far as they have been 'combined'" (334b17-18). As explained above, the four primary bodies, the "elements" of all compounds in the sublunary world, *must* interact in virtue of their contrary qualities, and either change into one another or combine to form a third substance (334b23-26):

For the actually-hot is potentially-cold and the actually-cold potentially-hot; so that hot and cold, *unless equally balanced*, are *transformed* into one another (and all the other contraries behave in a similar way). It is thus, then, that in the first place the "elements" are transformed; and that in the second place out of the "elements" there come-to-be flesh and bones and the like ...

Finally, and most significant for our purpose, Aristotle goes on to indicate the basic principle of this process in which contraries, more or less "equally balanced", combine to form the "homoeomerous" parts of all bodies (334b26-30):

... out of the "elements" there come to be flesh and bones and the like, — the hot becoming cold and the cold becoming hot *when they have been brought to the "mean"* (ὅταν πρὸς τὸ μέσον ἔλθῃ). For *at the "mean" is neither hot nor cold.* The "mean", however, *is of considerable extent and not indivisible.* Similarly, it is *qua reduced to a "mean condition"* (κατὰ μεσότητα) that the dry and the moist, as well as the contraries we have used as examples, *produce flesh and bone and the remaining compounds.*

For Aristotle, then, as for Plato, the concept of opposing elements *blending in a mean* provides the basis for the understanding, and explanation, of all changing material things formed from the four

[27] This phrase is omitted in the Oxford translation.

primary bodies, — "flesh and bone and the remaining compounds". The primary bodies must be conceived as opposites, therefore, not only to explain the constant change, destruction, and generation of sensible things, but their formal structure as well, according to the general principle which Aristotle enunciates elsewhere (*De Part. An.* 652b17-20): "Now everything has need of an *opposite as counterbalance* (δεῖται τῆς ἐναντίας ῥοπῆς) in order that they may achieve *moderation and a mean* (ἵνα τυγχάνῃ τοῦ μετρίου καὶ τοῦ μέσου); for it is the mean that contains the substance and proper proportion (τὴν γὰρ οὐσίαν ἔχει τοῦτο καὶ τὸν λόγον), and not either of the extremes apart from it".[28] And it is because they must join in a mean, where the "substance and proper proportion" of the compound body lies, that the opposing ingredients must be σύμμετρα, — balanced or proportionate — in order to form a new compound, as Aristotle explains in connection with human generation (*De Gen. An.* 767a16-20):

> ... for all things that come into being as products of art or of nature exist in virtue of a certain proportion (λόγῳ τινί). Now if the hot preponderates too much it dries up the liquid; and if it is very deficient it does not solidify it; whereas for the artistic or natural product it is necessary to have this proportion, — the proportion of the mean (δεῖ ... ἔχειν τοῦτον τὸν τοῦ μέσου λόγον)... Thus there is need of *due proportion* (δεῖ τῆς συμμετρίας) in the mixture of the male and female elements also.[29]

On the basis of this evidence we may conclude in summary, then, that Aristotle conceives the elemental structure of the animal body as a compound of opposing constituents, earth, air, fire, and water, which come together in proportionate strength (συμμετρία) to blend or combine (κρᾶσις), the extreme "power" (δύναμις) of each component moderating, and being moderated by, its opposite in a mean (μέσον) which links the extremes without being identical with any of them. This mean is not the same for all compounds,

[28] This is a departure from the Oxford translation. We have already noted the connection between form, essence, or substance and the λόγος, the proper proportion or formula of constituents. Cf. 335b8.

[29] This is a departure from the Oxford translation. Cf. also *De. Gen. An.* 743a29-31, where lack of συμμετρία between hot and cold (i.e. excess or deficiency of one or the other) is said to produce deformity in the homoeomerous parts.

but is "of considerable extent and not indivisible",[30] i.e. it varies, within certain limits, according to the relative strength or proportion (λόγος) of the contrary "powers" within each compound. This allows for the formation of bodies which differ qualitatively according to form and nature. Thus the same four elemental bodies, blended to this "middle state" or "mean condition" (μεσότης), but in different proportions, compose the various "homoeomerous" parts, — blood, bone, flesh, marrow, etc. — which in turn form the "non-homoeomerous" parts and, ultimately, the whole of the animal body.[31]

This matter is of first importance for an understanding of Aristotle's concept of health and disease. For it is obvious that if the homoeomerous parts are properly formed and retain their integrity the whole body will be healthy. But, from what we have seen, the homoeomerous parts will be true to their nature and remain integral as long as the proper proportion and blend of opposing constituents are maintained in each. Consequently, it becomes clear why Aristotle holds that health is "constituted in a blending and proper proportion" (ἐν κράσει καὶ συμμετρίᾳ) of hot, cold, moist, and dry within the body (*Physics* 246b6).[32]

Conversely, it follows that when one or other of the opposing "powers" becomes excessive and its contrary correspondingly deficient in the blend, the "mean condition" (μεσότης) or equilibrium among them will be disrupted, and the process of destruction of the weaker and generation of the dominant set off. This condition, the opposite of health, will be disease, and if allowed to run its

[30] 334b28; and cf. *De Caelo* 312b1-3. The Oxford translator (Joachim) comments on the first, that "the mean is a stretch, not a point". Cf. also his *Aristotle: On Coming-to-Be and Passing-Away*, p. 244 *ad* 334b28ff.

[31] On the distinction between "homoeomerous" and "non-homoeomerous" parts, and their composition in the total animal body cf. *De Part. An.* 646a13-24 and the remarks of A. L. Peck in his introduction to the Loeb edition of the *De Partibus Animalium* (London, 1945), pp. 28-30.

[32] The same notion is implied in the association of health with *good blend* (εὐκρασία) in *De Part. An.* 673b26. In general, therefore, it appears that the definition of health based upon συμμετρία found in the *Eudemus* and *Physics* VII is quite consistent with the evidence gathered from the treatises, *De Gen. et Cor.*, *De Part. An.*, *De Gen. An.*, etc. Cf. note 14 above.

course will result in the dissolution of the homoeomerous part or parts, to end in death.[33]

Thus Aristotle can say (*De Part. An.* 648b2-6) that the elemental qualities, — hot, cold, moist, dry — seem to be "practically the causes (αἴτια ταῦτα σχεδόν) of life and death, not to mention ... prime and age, disease and health." This is but a more specific form of the general principle enunciated elsewhere (*De Gen. An.* 777b28-30): "As we know, it is heat and cooling in their various manifestations which up to a certain due proportion (μέχρι συμμετρίας τινός) bring about the generation of things, and beyond that point their dissolution ..."[34] The hot, cold, moist, and dry are the "proper affections" (πάθη) of the body which, because of its material nature, "tend to promote or destroy its existence" (*Physics* 246b9; *Meteor.* 378b27-379a11). And the body attains its primary excellence, health, when these exist ἐν κράσει καὶ συμμετρίᾳ, that is, ἐν μεσότητι.

3. *Environment and Its Influence*

The importance of *environment* in Aristotle's physiology has already been indicated in the passage from the *Physics* (246b3-7) where health is said to be constituted in a blending and proper proportion of hot and cold, moist and dry "in relation either to one another within the body *or to the environment*" (ἢ πρὸς τὸ περιέχον). Aristotle, like Plato, denies the existence of the void

[33] For an example of disease, cf. *De Part. An.* 652b33-653a10, where catarrhs are said to begin when "the parts about the head are colder than the properly proportioned blend" (ψυχρότερα τῆς συμμέτρου κράσεως). That death is ultimately due to the loss of equilibrium between hot and cold is clear from *Parva Nat.* 479a8-9: "The source of life is lost to its possessors when the heat with which it is bound up is no longer tempered by cooling ..."

[34] The translation is that of A. L. Peck in the Loeb edition of *De Generatione Animalium* (London, 1953). In a note on this sentence (p. 180) Peck calls attention to the use of συμμετρία in the definition of health at *Physics* 246b4 and goes on to remark: "The language used in the context of this passage is very similar to that of *Eth. Nic.* Bk. II (dealing with the doctrine of the 'mean'), where it is stated that the *moral* ἀρεταί also are produced and preserved by τὰ σύμμετρα (1104a18), whereas they are destroyed by excess and defect, just as the corresponding physical ἀρεταί are."

(*Physics* 217b20-23) and consequently conceives the hot, cold, moist, and dry elements of the surrounding macrocosm as in constant contact with their opposites in the animal body. Being in contact, the opposites tend to interact, and unless the animal body can establish and maintain its blend of elements intact against the bulk and "power" of the surrounding elements, it is broken down and dissolved again into the primary bodies by the environment.[35]

Some further details on the constitution of the body and its relation to environment appear in the fourth book of the *Meteorology* (378a10 ff.), where the generation, change, and decay of plants, animals, and their parts are explained through the interaction of the "active" powers, hot and cold, with the "passive" powers, moist and dry. The living body is generated and maintained when its constituent active powers, tempering each other in proper proportion,[36] dominate (κρατεῖν) the passive so that "coction" (πέψις) and blend take place. The integrity and life of the body depend upon its retaining this blend, and the blend re-

[35] On the constant contact and effect of environment upon the animal body, see *Physics* 253a7-20; 259b1 ff.; *Parva Nat.* 465b1-33.

[36] On the necessity of συμμετρία between the male (hot) and female (cold) element for the generation of new life, see *De Gen. An.* 767a14-28 where, as we have noted above, it is said that they must be related in proper proportion (συμμετρία) because all that is made by art or nature exists in virtue of a proportion (λόγος), and so the hot and cold elements in generation must likewise have "the proportion of the mean" (ὁ τοῦ μέσου λόγος). If the male and female element are widely disproportionate, no conception takes place (766b15-16; 767a24-28); while if there is a certain proportion but the male element is unable to "dominate" the female element completely, then a female or some more defective offspring results (743a27-33; 765b7-766b27). Since, for Aristotle, the balance of hot and cold is the determining factor in conception and the differentiation of sex, it is clear why he attributes great influence in these matters to the environment. Thus wind direction, the phases of the moon, climate, and drinking water are all regarded as significant for generation (766b35-767a35). And the movements of the celestial bodies must be of interest to the physician and physiologist since these control the climate, the heating and cooling of the world around us, which in turn exerts a decisive influence on the generation, and dissolution, of things (777b24 ff.). The production of perfect offspring, the healthy male, demands that all these influences work together so that the opposing (hot and cold) elements contributed by the parents be in perfect proportion (συμμετρία) both with respect to each other and to the environment.

mains as long as the tempered heat[37] within the body continues to dominate and "concoct" its tempered moisture. If this fails, i.e. if the active powers are overcome by the passive, the body and its parts become "unconcocted", not firmly blended, and the way is open for destruction and decay. Decay increases as the passive powers grow progressively stronger; the body first becomes moist, and finally, dry. Aristotle remarks elsewhere (*Parva Nat.* 465b18-22): "We must remember that an animal is by nature humid and warm, and to live is to be of such a constitution; while old age is dry and cold, and so is a corpse."

Now the passive powers overcome the active within the body through the influence of the environment (διὰ τὸ περιέχον. *Meteor.* 379a12 ff.). For the external heat of the environment may exhaust the internal heat of the body by kindling it to excess, so that it quickly consumes all available nourishment and burns itself out; or the excessive cold of the environment may overwhelm the innate heat and extinguish it (*Parva Nat.* 469b20 — 470a4; 474b10-24). In either case the innate heat grows too weak to master the passive powers within the body, and as it dwindles the natural moisture of the body becomes dominant and is then evaporated by the heat of the environment until finally the parts crumble to dust (*Meteor.* 379a22-26).[38]

It is evident, then, that if the animal body is to remain in good condition its innate heat must strike some sort of balance with the heat and cold of the environment. This involves a double consideration. First of all, the animal can exist only in the region proper to the element preponderating in its particular constitution, — fish in water, birds in air, etc. (*Parva Nat.* 477a27-31; 477b14 ff.; *Hist. An.* 590a12-18). However, there is a difference between this natural material constitution of a particular animal and the state (ἕξις) or disposition (διάθεσις) of its body *with relation to* its normal constitution (*Parva Nat.* 477b17-19). And changes in environment have a strong influence in determining

[37] The nature of this "innate heat" (σύμφυτον θερμόν) and its importance for the generation and continuation of life will be discussed later. Cf. *De Gen. An.* 736b 33 ff., 742a14, 784a 34 ff.; *De Part. An.* 650a 2 ff.; *Parva Nat.* 469b7-20.

[38] Cf. *De Gen. An.* 784a 31-b13, where the same principle is applied to account for the greying of hair in old age.

these states, for better or for worse (*Parva Nat.* 465b27-30; 466b15-28). Thus, while Aristotle maintains that "the natural character (φύσις) of the material of objects is of the same nature as the region in which they exist, — the liquid ... in liquid, the dry on land, the warm in air", he adds that, on the other hand, "with regard to states of body, a cold situation has a beneficial effect on excess of heat, and a warm environment on excess of cold, for the region reduces to a mean (εἰς τὸ μέτριον)[39] the excess (ὑπερβολή) in the bodily condition" (*Parva Nat.* 477b30-478a3). Conversely, when the environment changes from one extreme to another, the body must respond proportionately to retain its normal or "mean" condition. Man, for example, must breathe faster in hot weather in order to refrigerate his innate heat and keep it from growing excessive, while in cold weather his breathing is retarded to conserve the internal warmth (*Parva Nat.* 472a31-b3). And the lower animals must compensate for external changes by some corresponding proportionate response.[40] Thus all living bodies must somehow maintain their proper internal balance of hot and cold, moist and dry, with the help of, or in spite of, the environment. Otherwise, they perish (*Parva Nat.* 474b25 ff.).

The health and continued existence of the body, therefore, is closely dependent upon the environment. Consequently Aristotle maintains (*De Gen. An.* 777b7-8) that "the real reason why any animal is long-lived is that its blend is about the same in comparison with the air surrounding it"[41] (κεκρᾶσθαι παραπλησίως

[39] For the use of τὸ μέτριον in the sense of τὸ μέσον as opposed to excess and deficiency, cf. also *Physics* 261b19; *Topics* 123b29; *Politics* 1295b4.

[40] Characteristic responses of the lower animals to changes in environment, e.g. migration, hibernation, etc., are treated at length in *Historia Animalium* 596b20ff., where they are evidently understood as efforts to maintain the animal's proper proportion of hot, cold, moist, and dry against the extremes of the environment. Cf. also the generalization in *Parva Naturalia* 463b26-33.

[41] This is a departure from the Oxford translation. The phraseology of this statement in the *De Generatione Animalium* is reminiscent of the definition of health in *Physics* VII (246b 3-7) as constituted ἐν κράσει ... πρὸς τὸ περιέχον, — a further indication that the *Physics* passage is entirely consistent with the physiological terminology and theory of the other treatises. It may be accepted, therefore, as a fair formulation of Aristotle's mature doctrine on the subject. Cf. note 14 above.

πρὸς τὸν περιέχοντα ἀέρα). For when the blend of hot and cold, moist and dry in the environment matches the normal internal blend of the animal body, it tends to preserve the balance of opposites within the body and to "reduce to a mean" any excess in the bodily condition. This, as we have seen, is to preserve the integrity and insure the continued health of the body. It is clear, then, why in defining health as constituted in a κρᾶσις καὶ συμμετρία of hot, cold, moist, and dry (*Physics* 246b3-7) Aristotle is careful to indicate that this relationship involves not only the opposites within the body but those of the environment (πρὸς τὸ περιέχον) as well.

Thus, through an analysis of the elemental structure of the organism and its relation to environment, we have found that the notion of opposing constituents blending in a mean is basic to Aristotle's explanation of the formation of the animal body from earth, air, fire, and water. Consequently health, the best condition of the animal body, is defined as an habitual disposition of the body dependent upon the συμμετρία and κρᾶσις of its opposing elements, both in itself and with relation to the environment. Clearly, then, Aristotle's basic concept of health agrees substantially with that of Plato and the medical writers.

B. PROPORTION AND THE MEAN IN THE LARGER ASPECTS OF THE ORGANISM

For Aristotle, as for Plato and the medical writers, the concepts of balance, proportion, symmetry, and of opposites uniting in a mean explain not only the elemental formation of the animal body but the larger aspects of anatomy and physiology as well. This is not the place to attempt a complete and systematic exposition of Aristotle's anatomical and physiological doctrines; but a few of the more obvious and important applications of these principles may be indicated.

1. *Ultimate Basis of the Vital Processes and Organs*

Aristotle's discussion of particular organs and functions can only be understood in the light of the general principles which he detects behind all the varied manifestations of organic life. One of the most important of these is that the life of plants, animals, and man depends upon the presence of a properly tempered *heat* within the organism, a heat that is not communicated from outside but *innate* (σύμφυτον θερμόν).

One fundamental reason for this we have already seen, namely, that unless the passive powers (moist and dry) are properly balanced and "mastered" by the active (hot and cold) the parts formed by the blending of these opposites tend to disintegrate. Consequently the hot, tempered by the cold, must always be present if the body is to retain its integrity.

Moreover, without innate heat the homoeomerous parts could never come into existence in the first place, since heat is necessary for all growth. Nor could they remain in existence, since heat is necessary for nutrition, by which the body repairs the losses suffered through environment. "For without food, growth and even existence are impossible" (*De Part. An.* 655b3; and cf. *De Anima* 416b8-15). And food must be transformed and assimilated by the living body through the action of innate heat (*De Part. An.* 650a3-8):

Now since everything that grows must take nourishment, and nutriment in all cases consists of fluid [moist] and solid [dry] substances, and since it is by the power of heat that these are concocted and changed, it follows that all living things, animals and plants alike, must on this account, if on no other, have a natural source of heat.

The innate heat of the living body is, in fact, the material medium in which soul manifests itself most immediately, and the principal means by which it carries on the vital functions (*De Part. An.* 652b8-17):[42]

Some writers assert that the soul is fire or some such power (δύναμις). This, however, is but a rough and inaccurate assertion, and it would

[42] This is a departure from the Oxford translation. The reason why *soul*, rather than some material agent like heat or fire, must be the principal cause in nutrition and growth is given in *De Anima* 416a 8-18. The way in which soul uses heat to form the various parts, each according to its proper λόγος, is discussed in *De Generatione Animalium* 740b26-741a3, 742a25-744b28; and cf. *ibid.* 783a 33-b2, *Meteor.* 328a 22 ff.

perhaps be better to say that soul subsists in some body of a fiery character. The reason for this is that of all bodies the hot is most serviceable for ministering to the functions of the soul. For nutrition and the imparting of motion are the function of the soul, and these are brought about through the power of heat especially. To say, then, that the soul is fire is much the same thing as to confound the auger or the saw with the carpenter or his craft simply because the work is accomplished by the two in conjunction. So far, then, this much is plain, that all animals must necessarily have a certain amount of heat.

Obviously, then, Aristotle considers innate heat utterly indispensable for life in all its forms. "Everything living has soul, and it, as we have said, cannot exist without the presence of heat in the constitution" (ἄνευ φυσικῆς θερμότητος; *Parva Nat.* 470a20-21). "Generation is the initial participation, mediated by warmth (ἐν τῷ θερμῷ), in the nutritive soul, and life is the maintenance of this participation" (*ibid.*, 479a28-30). "Now the other psychical faculties (δυνάμεις) cannot exist apart from the power of nutrition ... nor this apart from the natural fire (ἄνευ τοῦ φυσικοῦ πυρός) by the union with which Nature has set it aglow" (*ibid.*, 474b10-13).

This innate heat or "fire", which is the immediate vehicle and instrument of soul, is not fire in the ordinary sense but "analogous to the element of the stars" (*De Gen. An.* 737a1-8),[43] i.e. to the so-

[43] This is, of course, a very summary treatment of the nature of σύμφυτον θερμόν. It must, however, suffice for the present to remark that I believe Ross is correct in distinguishing between σύμφυτον θερμόν and πνεῦμα (*Aristotle: Parva Naturalia* [Oxford, 1955], pp. 40-43), in contrast to Peck who identifies them (*Aristotle: Generation of Animals*, Loeb Library, 1951, p. 582, Appendix B, No. 13). Aristotle defines pneuma as *hot air* (*De Gen. An.* 736a1), air being conceived as "a sort of aquaeous vapor" (*De Gen. et Cor.* 330b5). Hence pneuma is really a combination of the hot and the moist, the first being "active", the second "passive" (cf. *De Gen. An.* 742a15). Consequently pneuma must be distinguished from "innate heat" as effect from cause or whole from part.

It is surprising, in view of his recognition of this distinction, to find Ross referring to *pneuma* as "analogous to the element of which the stars are formed" (*op. cit.*, p. 43). In the passage concerned (*De Gen. An.* 736b30-737a8) Aristotle is describing the nature of the particular material thing (σῶμα) which serves as the immediate instrument and vehicle of soul in the process of generation. We have already seen that he regards innate *heat* as such an instrument in all living organisms (cf. *De Part. An.* 652b8-17; *Parva Nat.* 470a20-21). In the present passage he says that all soul acts in conjunction with a material thing "differing from, and more divine than, the so-called elements", earth, air, fire, and water. This thing, present in the seed and making it fertile, is called fifth element or αἰθήρ (cf. *De Caelo* 269a31 ff.). However, it does conform to the nature of ordinary fire in that it can grow feeble and be extinguished or grow excessive and exhaust itself, either of which is the immediate cause of death (*Parva Nat.* 474b13-23; and cf. 469b20 ff.):

> But fire ... is destroyed in two ways, either by extinction or by exhaustion. It suffers extinction from its opposite. Hence it can be extinguished by the surrounding cold ... But exhaustion is due to excess of heat; for if there is too much heat surrounding it and the thing burning does not have a fresh supply of fuel added to it, it goes out by exhaustion, not by the action of cold.

It is clear, then, that if the animal is to survive, the vital heat must be maintained in a "mean state". It must be fed constantly through the nutritive process so that it does not become deficient; yet, lest it become excessive "there must be some way of cooling the heat resident in the source of warmth" (*Parva Nat.* 470a5-7). Hence Aristotle says (*ibid.* 476a16): "Every animal in order to exist requires nourishment; in order to prevent itself from dying, refrigeration." These two processes, then, are necessary to maintain the right balance or equilibrium between hot and cold, so that the properly tempered hot may continue to dominate the moist and

---

what we call *heat* (τὸ καλούμενον θερμόν). It is not fire or any such substance, "but the pneuma contained in the seed ... *or rather* (καί: cf. H. Smyth, *Greek Grammar for Colleges*, [New York, 1920] No. 2870) *that substance* in the pneuma, analogous to the element of the stars" (ἡ ἐν πνεύματι φύσις, ἀνάλογον οὖσα τῷ τῶν ἄστρων στοιχείῳ). The rest of the context leaves little doubt that by "the substance in the pneuma" Aristotle means innate *heat*, and that *this* is what he considers to be "analogous to the element of the stars", for it is the heat (θερμότης) of animals and of the sun which is contrasted to fire as being capable of producing life. Cf. *De Gen. An.*, 762a 20 ff., where "the *soul-heat* (θερμότης ψυχική) *present in* all pneuma" is named as the ultimate physical cause (ἀρχή) of spontaneous generation: and cf. *ibid.* 743a 27-36, where the "heat in the seminal residue" is the determining factor in embryonic development, being *proportionate* (σύμμετρος) to each part, and causing malformation when excessive or deficient. A brief history of scholarly opinion on the problem of connate pneuma appears in H. A. T. Reiche, *Empedocles' Mixture, Eudoxan Astronomy, and Aristotle's Connate Pneuma* (Amsterdam, 1960), pp. 78-80. Add: A. L. Peck, "The Connate Pneuma", in *Science, Medicine, and History*, edited by E. A. Underwood (Oxford, 1953), vol. I, pp. 111-121.

dry. The basic organic processes, therefore, are directed to a mean.

But nutrition and refrigeration can only be carried on through suitable organs. Consequently man, among other higher animals, is equipped with heart, digestive and vascular systems, brain, lungs, etc. For "the body exists in some way for the sake of the soul, and the parts for the sake of the functions for which each has been formed by nature" (*De Part. An.* 645b19-20).[44] Even the senses, with their accompanying appetites and emotions, serve the nutritive process, though in man this purpose is subordinated to the higher functions of the rational soul (*Parva Nat.* 436b19-437a2; *De Anima* 434a30-b2):

> To all that possess them [the senses] they are a means of preservation; their final cause being that such creatures may ... both pursue their food and shun things that are bad or destructive. But in animals which have also intelligence they serve for the attaining of a higher perfection.

Basic, then, to Aristotle's physiology is the notion that the union of soul and body, constituting the living organism, is conditioned by the presence of innate heat, properly balanced or tempered by cold, which "concocts" the moist and the dry and keeps intact the λόγος or proportion of elements proper to each of the bodily parts. The parts, in turn, at least those concerned with nutrition and refrigeration, are explained teleologically as directed toward maintaining this balanced blend or "middle state" of hot, cold, moist, and dry which is, as we have seen, the cause of the body's primary perfection or excellence, — health.

2. *The Principal Organs and Their Functions*

The principle of symmetry, proportion, and the mean finds constant application throughout Aristotle's discussion of particular bodily parts, systems, and functions. This can be illustrated adequately by considering particularly two specific parts which, Aristotle holds (*De Part. An.* 673b11-12), "possess supreme power over the life of the body", namely, the heart and the brain.

[44] This is a departure from the Oxford translation. For soul as the final cause of the living body, cf. *De Anima* 415b15-22.

a. *The Heart: Central Organ of the Body*

The heart is, for Aristotle, the most important organ of the body. He maintains (*Parva Nat.* 467b13-16) that while the soul in its essential reality (οὐσία) cannot be corporeal, yet "it must exist in some bodily part, which must be one of those possessing control over the members". Since any animal soul, including man's, must have at least the double power of nutrition and sensation, its proper bodily organ "must be numerically one and the same and yet possess multiple and disparate aspects" (*ibid.* 467b26-28). And this organ, to be appropriately located for its double function, must be central (*ibid.* 467b28-468a1):

> Since then the organs of special *sensation* have one common organ (αἰσθητήριον) in which the senses when functioning must meet, this must be situated midway (μέσον) between what is called "before" and "behind" (we call "before" the direction from which sensation comes, "behind" the opposite). Further, since in all living beings the body is divided into upper and lower ... clearly the *nutritive* principle must be situated midway (ἐν μέσῳ) between these regions.

That the heart is this central organ of nutrition and sensation can be shown, Aristotle believes, by both observation and inference (*ibid.* 468a13 ff.). He argues from observation that the heart must be the center of nutritive life because it is the first organ formed in the embryonic development of blooded animals (*ibid.* 468b28-32; *De Gen. An.* 739b34 ff., 742a17 ff.) and is the source of the vascular system which supplies the rest of the body with nourishment in its final form as blood (*Parva Nat.* 468b31 ff.; *De Part. An.* 665b10 ff.). Furthermore, since the organs of taste and touch can be traced to the heart (*Parva Nat.* 469a10-17; 438b25 ff.) and since nutritive life and sensation cannot be separate in animals (*ibid.* 469a18-19), it is clear that "the source of the sensitive soul, together with that connected with growth and nutrition, is situated in this organ and in the central one (ἐν μέσῳ) of the three divisions of the body" (*ibid.* 469a23-26).

Deduction leads Aristotle to the same conclusion (*ibid.* 469a26ff.). For "when several results are open to her, Nature always brings to pass the best." Now if the principle of nutrition and sensation is located centrally, joining opposite parts of the body, they will function best, for the principle will then be close to both and thereby best able to mediate the vital processes to the extremities.[45] The center (μέσον) is, therefore, the proper place for the ruling principle and source (ἀρχή), the position "of a primary or dominating part. For Nature, when no other more important purpose stands in her way, places the more honorable part in the more honorable position; and the heart lies about the center (περὶ μέσον) of the body, but rather in its upper than its lower half, and also more in front than behind" (*De Part. An.* 665b18-22; cf. also 665a11-25, 655b29 ff.).

Nature, then, like an intelligent craftsman, operates according to the principle of symmetry, placing the most important organ, the heart, in the center and forming the rest of the body proportionately about it.[46] "In reality ... all the organs are double. The reason for this is that the body itself is double, consisting of two halves, which are, however, combined together under one supreme center. For there is an upper and a lower half, a front and rear, a right side and a left" (*De Part. An.* 669b18-22). Thus the body, in its gross aspect, is conceived by Aristotle as divisible into equal but opposite halves grouped symmetrically about a central organ, the heart.

b. *The Heart: Center of Innate Heat and Activity*

The fundamental reason why the heart alone can be, for Aristotle, the primary organ of the soul in nutrition and the other vital functions is that he regards it, above all, as the center and source of innate heat in the living body (*Parva Nat.* 474a25-b3):

> We have already stated that life and the presence of soul involve a certain heat. Not even the digesting process to which is due the nutrition

[45] Cf. also *De Part. An.* 666a13-16: "... of all places the best suited for a source is the center. For the center is one, and is equally or almost equally within reach of every part."

[46] Nature is frequently compared to a skilled craftsman; cf. *De Part. An.* 645a10, 654b29ff., 687a12; *De Gen. An.* 731a24; etc. The importance which Aristotle attaches to symmetrical position may be seen in the fact that he rules out the liver as the primary organ because its position is *not* central, and because it is counterbalanced by the spleen (*De Part. An.* 666a26-28).

> of animals occurs apart from soul and warmth, for it is to fire that in all cases elaboration is due. It is for this reason, precisely, that the primary nutritive soul also must be located in that part of the body and in that division of this region which is the immediate vehicle of this principle ... intermediate (ὁ μέσος τόπος) ... in the sanguineous animals ... the heart.

Although "in animals all the members and the whole body possess some connate warmth", still "the source of this warmth must be the heart in sanguineous animals" (*ibid.* 469b6-11), for "the source of heat in all the other members depends on this, and the soul is, as it were, set aglow with fire in this part ... " (*ibid.* 469b14-16; cf. 474b12, 478a29). The heart is, for this reason, essential to every blooded animal, "that it may lodge the central source of heat. For some part or other there must be which, like a hearth, shall hold that which kindles the organism" (*De Part. An.* 670a24-26).

Since, as we have seen, the soul manifests itself most immediately through the vital heat, using it as an instrument in the life processes by which the body is formed and conserved, it follows that the heart, as center and source of the vital heat, must be the center of the soul's activity wherever that activity involves the body, i.e. in nutrition, sensation, autonomic movement, and the emotional responses leading to external action. The heart's function in sensation, emotion, and local movement will be considered later; our present discussion will be confined largely to the heart as center of *nutrition*, the basic vital activity upon which all others are contingent.

For as center of growth and nutrition the heart governs life and death. In the generation of blooded animals it is the first organ formed, and in this sense is "the source of their nature" (ἡ ἀρχὴ τῆς φύσεως: *De Part. An.* 666a20-22; *De Gen. An.* 741b15-18; cf. *Metaph.* 1013a4-7). It is also the organ in which life finally fails: "It is always to some lack of heat that death is due, and in perfect creatures the cause is its failure in the organ containing the source of the creature's essential nature" (*Parva Nat.* 478b32-33; *De Gen. An.* 741b18-24). The life of the animal, therefore, begins at the center and its body builds out from the center (cf. *De Gen. An.* 741b25 ff.); the animal withers and dies as the vital heat is exhausted and finally extinguished at the center (*Parva Nat.* 478b32-479b6).

c. *The Heart: Center of Nutritive and Vascular Systems*

As center of the vital heat, the heart is the most important organ of the soul in carrying on the nutritive process, for, "though all parts of the body by their natural heat elaborate and concoct the nutriment, the governing organ takes the chief share in this process" (*Parva Nat.* 469b12-13).[47] By the "chief share" in the nutritive process Aristotle has in mind the concoction of the blood, which takes place in the heart (*De Part. An.* 647a25-b8, 650a3-b13, 667b27-32; *De Gen. An.* 739b34 ff.; *Parva Nat.* 456a30 ff.). Here the useful elements of food are formed into blood through the action of the vital heat, and then pumped out through the vascular system to develope and nourish the rest of the body. Hence the heart is also the primary container of the blood, and the chief organ and source of the vascular system (*De Part. An.* 650a32 ff., 665b10 ff., 666a6 ff.; *De Gen. An.* 743a1 ff.; *Parva Nat.* 474a25 ff.).

The principle of proportion and the mean appears in several important details of the nutritive-vascular processes. For example, the food which a particular animal takes must be proportionate to its innate heat, neither excessive nor deficient. If food is deficient, "the natural warmth which is the primary cause of digestion consumes the material in which it is located" (*Parva Nat.* 466b31-32), i.e. it consumes the body, and the animal perishes. Or food may be excessive, either because the innate heat is too feeble to concoct it or because the food itself is excessive in quantity or indigestible in quality (*De Part. An.* 668b12-16). Thus, in the case of athletes, who over-eat, "owing to the quantity of their food their nature is not able to master it all so as to increase and arrange their form symmetrically; therefore their limbs develop irregu-

[47] Aristotle discusses the process of nutrition and growth on various levels throughout the treatises, from its metaphysical aspects to the concrete details of digestion and assimilation. For some of the general principles cf. *Physics* 260a 29 ff.; *De Caelo* 321b18-322a27; *De Gen. et Cor.* 321b18-322a27, 328a 33 ff., 334b 31 ff.; *De Anima* 416a18 ff. The concrete details of the process may be found neatly summarized, with references, by the Oxford translator (Ogle) in a note to *De Partibus Animalium* 668a 8.

larly ..." (*De Gen. An.* 768b29-33). And lack of proportion between food and vital heat brings on disease. For example, the function of the spleen, according to Aristotle, is to draw off the residues of food from the stomach and assist in concocting them. "Should, however, this residual fluid be too abundant, or the heat of the spleen too scanty, the body becomes sickly from over-repletion with nutriment" (*De Part. An.* 670b5-8). Aristotle generalizes elsewhere as follows (*De Gen. An.* 784a34-b2):

> We must remember that the nutriment coming to each part of the body is concocted by the heat appropriate to the part; if the heat is inadequate the part loses its efficiency, and destruction or disease results.

Moreover, the quantity and quality of the blood is proportionate to that of the food consumed by the animal. For the food is broken down into its elements by the various digestive organs and the useful part concocted into blood by the heart. The amount and quality of blood, then, depends upon that of the food, since blood is simply food reduced to a form in which it can be immediately assimilated by the elements of the living body. Consequently, "the blood diminishes in quantity when no food is taken and increases when much is consumed" and "becomes healthy and unhealthy according as the food is of one or the other character" (*De Part. An.* 650a37-b3).

The principle that excellence lies in a "middle state" or "mean condition" is applied by Aristotle in determining the best type and amount of blood. "Blood at its best, before it undergoes deterioration from either natural decay or from disease, is neither very thick nor very thin" (*Hist. An.* 520b22-23). "Animals that are in good condition ... have the blood neither too abundant ... nor again very scanty ..." (*ibid.* 29-32). Accordingly, blood is said to be in its best or "mean" condition in middle age, when the animal is at its prime: "In very young animals it resembles ichor and is abundant; in the old it is thick and black and scarce; and in middle aged animals its qualities are intermediate" (μέσως: *ibid.* 521a32-34).

Obviously, then, Aristotle would agree with the author of *Ancient Medicine* (Ch. 9) that in prescribing proper diet one must "aim at the mean". For food must be proportionate in quantity and quality to the vital heat, so that it will be concocted into the mean amount and quality of blood. Then it will nourish the various parts of the body proportionately, and maintain, or restore, the internal equilibrium of health.

d. *The Heart's Structure and the Principle of the Mean*

Finally, the principle of the mean is evidently assumed in Aristotle's explanation of the internal structure of the heart itself. He insists, contrary to the facts, that the heart in the largest and most perfect animals has *three* chambers or cavities (*Hist. An.* 496a4 and 17ff., 513a27ff.; *De Part. An.* 666b22ff.; *Parva Nat.* 458a10ff. and 19).[48] The reason offered (*De Part. An.* 666b25ff.) is that there are two main blood vessels from which all others stem, differing from one another, so that it is "better that they have separate sources". Hence, in larger animals, the heart will have at least two separate cavities or receptacles to serve as sources for these. However, Aristotle believes, "it is better still that there be three chambers, in order that there may be a single common center and source

[48] This point has exercized the commentators from Galen to D'Arcy Thompson. Cf. the latter's long note in the Oxford translation of *Historia Animalium* at 513a35, where it is suggested that Aristotle's assertion of three chambers may be explained as an example of "superstition ... in conflict with the results of observation", due to the influence of "tradition or mysticism", the "persistence of archaic and traditional views in regard to this central organ". To invoke "superstition", "mysticism", and "tradition" seems no satisfactory explanation but a measure of desperation. In the light of the evidence we have seen, I propose instead that Aristotle's three-chambered heart is another application of the rational principle of proportion, symmetry, and the mean, which pervades his physiological thought as it does that of Plato and the medical writers. This is suggested by his explanation in *De Partibus Animalium* 666b22-667a7. It is logical that Aristotle should find this basic principle embodied in the most perfect organ of the most perfect animal, and we need not be surprised to find his judgement in the matter influenced at least as much by deduction as by observation, since he employs both methods in physiological investigation (for example, in arriving at the *heart* as the central organ of soul and body: *Parva Nat.* 468a13ff.). Cf. his rigorous application of the principle of symmetry to the internal organs, including the heart, in *De Partibus Animalium* 669b13ff. In this connection see the illuminating essay on "Aristotle's Method in Biology" by I. Düring in *Aristote et les problèmes de méthode* (Louvain, 1961), pp. 213-221.

(ἀρχή). But the middle and odd-numbered is the source" (τὸ δὲ μέσον καὶ περιττὸν ἀρχή).[49]

Accordingly, the middle chamber is related in size to the others as mean to extremes (*Hist. An.* 496a20-22; cf. 513a32-35), "the right-hand one the largest of the three, the left-hand one the smallest, and the middle one intermediate in size" (μέσην δὲ μεγέθει τὴν ἀνὰ μέσον).

Likewise, the *blood* contained in the central chamber is correspondingly related to that of the other chambers as mean to extremes (*De Part. An.* 667a1-4):

Of these three cavities it is the right that has the most abundant and the hottest blood ... The left cavity has the least blood of all and the coldest; while in the middle (μέσον) cavity the blood, as regards quantity and heat, is intermediate (μέσαι) to the other two, being however of purer quality than either (καθαρώτατον).[50]

The purpose which Aristotle sees behind this precise anatomical structure is that the source and center of the living body be kept calm and stable, a state which evidently he, like Plato and the medical writers, associates with a "middle" or "mean" condition. "For it behoves the supreme part (ἀρχή) to be as tranquil (ἠρεμεῖν) as possible, and this tranquility can be ensured by the blood being pure, and of moderate (μέσος) amount and warmth" (*De Part. An.* 667a4-7).

The middle chamber of the central organ, then, is regarded as the true center and ultimate source of the living body. Here, we are told (*Parva Nat.* 458a10ff.), the finest and purest blood is separated from the thick and turbid and sent into the upper parts of the body, where it nourishes the nobler organs (heart, blood-vessels, flesh, sense organs; cf. *De Gen. An.* 744b12-28), while the thick and turbid flows to the lower parts (cf. *De Part. An.* 647b29-36). Hence the central cavity of the heart may properly be considered the center of the nutritive-vascular system, the *mean* which unites extremes in the structure of the heart itself and in the whole body.

[49] This is a departure from the Oxford translation.

[50] Aristotle explains the off-center position of the heart in man as compensation for the excessive size and heat of the right-hand chamber, thus maintaining the principle of symmetry: *De Part. An.* 666b7-11; 667a2-3.

e. *The Brain: Counterpoise of the Heart*

The generation, development, and continued existence of every animal requires, as we have seen, that the vital heat essential to life be tempered, i.e. be properly balanced and moderated by cold. A new organism cannot even be conceived until hot and cold are brought together in a mean (*De Gen. An.* 767a17): "For all things that come into being as products of art or nature exist in virtue of a certain ratio (λόγῳ τινί). Now if the hot preponderate too much it dries up the liquid; and if it is very deficient, it does not solidify it; for the artistic or natural product we need this ratio, — the ratio of the mean" (τοῦτον τὸν τοῦ μέσου λόγον).

In the embryonic development of blooded animals the heart, as center of the vital heat, must be the first organ formed. But in order to counterbalance the heart and its heat, nature proceeds at once to establish a center of cold in the organism by forming the brain (*ibid.* 743b27-33):[51]

Because of the heat of the heart, the cold causes the brain to form in the place where the blood-vessels terminate above, countering (ἀντίστροφον) the heat surrounding the heart. Hence the parts about the head begin to form next in order after the heart and surpass the other parts in size, for the brain is from the first large and fluid.

Thus Aristotle regards the brain as essentially related to the heart. He conceives both organs as necessary, demanded by the principle that the existence of the whole organism, like that of its parts, derives from opposites combining in a mean according to a proportion proper to the essential nature of the organism. His formulation of the general law has already been noted (*De Part. An.* 652b7-23):[52]

The brain is present in order to preserve the animal organism as a whole ... So far this much is plain, that animals must necessarily have

[51] This is a departure from the Oxford translation. It should be noted here that Aristotle regards *cold* not merely as a privation of heat (cf. e.g. *De Gen. An.* 743a 37) but as a positive or active δύναμις (cf. *De Part. An.* 649a19; *Meteor.* 382b 5 ff.; 347b 2-7; 348b 2-8).

[52] *Supra*, p. 172. The translation here is an adaptation of that of A. L. Peck in the Loeb Library edition of *De Partibus Animalium*. In connection with the present passage cp. *Parva Naturalia* 439a 3, where the heart, as the hottest of all bodily organs, is called the "counterpoise of the brain".

a certain amount of heat. Now everything has need of an opposite as counterbalance in order that they may achieve moderation and a mean; for it is the mean that contains the substance and proper proportion, and not either of the extremes apart from it. For this reason nature has contrived the brain to counterbalance the region of the heart and the heat in it; and that is why animals have a brain, the composition of which is a combination of water and earth.

Water and earth are the coldest of the elemental bodies (*De Gen. et Cor.* 330b5-6) and the brain the coldest part in the living body (*De Part. An.* 652a28; *Parva Nat.* 457b30), designed thus by nature for its function of cooling the heat in heart and blood. "The brain, then, tempers the heat and seething of the heart" (*De Part. An.* 652b26-27) and "chills the blood that streams upward after food" (*ibid.* 653a14).

To perform its function the brain must be proportionate in quantity and quality to the heart and its heat. Hence, "of all animals, man has the largest brain in proportion to his size; and it is larger in men than in women. This is because the region of the heart and of the lung is hotter and richer in blood in man than in any other animal; and in men than in women ... It is then as a counterpoise to his excessive heat that in a man's brain there is this superabundant fluidity and coldness" (*ibid.* 653a29-34). Because brain and heart are thus properly proportioned they "achieve moderation and a mean" (*ibid.* 652b18), the cold of the brain tempering the heat of the heart so that it is "well-blended" (εὔκρατον: *ibid.* 27), both contributing to the superior condition of man which Aristotle characterizes as "excellence of blend" (εὐκρασία: *De Gen. An.* 744a31).

Because of the essential connection and delicate balance between brain and heart as centers of cold and heat, any change in the brain brings an immediate reaction in the center of life at the heart. "For the cardiac heat and the center of life is most delicate in its sympathies, and it is immediately sensitive to the slightest change or affection of the blood on the outer surface of the brain" (*De Part. An.* 653b5-8).

f. *The Brain: Effect on Sense Life; Location of Senses*

Through its close connection with the heart and its function in cooling the blood, the brain plays an important role in sensation, which we mention here because it helps Aristotle to explain why nature has located both brain and sense organs in the head. Though the brain itself is not a sensorium (*De Part. An.* 652a24 ff.; 656a15), it affects the sense life of the animal through its influence upon the center of sensation in the heart. Thus it plays an essential part in Aristotle's explanation of the phenomena of sleeping and waking (*ibid.* 653a12-18; and cf. *Parva Nat.* 455b28 ff.):

It is the brain again ... which is the cause of sleep. For either by chilling the blood that streams upwards after food, or by some other similar influences, it produces heaviness in the region in which it lies ... and causes the heat to escape downwards in company with the blood. It is the accumulation of this in excess in the lower region that produces complete sleep ...

We have already noted, moreover, that for Aristotle, as for Plato, the blood stream is the medium which carries the sense impulses from the organs to the center of sensation. The quality of the blood, therefore, will be of considerable importance in the sense process. The brain, then, affects the sense life of the animal through its influence in determining the quality of the blood. Aristotle holds that when blood is thin and cold it conveys sense impressions quickly and accurately, though not so much from its coldness as from its thinness (*De Part. An.* 648a3-4; 650b19-23). Now the brain by its coldness imparts this desirable quality to the blood passing through its blood-vessels, and so nature locates the organs of the majority of senses, — sight, hearing, taste, and smell — in the head, in close proximity to the brain (*ibid.* 686a7-12):[53]

The head exists mainly for the sake of the brain, for blooded animals must have this part, and in a place counterbalancing that of the heart, for the reasons mentioned earlier. But nature has also located some of the senses up in the head, because the blend (κρᾶσις) of the blood there is properly proportioned (σύμμετρος) and adapted both to the warming of the brain and to the tranquillity and accuracy of the senses.

[53] This is a departure from the Oxford translation. The physiological aspects of sensation will be treated later. Some of Aristotle's reasons for ruling out the brain as the center of sensation may be found at *De Part. An.* 656a14ff.; and cf. the Oxford translator's note to 656a24.

Elsewhere Aristotle observes (*ibid.* 656b3-8):

Moreover, it cannot but necessarily be that the more precise senses will have their precision rendered still greater if ministered to by the parts that have the purest blood. For the motion of the heat of blood destroys sensory activity. For these reasons the organs of the precise senses are lodged in the head.

The brain, therefore, performs a function of the highest importance for the sensitive, as well as for the nutritive, life of the animal. Hence it conditions also, though indirectly, the intellectual activity of man, which depends upon sensation.

g. *The Brain: Its Proper Blend and the Principle of the Mean*

The blood-vessels in the brain, as Aristotle mentions above, also serve to moderate the cold of that organ. For though the brain is the center of cold in the body, even here the extreme must be tempered (*De Part. An.* 652b27-33):[54]

In order, however, that it may not itself be absolutely without heat, but may have a moderate amount, branches run from both blood-vessels ... and end in the membrane which surrounds the brain; while at the same time, in order to prevent any injury from the heat, these encompassing vessels, instead of being few and large, are numerous and small, and their blood scanty and clear instead of being abundant and thick.

The brain, therefore, must also have its own proper proportion of heat and cold. The heat which tempers its cold is supplied not only by the blood-vessels on its surface, but also by the spinal-marrow. It is precisely in this that Aristotle sees the reason for the brain being continuous with the spinal marrow (*ibid.* 652a27-33):

In reality the two may be said to be utterly opposite to each other. For of all the parts of the body there is none so cold as the brain; whereas the marrow is of a hot nature ... Indeed this is the very reason why the brain and spinal marrow are continuous with each other. For wherever the action of any part is in excess, nature so contrives as to set by it another part with an excess of contrary action, so that the excesses of

[54] Cf. *De Part. An.* 658b1-10, where Aristotle explains that man has more hair on his head than any other animal because he has the largest and most fluid brain, which requires proportionately heavy protection against the extremes of heat and cold.

the two may counterbalance each other (ἵνα ἀνισάζῃ τὴν θατέρου ὑπερβολὴν θάτερον).

The last statement is important as representing another formulation of the general law amounting in effect to this, that nature, operating for the best, always moderates extremes by joining them in a mean, so that they counterbalance each other in a dynamic equilibrium of opposites. Hence that is, for Aristotle, the natural or normal condition, and when it prevails each part performs its proper function and the whole organism is healthy. But when the proper proportion or balance of opposites is disrupted the equilibrium is broken, malfunction and disease result.

So it is with the brain. Disturbing its proper internal proportion of hot, cold, moist, and dry brings on malfunction and disease. Thus Aristotle explains catarrhs or "colds" as arising in the head whenever the parts about the brain grow "colder than the properly proportionate blend" (ψυχρότερα τῆς συμμέτρου κράσεως: *De Part. An.* 652b36). Elsewhere he says (*ibid.* 653b3-5) that "if the brain be either too fluid (moist) or too solid (dry), it will not perform its function, but in the one case will freeze the blood, and in the other will not cool it at all; and thus will cause disease, madness, and death."

This illustrates well the principle that the proper *function* of the part, as of the whole organism, depends upon the proper *internal disposition* or proportion of its opposing constituents. When this latter is disturbed the part, and the whole organism, are unable to perform their function properly, and pain, disease, destruction are the result.

It must be clear now why Aristotle considers heart and brain "the main governing powers of life" (*ibid.* 673b12). For he conceives them primarily as coordinate opposites, the centers of heat and cold, which produce and control that proportionate blend or "middle state" essential to the existence and well-being of the animal organism.

3. *Summary*

The evidence presented even in this brief discussion has, it may be hoped, shown clearly enough that the notion of properly proportionate or symmetrical opposites blending in a mean is fundamental to Aristotle's analysis of man's physical or bodily aspect at every level.

He defines health as an excellence arising when the hot, cold, moist, and dry blend in due proportion both within the body and in relation to the environment. We have found this definition consistent with his explanation of the elemental structure of the homoeomerous parts, which form when earth, air, fire, and water join in a proportion proper to the nature of each part and blend to a mean or "middle state" through the action of heat. The body, to retain its integrity and well-being, must maintain this blend or "mean condition" against the environment, compensating for changes in the environment by proportionate responses. Otherwise disease and destruction result.

The critical factor in all this, we have seen, is the presence of a properly moderated innate heat, the immediate vehicle and instrument of soul in the vital processes by which the parts and the body as a whole are formed and preserved in existence. When the vital heat grows excessive or deficient the body disintegrates, its matter no longer disposed to retain its form, and the union of soul and body is dissolved. Consequently, to maintain the vital heat in a "mean state", properly proportioned or tempered by cold, nature designs the processes of nutrition and refrigeration, and even of sensation and emotion, with the organs by which they are carried on.

Aristotle postulates, therefore, a center and source of vital heat in the organism, and locates this, according to the principle of symmetry, in the heart, the organ closest to the mid-point of the body.[55] Here, in the "mean" position, the place of honor, the

[55] Much more could be said of the symmetry or balance which Aristotle finds in the organism as a whole, and especially in the body of man, whom he regards as the most complex and divine among the animals (*De Part. An.* 656a3-13; 686a25ff.). We have noted his formulation of the principle that Nature always strives to balance one extreme with its opposite (*ibid.* 652a31-33) in the body, an ideal which is best realized in the "most highly-finished" animals (cf. 666a28). Accordingly he views the structure of the whole body as a combination of equal but opposite halves, upper and lower, right and left, front and back, joined at a single source and center (669b18ff.). Consequently the viscera,

heart mediates the nutritive and sensitive processes to the extremes. As center of the nutritive process, it receives the useful residue of food, concocts it into blood, and sends it out through the vascular system to nourish the rest of the body.[56] Consistently, Aristotle divides the human heart into three chambers, of which the center is related to the others as mean to extremes and contains the best blood, i.e. blood of mean quantity and quality, not too abundant or scanty, thick or thin, hot or cold.

Finally, Aristotle conceives the brain as a center of cold, set by nature in the head to counterbalance the heat of the heart and thus to establish the proper middle state or mean condition in the organism as a whole. To fulfill this function the brain itself must be properly proportioned, comparatively large and fluid in man to balance his heart, yet not too hot or cold, moist or dry. For any disproportion in the brain throws off the balance at the center of life in the heart, and so disrupts the equilibrium of the whole organism, bringing disease and death.

Thus, from its elemental structure up through the great processes that govern life and the chief organs by which these processes are carried on, Aristotle sees the living body as a dynamic equilibrium of opposites blending in a mean.

— brain, heart, lungs, kidneys, liver and spleen, etc. — are correspondingly divided (*ibid.*), as are the sense organs (*ibid.*; and cf. 656b33-34). "The reason, then, why the viscera are bilateral is ... that there are two sides to the body, a right and a left. For each of these sides aims at similarity with the other, and so likewise do their several viscera; and as the sides, though dual, are knit together into unity, also so do the viscera tend to be bilateral and yet one by unity of constitution" (670a3-7). For all that they may "aim at similarity", however, the halves are conceived essentially as opposites, conforming in some degree to the traditional "column of opposites" mentioned in *Metaphysics* 986a21ff. Thus Aristotle remarks that "the left side of the body is generally more watery and colder than the right. As we know, the opposites are divided into two columns, so that each is classed with those akin to it, e.g. right is in the opposite column to left, and hot to cold; and thus some of them stand together in the same column, as I have just indicated" (*De Part. An.* 670b19-23, translated by A. L. Peck in the Loeb Library).

[56] The heart as center of sense and emotional activity will be considered hereafter, as remarked above.

C. PROPORTION AND THE MEAN IN SENSE PERCEPTION

From Aristotle's point of view our discussion so far has been one-sided, concerned largely with the material aspect of the living organism and neglecting its most important aspect, — the form or soul. Yet what we have seen provides a natural and necessary complement to any discussion of Aristotelian psychology. For the first positive point made by Aristotle in the *De Anima* is that the soul in most of its experiences (πάθη) cannot act or be acted upon without involving the body (403a5-7).[57] Examples are "anger, courage, appetite, and sensation generally" (*ibid.*) and again, "passion, gentleness, fear, pity, courage, joy, loving and hating", in all of which "there is a concurrent affection of the body" (403a16-18). He warns against discussing or defining any one of these from the psychic or somatic aspect alone, as, for example, the dialectician might define anger as "the appetite for returning pain for pain", while the pure physiologist would call it rather "a boiling of the blood or warm substance surrounding the heart" (403a29-33). In this the one states only the form or formulable essence of anger and the other its material conditions, whereas the two aspects are *combined* in the affection of anger as it exists in reality. The "genuine physicist" must therefore keep in mind *both* the psychic and physical aspects of sensation and emotion (403b7-8, 18-19). And this we must try to do in understanding Aristotle's psychology.

1. *Body, Soul, and the Living Composite*

This close interdependence of the psychic and somatic is explained ultimately through Aristotle's concept of soul and body as the form and matter of the living organism. He first defines soul in the *De Anima* (412a19-20) as "substance in the sense that it is the form of a natural body having in it the capacity of life" (412a19-

[57] In the following section references to the *De Anima* will be given by Bekker number *alone*. The Greek text of R. D. Hicks (*Aristotle*: *De Anima*, [Cambridge, 1907]) and the Oxford translation will be used, though the latter will be altered at times where details require a more literal version.

20), relating it to the matter of the organism as act to potency, and as "first act" upon which follow all the "second acts" of the life functions (412a21-28). However, since a body is capable of life only in so far as it possesses the organic structures necessary for the functions of nutrition, sensation, etc., soul is later defined more precisely as "the first grade of actuality of a natural body furnished with organs" (412a28-b6).

Soul, as the form or intelligible element, is the very essence of the living organism, like axness to the ax or eyesight to the eye. As the pupil joined to eyesight constitute the eye in its proper sense, so body and soul constitute the living organism (412b10-413a5). They are the two principles or causes of the one reality, really distinct, but incomplete in themselves. The body can no more live and function apart from the soul than soul can activate a body lacking the organic structure necessary for the various life functions (414a16-27; *Met.* 1036b21-32).

Aristotle classifies living organisms according to their capacity for characteristic life activities on various levels (414a29 ff.). Plants display the activities of nutrition, growth, and decay. Animals are characterized by sense perception and appetency, at least through the sense of touch. Higher animals possess the other senses and the power of self-movement. Man, finally, is capable of rational activity, of forming and contemplating the universal. Soul is the principle for all these functions and distinguished accordingly as nutritive, sensitive, movent, and rational.

These various types of soul, then, exist separately in the different classes of organisms; but in man and animals the higher functions presuppose and include capacity for the lower functions as well. Thus Aristotle comes to speak of the nutritive, sensitive, and rational "souls" or "parts of soul" in man, though he maintains that these are but separate functions or faculties (δυνάμεις, "powers") of the one soul, and that they cannot be really distinct, nor each assigned to a different part of the body (411a26-b27; 413b11-32).

The soul of the organism, however, is no mere static form, like that of the statue or the bronze sphere. Rather it is the source of all the change and movement involved in the life functions, the immanent principle behind the ordered processes of growth and development from conception to maturity (415b22-27). As such, it is the *efficient* cause of the organism. And since the body exists for the sake of the soul, the soul is also the *final* cause of the composite (415b15-22).

Yet the soul can carry on the functions of nutrition and sensation only by reason of informing a body, which it moves, and by reason of which it can be moved. Hence the activities of soul on these levels can never be understood apart from the body and, more particularily, apart from the specific bodily organs which these various functions require for their operation. To say that the *soul* sees or is angry, then, is for Aristotle as inaccurate as to say that it weaves or builds (408b13-18):

It is doubtless better to avoid saying that the soul pities or learns or thinks and rather to say that it is the man who does this with the soul. What we mean is not that the movement is in the soul, but that sometimes it terminates in the soul and sometimes starts in it ...

The soul as the "first grade of actuality of a body *equipped with organs*", then, must always be understood as functioning, on the nutritive and sense levels at least, through and in appropriate bodily organs. It is not the soul that carries on these life functions, but the compound, the ensouled organ, and, ultimately, the man.

Of the three great levels of activity in man — nutritive, sentient, and rational — we have already considered in brief the organs by which the nutritive "part" of his soul operates, the part by which he is said primarily to "live". While man's soul activates or informs all the organs and carries on, necessarily, its nutritive function in *all* parts, still, as we have seen, this is possible only because some organs are specifically designed to receive food, change it into an assimilable form as blood, and send it out to all the other parts. Hence the nutritive "part" of man's soul is thought to operate through these organs preeminently, and particularily through that organ whose unique position and structure qualify it as the supreme center and source of the nutritive-vascular system, — the heart.

In this organ, as we have seen, is centered the innate heat which is the chief instrument of soul in the vital functions. Hence the soul itself is conceived as operating principally in and through the heart. For soul (with the possible exception of intellect) can exist and operate only as activating an appropriate *organ.* The heart is the one organ designed to serve as the center and source of the entire organism. Hence the heart can be said to be the center and primary seat of the soul in its nutritive functions at least. This is not to deny that the soul activates and operates in the other organs of the nutritive-vascular system and the whole of man's body. However, it can activate and operate in them only according to their structure and function in the whole; and in this, as we have seen, they are all subordinate to the heart.[58]

Finally, we have seen that the formation of the homoeomerous parts as well as the larger structure and function of the organs concerned with nutrition, etc. are determined by the principle of proportion, symmetry, and the mean. Now since the soul as the formal-final-efficient cause of the body is ultimately responsible for this formation and function, we may conclude in general that the soul by nature operates according to this principle, at least on the nutritive level, and that when a "middle state" (μεσότης) is achieved in all respects the bodily organism is fully formed, attains its end, and functions most perfectly. "For it is the mean that contains the substance and proper proportion, and not either of the extremes apart from it" (*De Part. An.* 652b19-20). And the soul, as form of the living organism, *is* its "substance and proper proportion" (*Met.* 1035b14 ff.).

We must now examine briefly Aristotle's notion of the operation of soul and body in sense perception.[59]

[58] The theory of Francois Nuyens on the development of Aristotle's psychology must be considered at this point. A brief discussion of his position will be found in Appendix II. Cf. also note 75 below.

[59] There is place here for only the merest sketch of this large aspect of Aristotle's thought, enough to render somewhat intelligible his application of the notion of μεσότης in explaining sensation. For further detail, J. I. Beare, *Greek Theories of Elementary Cognition,* (Oxford, 1906), affords a classic discussion of sensation, which is supplemented, and to a certain extent corrected, by Paul Siwek, *La Psychophysique humaine d'aprés Aristote* (Paris, 1930), ch. III.

2. *The Problem and Conditions of Sense Perception*

The animal soul, Aristotle holds, must share with the soul of a plant the functions of nutrition, growth, and reproduction (411b27-30; 415a1-2). However, even the simplest animal is superior to the plant in having a further "power" or faculty by which it becomes *aware* of the objects around it, distinguishing one from another, so as to seek what is nourishing and avoid what is harmful (*Parva Nat.* 436b10-22). This is the faculty of *sense perception* (αἴσθησις), upon which the related faculties of appetite and locomotion depend. The animal body is capable not only of the mechanical reaction involved in nutrition and physical response to environment, in which both the matter and form of surrounding objects affect the organism; but the animal is distinguished by its power to assimilate the forms of surrounding material objects *without* their matter, so that the form alone of the object is reproduced within the animal (424a17-19, b1-3; 425b23; 434a29). This basic notion lies behind Aristotle's whole theory of sense perception.

The material objects surrounding the animal and constituting its environment are, as we have seen, composed of earth, air, fire, and water blended in the fixed proportions demanded by the nature or form of each object. The form of each object, then, is contained or embodied in the material elements so blended, since it is the principle which fixes their proportions and determines their structure in the total blend. The nature or form of each object, therefore, is manifest materially in the elemental structure of its body. Aristotle's problem is to explain how the animal can assimilate the forms of these objects without their proper matter, so that it becomes aware of the objects and can distinguish one from the other.

The first step toward a solution Aristotle finds in certain properties of the elemental bodies themselves, — earth, air, fire, and water. These, embodying as they do the forms of external objects in blends of various proportions, must be capable of stimulating characteristic reactions in the animal, either in the whole of it or in certain specialized parts. And so the elemental bodies are conceived as combinations of four causal factors, — the hot, the cold, the moist (fluid), and the dry (solid) — which have this active power of stimulating characteristic responses in the animal (cf. 423b27-30; 429b13-16). These retain their power in the blend, as we have seen, thus imparting to the material bodies which they constitute their particular *sensible qualities* (πάθη), — hot, cold, moist, dry, heavy, rough, smooth, bitter, sweet, black, white, etc. (*Meteor.* 384b28-385a4; *Parva Nat.* 440b10-22, 445b4 ff.; *De Part. An.* 646a15-21). These sensible qualities of the object, then, are the active cause of sensation. And since the particular combination of sensible qualities possessed by each material object depends upon the proportion in which its ingredients are blended, i.e. upon its form, this particular combination of sensible qualities, when assimilated by the animal, will at the same time convey the form of the object, enabling the animal to distinguish it from other objects. Thus far the objective foundation of sense perception (cf. *Physics* 244b5-245a4, as in the Oxford translation, following Prantl's text; and *Categories* 9a28-b9).

In regard to the perceiving subject, the animal, Aristotle holds that the (sensitive) soul forms and activates specific organs suitable for the assimilation of the sensible qualities of the object. These sense organs, in general, are of such a nature that they react characteristically to the different sensible qualities of the object, presented in the proportion in which the elements are blended in the object, and so receive the sensible form of the object without its matter. This is the passive aspect of sensation, in so far as the sense organs are determined by the sensible qualities of the object.

Aristotle, then, sees the whole process of sense perception as a sort of *qualitative change* (ἀλλοίωσίς τις),[60] a κίνησις set up by the material object and received in the sense organ, and so assimilated by the animal, both body and soul (415b24; 416b32 ff.; *Parva Nat.* 436b6). The process from beginning to end is an activity of the *compound*, never of the soul alone (408b1-18). He compares it to the process by which a seal is impressed upon wax with a signet ring (424a17-24). The sense organ assimilates the form of the object without its matter as the wax assimilates the form of the

[60] But cf. the qualification of ἀλλοίωσις as applied to sensation: 417b3-418a6, 431a1-8.

seal without the bronze or gold in which it is cut. Like any other change, therefore, the process of sense perception has its active and passive aspects and takes place formally in the subject (426a8-12).

Moreover, the sense process must conform to the conditions of all change, namely, 1) that there be *contact* between agent and patient, and 2) that agent and patient be *opposites*, since change can take place only between opposites (416b33 ff.). In conformity with the first principle, Aristotle postulates a special *medium* for each sense, by which contact is established between the sense organ and its proper sensible object. For example, the colored object affects the eye through "the transparent" in air and water, the sounding object affects the ear through air, the hard object affects touch through the flesh, etc. (cf. 419a17 ff.). In every case the sensible qualities impress themselves on the medium through a sort of qualitative change which in turn sets up a corresponding change in the sense organ. Without the medium there would be no contact between object and organ, no change, and no sensation.

However, it is in explaining the application of the second principle to the sense process that Aristotle comes to grips with the central problem of sensation. How can the sense power, embodied as it is in the matter of its own proper organ, be at the same time the *opposite* of all the innumerable sensible qualities that distinguish material objects, so that it can respond to the stimulus of each quality and change to conform with any combination of qualities? How can the sense be a single, unified faculty and yet be the opposite of sensible qualities which are themselves opposites, like hot and cold, black and white? Furthermore, how can the sense retain its identity and yet become like its opposites, reproducing them? The answer to these questions must be found in the very essence of the sense power.

First of all, Aristotle recognizes the obvious fact that not all the sensible qualities in the object affect all sense organs equally, but that the senses are distinguished by their response to *different sets* of qualities attributable to the same object. Thus touch perceives the object as hard or soft, hot or cold, rough or smooth; taste, as bitter or sweet; sight, as black or white; hearing, as low sounding or high sounding, etc. Each sense is specified by its response to a certain type of sensible quality in the object, its own "proper sensible" (426b8-11).[61]

Next, Aristotle conceives the sum total of objective qualities *proper* to each sense as falling into a continuous stretch or range of temperature, color, sound, etc. The various qualities in each class form, as it were, an even scale which gradually advances by degrees from one extreme to the opposite, after the manner of "the more and the less" or of "excess and defect". Thus the sense of touch is sensitive to the variations of temperature ranging from extreme heat to extreme cold, the sense of sight to the variety of colors ranging (as Aristotle believes) between black and white, the sense of taste to every flavor from bitter to sweet, etc. (422b23-31; *Parva Nat.* 445b21 ff.). And each one of the intermediate qualities is the result of the blending, in greater or less proportion, of the extremes. For the intensity of one extreme can be moderated only by its opposite.[62]

However, though Aristotle speaks of the proper sensibles as ranging between extremes, he does not think of the range of each sense as *unlimited*. Rather, he recognizes that each sense can respond only to a limited range of sensible qualities, and that beyond those limits there exist further extremes of objective quality, some of which are too deficient to affect the normal organ (cf. *Parva Nat.* 464a6 ff.), others so excessive that they can overwhelm the

[61] Aristotle considers touch the basic sense, not however because the others are reducible to touch, but because it is present in *all* animals, even those who do not have any of the others. The reason for this is that touch is the minimum and indispensible sense required for the nutrition and preservation of the animal; without it the others would be useless or dangerous; and when touch is destroyed the animal is destroyed (413b4-9; 434b11-24; 435b3-25).

[62] Cf. 426b5; *Parva Nat.* 440a33 ff., 442a12-25; *Met.* 1043a10. The particular proportion in which the sensible qualities blend in any one object is, as explained above, the same as the proportion in the blend of its elements (*Parva Nat.* 440b13-20), since these are the foundation possessing the sensible qualities. This allows the particular blend of qualities in the object to convey the *form* of that object; for it is the form which ultimately determines the proportions of the elements, and their sensible qualities, in the blend. The sensible qualities associated with each elemental body are clear enough with regard to the sense of *touch* (i.e. hot, cold, moist, dry). With reference to *sight*, fire in its purest form appears white, earth, its opposite, black. Cf. Beare, p. 65.

organ and destroy it (422a20ff., 424a10-15). Hence the range of sensibles proper to each organ is limited and its extremes determined by the limitations of the organ.

In general, then, Aristotle conceives the senses as several faculties or powers of the animal soul by which it responds to the different ranges of sensible qualities. But since the animal soul can exist only as informing and operating through bodily organs, it must develop and inform organs suitable to these different powers of response. Hence each sense must be embodied in an organ materially constituted so as to respond to the variations within its proper range of sensible qualities, when these are presented by the material objects surrounding the animal. A sense organ, therefore, adequately conceived and properly so-called, must contain both a material and a formal aspect, i.e. a particular sense power of the soul informing its material organ (424a24ff.). Thus the eye is conceived as the power of sight activating a pupil; and the eye without sight is no more an eye than a body without a soul is a man. For both have lost their form. This double aspect must be kept in mind in all of Aristotle's discussions of the senses. For him the power of sense can never exist except as organizing and informing a proper bodily organ, though this power or form cannot be reduced to the matter of its organ any more than the entire soul can be reduced to the matter of the body it informs (412b10-413a6, 416a8-18; *Parva Nat.* 454a7-12).

All this, however, is largely descriptive and leaves the essential problems untouched. How can any one sense, embodied in its bodily organ, be unlike or opposite its whole range of proper sensibles, so that it can be moved by either extreme or by their intermediate qualities? Aristotle arrives at the ultimate stage in his analysis of the sense power when he proposes that the form of each sense, and the sensitive organism as a whole, is in the nature of a μεσότης.

3. *The Special Senses and the Mean*

Aristotle's use of the term μεσότης in the contexts already examined indicates that for him it signified, in general, the "middle state" or "mean disposition" of a compound in relation to extremes of "greater or less", "excess and defect", achieved in the proper proportion or balance of its opposing constituents. When he calls the *sense* a type of μεσότης,[63] therefore, he must be thinking of each sense as form disposing and activating its matter in a state of equilibrium, i.e. a *state of balance about a point* (μέσον), by nature constituting its proper organ in a condition of poise *midway between the extremes of its proper range of sensible qualities.* This "middle state" or equilibrium, then, is of the essence of the sense organ, for it is this precisely which enables it to perform its function of discriminating (κρίνειν) or distinguishing one quality from another (424a5-7).

The form which imparts this equilibrium, be it remembered, is the "first actuality" of the organ, holding it poised, ready to respond one way or the other, in the direction of excess or defect, depending upon the particular sensible quality presented by the object. The objective quality stimulates the sense organ by setting up a kind of qualitative change in the medium. This in turn affects the organ, tipping its natural balance to conform to the nature and intensity of the stimulus, thereby throwing the organ into "second act". Before stimulation, however, it is the nature of the organ to be neutral, indifferent to any specific quality, balanced between both extremes and ready to respond to either in whatever degree it is presented. In this way the organ has the capacity to conform to either extreme or their intermediates in "second act", so that it is potentially all of them; in the actuality of its "first act", however, it is like none of them (424a7-10).

In this way, then, Aristotle solves the difficulty presented by his second principle, i.e. that change can only take place between *opposites.* Being in a neutral or "mean" state with regard to the varying degrees of its proper objective quality, the sense organ will be *opposite* or unlike the *particular* objective quality presented. For the range of each type of quality is conceived as constituted of its two extremes and their intermediates, and the mean is equally opposed to either extreme. In this way, too, the sense organ, being

[63] Cf. 424a 4 (μεσότης τις), 435a 21 (ὥσπερ μεσότης); *Meteor.* 382a19 (ὡς μεσότης).

in a "mean state", can retain its identity and yet be equally opposed to extremes which are themselves opposites.

Aristotle holds, as we have seen, that opposites react when contact is established between them, provided that one be stronger than the other. Therefore, if the objective quality is strong enough, the qualitative change which it sets up in the medium evokes a corresponding change in the sense organ, i.e. the organ responds in the direction of that quality in proportion to its intensity. In doing so, the organ becomes *like* the objective quality, thus taking on the form of the object without its matter. Thus through the organ, a living compound of appropriately organized matter informed by the animal soul, the animal assimilates the form of the external object.[64]

Because the sense is in form a μεσότης, a single equilibrium which responds in one direction or another according to the quality presented and then returns to its original "middle state" when the stimulating cause is removed, the sense organ is capable of "judging" between one quality and another, e.g. between a high and low sound, between light yellow and dark yellow, etc. By the very fact it responds differently to different qualities in the same range it distinguishes one from the other (424a5-7). Like the balance arm of a scale, it constitutes one and the same standard that responds now one way and now another, thus registering the differences in the objects that move it. The sense can unite and compare different objective qualities, then, precisely because, as a μεσότης, it is a *single* power capable of *various* relationships or modes of existence.[65] Thus the concept of μεσότης solves for Aristotle the basic problems of sensation.

[64] Because he conceives the sense as a μεσότης Aristotle can say both that "like is known by unlike" (i.e. the object and the organ *before* the change) and that "like is known by like" (the object and organ *after* the change), — his solution to a point of controversy among his predecessors (cf. 416b35ff.; *De Gen. et Cor.* 323b1 ff.).

[65] This aspect will be examined at greater length in the discussion of the central sense power. The importance of the concept of μεσότης in Aristotelian psychology has been brought out clearly by W. J. Oates ("The Doctrine of the Mean", *The Philosophical Review*, XLV [1936], 393ff.) who concludes (p. 394): "It is perfectly obvious, if the mean is not only a principle which possesses the power of judging but also is that which enables the sense organ to perceive

Aristotle apparently finds his theory of sense as a μεσότης most clearly confirmed by the response of *touch* to variations in its proper qualities. For he first mentions and illustrates the theory in the *De Anima* in this connection (423b27-424a10):

What can be touched are distinctive qualities of body *as* body; by such differences I mean those which characterize the elements, viz. hot, cold, dry, moist, of which we have spoken earlier ... The organ for the perception of these is that of touch, — that part of the body in which primarily the sense of touch resides. This is that part which is potentially such as its object is actually; for all sense perception is a process of being so affected; so that that which makes something such as it itself actually is makes the other such because the other is already potentially such. That is why when an object of touch is equally hot and cold or hard and soft we cannot perceive; what we perceive must have a degree of the sensible quality lying beyond the neutral point. This implies that the sense itself is a μεσότης between any two opposite qualities which determine the field of that sense. It is to this that it owes its power of discerning (κρίνει) the objects in that field. What is "in the middle" is fitted to discern (τὸ γὰρ μέσον κριτικόν); relatively to either extreme it can put itself in the place of the other. As what is to perceive both white and black must, to begin with, be actually neither but potentially either (and so with all the other sense organs), so that the organ of touch must be neither hot nor cold.

Touch, then, seems to be for Aristotle not only the basic sense,[66] but the *typical* sense as well, since he assumes that the other senses operate in an analogous manner. Each is conceived as a μεσότης with regard to the range of its proper objective quality, so that it can be stimulated only if the quality presented exceeds the "mean" in the direction of one extreme or the other, just as the object must be hotter or colder than the organ of touch if that sense is to be

form without matter, that it constitutes the very core of Aristotle's psychology". Professor Oates underscores the ontological reality of the μεσότης in Aristotle's psychology, which distinguishes it from mere mathematical imagery.

[66] Touch is the basic sense as explained above, note 61. It is noteworthy that Aristotle uses the term μεσότης of a *special* sense in contexts which deal specifically with *touch* (424a4, 424b1, 435a21; *Meteor*. 382a19). However, that Aristotle regards the other special senses analogously as μεσότητες is clear from 424a7-9 (and cf. 426a27-b7). Furthermore, he applies the notion to explain the nature of the central sense power, as we shall see.

moved from its normal equilibrium and register or distinguish the difference between itself and the object, and between one object and another of different temperature. So much for the *formal* aspect of the particular senses.

Further, as we have mentioned, each sense power informs its proper *matter*, joining with it and disposing it to form one actual reality, — the living sense organ. Aristotle explains (424a25-28):

The sense and its organ are the same in fact, but their essence is not the same. What perceives is, of course, a spatial magnitude (μέγεθος), but we must not admit that either the having the power to perceive or the sense itself is a magnitude; what they are is a certain ratio and power *in* a magnitude (λόγος τις καὶ δύναμις ἐκείνου).

The sense, then, is a *power* (δύναμις) which disposes its matter according to a definite *proportion or ratio* (λόγος), constituting it in a "middle state" or equilibrium so that the quantified substance (μέγεθος), the organ composed of matter and form, is capable of responding one way or another to its proper object. The organ, then, must be materially constituted so as to admit of a certain flexibility and ready response to external stimuli. This demands, therefore, a blending of opposing elements in the organ,[67]

[67] It should be recalled that Aristotle regards even the elemental bodies as being a kind of *mixture*, at least as they exist in the sublunary world (cf. *De Gen. et Cor.* 330b 22ff.). Hence when he speaks of the organ of hearing as constituted of air, and sight as constituted of water (e.g. 425a 3-9, 435a14-15) he is not excluding the presence of the opposites entirely, but designating that which is dominant and characterizes the mixture. The fact, for instance, that he accounts for defects in vision by deviations of thickness or scantiness of the fluid in the eye (*De Gen. An.* 779b 26-780a 25) shows that he regards the fluid as *composed*, since he explains differences of "greater and less" in material substances by the presence of opposites blended in different proportions. Thus Simplicius, commenting on Aristotle's designation of αἴσθησις as a μεσότης τις at 424a 4, explains (*Commentaria in Aristotelem Graeca* XI, 165) that Aristotle is using αἴσθησις (sense) for αἰσθητήριον (sense organ), or that he speaks of the *perceiving substance* (ἡ αἰσθητικὴ οὐσία) as a "kind of μεσότης" because of the fact that it employs *the fine median blend of the sense organ* as its matter or substrate. Philoponus points out (*ibid.* XV, 435-436 ad 424a 2) that, of course, only in the case of touch is the organ actually *constituted* by a median blend of the *opposites which are its proper object* (i.e. the constituents of all bodies, the hot, cold, moist, and dry). The organs of the other senses must be

and particularily the presence of the two "intermediate" elemental bodies, air (vapor) and water (cf. *De Gen. et Cor.* 330b31 ff.), which are most capable of expansion and contraction.

Aristotle explains that plants — and parts of the animal body like bone, sinew, and hair — do not have even the minimum power of touch precisely because they are composed predominantly of *earth* and lack the balance of material constituents needed for response to either extreme (435a21-b3):

For touch is as it were a μεσότης between all tangible qualities, and its organ is capable of receiving not only all the specific qualities which characterize earth, but also the hot and the cold and all other tangible qualities whatsoever. That is why we have no sensation by means of bones, hair, etc., because they consist of earth. So too plants, because they consist of earth, have no sensation. Without touch there can be no other sense, and the organ of touch cannot consist of earth or of any other single element.

Thus, though plants have souls and are affected by tangible objects, Aristotle contends that they have no power of touch because they lack the proper material composition (424b1-3): "The explanation is that they have no mean condition (μεσότης), and so no principle (ἀρχή) in them capable of taking on the forms of sensible objects without their matter."

Since each sense power demands appropriate matter, Aristotle does not regard, for example, the entire eye and ear as the sense organ proper, but only that part of each immediately activated by the power of sight and hearing to form the embodied μεσότης responsive to color and sound, i.e. the pupil of the eye, and the air enclosed within the ear (413a2-3; 420a4-11). Furthermore, the material of the organ proper must be allied to the *medium* if the organ is to respond to changes in the medium (cf. *De Gen. et Cor.* 323b30-324a9). For example, since the "actual transparent" in air and water is the medium of color, and air the ordinary medium of sound, the organ of sight, the pupil, is watery, while that of

constituted of these also, but blended in such a balance as to respond to either extreme of their own proper objects (e.g. white or black color, high or low sound). Cf. Beare, pp. 239-241, and 248, note 2.

hearing is air (418a31 ff., 420a33 ff., 424b31 ff.).[68] Touch, whose range includes two basic sets of contraries (hot-cold, moist-dry), must be composed of all four elements in almost equal proportions (422b24-423a17, 423b25-424a2, 425a2-7; *De Part. An.* 653b23-29 and the Oxford translator's note *ad* 29). In every case the sense organ proper must be homoeomerous, of material suitable to its particular range of objective qualities (*De Part. An.* 647a3-24).

Thus Aristotle conceives the sense power as essentially united with its appropriate matter to form a single existing reality, the sense organ. Because anything that affects the organ affects the compound, material changes conveyed to the organ affect both body and soul. If these are inaugurated in the medium by objective qualities proper to the sense, and fall within the range and intensity to which the material organ naturally and easily responds, the result is normal sensation. However, if the stimulus is weak it may affect the organ slightly but will not cause a clearly defined sensation. On the other hand, the stimulus may be so powerful that by violently disrupting the essential balance of the material organ it injures or destroys the sense permanently (424a30-33; cf. also 426a27-b7): "If the movement set up by an object is too strong for the organ, the equipoise of contrary qualities (λόγος) in the organ, which just *is* its sensory power, is disturbed; it is precisely as concord and tone are destroyed by too violently twanging the strings of a lyre." The animal may survive the loss of any other sense but touch, since touch is the minimum sense required for life; hence "excess in intensity of tangible qualities destroys not merely the organ, but the animal itself" (cf. 435b8-18).

[68] The nature of the physical impulse which affects eye and ear in one way or another so as to produce opposite sensations is not very clearly explained. However, there is some information offered in 420a16-b3, where the *high* tone is said to be produced by an impulse which moves the organ of hearing "much in a little time", while the *low* tone moves it "little in much time". The *high* is analogous to what is *sharp* to the touch in that it *pierces*, while the *low* is like the *blunt* in that it *pushes*. Elsewhere (*Met.* 1057b8-10; *Topics* 107b27-32, 119a30-32) the color contraries *white* and *black* are distinguished as "sight-*piercing*" and "sight-*compressing*", and this is consistent with the notion that *fire* causes *white* and *earth*, being solid, causes *black*, though it is difficult to tell from the contexts whether Aristotle holds this or merely proposes it for the sake of example, borrowing from Plato, *Timaeus*, 67 D, E.

Finally, since the sense organ is by nature a μεσότης and its function dependent upon this essential balance, any deviation of excess or deficiency in its material structure results in malfunction, i.e. faulty response and distorted perception. Aristotle maintains, for example, that hard-skinned men are less sensitive to tangible qualities, and so less intelligent (421a25). The tongue has no perception of taste if too moist or too dry (422b5-6), and when sickness disturbs its normal elemental structure, sweet objects taste bitter (422b8-10). Thin or scanty fluid in the eye reacts too violently to stimulus, while thick or over-abundant fluid is not responsive enough, so that in either case faulty vision results. Vision is best when the fluid reacts proportionately to the stimulus, being "intermediate (μέση) between too much and too little" (*De Gen. An.* 779a26-780a25). Even the size of the total organ affects its function and the "medium" eye and ear are pronounced best (*Hist. An.* 492a7, b1). With the sense organs, then, as with the animal organism as a whole, proper function and response are dependent upon internal proportion, symmetry, equilibrium. If the internal equilibrium is thrown off center, the organ cannot function properly, i.e. it cannot respond proportionately to the objects in its environment, so that perception is faulty. The best state of the whole organism (health) and of the particular sense organ (the capacity for clear, accurate sensation) results when their opposing constituents exist in a state of correct balance, i.e. ἐν μεσότητι.

4. *The Central Sense Power and the Mean*

Aristotle's account of sense perception would be inadequate, however, if it went no further than the five special senses. For the various types of objective quality — color, sound, odor, temperature, etc. — existing united in the object and affecting the perceiving subject through one or more of the five different avenues of sense must be *reunited* again within the subject if it is to receive a fair sense impression of the whole object. This cannot be done by any one of the five special senses because of their specialized nature (426b8-11); nor can two or more distinct senses working in conjunction effect a single unified impression (426b17 ff.); nor again, can the function be assigned to a power of reason higher than sense, since it is *sense* impressions that must be unified (426b14-15). Hence, in addition to the five special senses, Aristotle postulates another power of the sensitive soul, a central, unifying power which receives the data of the special senses and, while *distinguishing* among them, *at the same time reunites* them in a single impression reproducing the sensible form of the object as a whole. This is the so-called common or central sense power.

In establishing the existence and nature of this central sense Aristotle appeals to the testimony of consciousness (426b8 ff.) that we perceive distinctions (κρίνειν) not only between objects of each individual sense (e.g. between black and white, bitter and sweet) but also between the objects of *different* senses (e.g. between black and bitter, white and sweet). Now distinction implies comparison, and comparison demands that some *one* sense power unite the objects of different senses (e.g. white and sweet) within itself; for "what asserts that two things are different must be one" (426b20-21). Moreover, the comparison or distinction must be effected in a single moment of time, i.e. in a single act of this unifying power, which asserts that "the objects are different *now*" (426b28). Hence, "both the discriminating power and the time of its exercise must be one and undivided" (426b28-29).

However, it seems impossible that a single power should be moved simultaneously by the forms of different, and opposite, sensible objects and yet retain its identity (426b29 ff.). Sweet moves sense in one determinate way, bitter in the opposite, white in another. How can these different qualities, then, be *united*, yet *distinguished*, by one sense power in the same act or instant? For it seems that to distinguish them the power must be divided between (or among) them. And the problem is as acute for each of the *special* senses as for the central sense power.

The possibility of a solution is found by Aristotle in the analogy between the sense power and the *point* (στιγμή) that divides a line (427a9-14).[69] For the point is one in place and number, yet dual

[69] This passage of the *De Anima* is best understood against the background of the *Physics*, where the nature of the point is treated in various contexts.

in its relationship to the opposite parts of the line which it divides. As a single point it unites the opposite parts by serving simultaneously as the end of one and the beginning of the other. Yet at the same time it is dual precisely because it serves as the boundary between them, being an *end* with relation to one and a *beginning* with relation to the other. Thus it unites the opposite parts of the line in the same act by which it divides or distinguishes them. "For the point ... both connects and marks off the length; it is the beginning of one and the end of the other" (*Physics* 220a10-11). Aristotle applies the analogy to the sense power as follows (427a11-14):[70]

In so far, therefore, as that which judges (τὸ κρῖνον) is undivided, it is one and judges instantaneously; but in so far as it is divided it is not present as one, for it uses the same distinguishing mark (σημεῖον) twice simultaneously. In so far as it uses the boundary point (πέρας) as *two*, it distinguishes two objects and these are separated by what is, as it were, itself separated; but in so far as it is *one*, it uses the point as *single*, and instantaneously.

Now it is evident that the *point* which Aristotle has in mind here is of the nature of a μέσον, for only a point *between extremes* can have the attributes described, as explained in the *Physics* (262a19-20):

We have three points, starting-point, middle-point (μέσον), and finish-

---

Cf. 219b12-220a26, 222a12ff., 253a7ff., 259b1ff., 262a21-26, b19, etc. This appears to be no mere mathematical imagery but a serious attempt to analyze the ultimate form of the sense informing the material organ. Aristotle declares elsewhere in the *De Anima* (431b15-16) that we conceive the objects of mathematics as separate from matter, *though in fact they are not*. He holds that "geometrical subjects are predicable as properties of a perceptible substratum" (*Post. Anal.* 79a8-10) and that "physical bodies contain surfaces, volumes, lines, and *points*" (*Physics* 193b24-25), though the mathematician as such "does not treat them as the limits of a physical body, nor does he consider the attributes indicated as the attributes of such bodies" (*ibid.* 31-32). Yet in abstracting from their material substratum, the mathematician does not distort the realities of point, line, surface, etc., "for in thought they are separable from motion, and it makes no difference, nor does any falsity result if they are separated" (*ibid.* 34-35). Aristotle treats the concrete moving body (e.g. a stone, or Coriscus) as a *point* in his explanation of time: cf. *Physics* 219b13-32.

70 This is a departure from the Oxford translation, and adopts the traditional reading of 427a14 (... ᾗ δὲ ἕν, ἑνὶ ...).

ing-point, of which the middle-point in virtue of the relations in which it stands severally to the other two is both a starting-point and a finishing-point, that is, it is *one* in number (ἀριθμῷ) but *two* in its relationship (λόγῳ).[71]

This is consistent with Aristotle's explanation that the special senses are capable of distinguishing between opposing qualities precisely because of their character as a μεσότης τις, — a proportionate distribution of parts about a μέσον, constituting a state of balance or equilibrium responsive to whatever extreme is presented by the object. The whole dynamics of sensation as a κίνησις τις διὰ τοῦ σώματος τῆς ψυχῆς (*Parva Nat.* 454a9-10) depends upon this concept of the sense organ as balanced at a point midway between extremes (cf. *Physics* 224b28-34, 226b1-9, 229b15-20, 262a19-23). Upon this the essential discriminatory function of sensation depends: τὸ γὰρ μέσον κριτικόν. And now the point of balance, the μέσον, explains for Aristotle the simultaneous unity and plurality of the sense power.[72]

It is the same concept of μεσότης, then, that Aristotle has in mind in explaining the common or central sense power, as is evident in the following summary of the sense process (431a17-19):[73]

The air makes the pupil of the eye such or such (τοιανδί); the pupil makes something else in turn such and such correspondingly; the organ

[71] On Aristotle's use of τῷ λόγῳ and τῷ εἶναι, cf. *Parva Nat.* 449a 20 and the Oxford translator's note *ad loc.*; also, Beare, p. 279, note 2.

[72] We may venture the following explanation: The sensitive soul, unmoved and undivided like the point upon which is balanced the sense organ it informs, unites and distinguishes by this point the opposite qualities affecting the organ. The sense organ is by its *nature* poised at this μέσον between contrary sensible qualities. It can be moved, therefore, only by a sensible quality which is in *excess* of the mean in one direction or the other. Therefore in responding proportionately to the quality presented, it is moved in the direction of an excess of one contrary quality, and measures the degree of excess by its response. But in the very act of responding in one direction, like the balance arm of a scale, it registers at the same time a *corresponding deficiency of* the *opposite quality* in the object. Thus both the stimulating quality and its opposite are present to the sense simultaneously, the one by way of act, the other by way of corresponding privation, because both are related to the sense as to a common *point* or μέσον.

[73] This is a departure from the Oxford translation in the direction of greater literalness.

of hearing reacts similarly; and the ultimate is *one* thing (ἕν), that is, a *single* μεσότης, but with a *plurality of relationships* (τὸ δ' εἶναι αὐτῇ πλείω).

Aristotle refers immediately to the passage previously considered (427a11-14), remarking that the sense power that distinguishes sweet and hot is "one thing, but in the manner of a boundary mark" (431a21-22). The same solution is proposed again in the *De Sensu* (*Parva Nat.* 449a5 ff.), where Aristotle concludes that there is "some one faculty in the soul with which the latter perceives all its percepts, though it perceives each different genus of sensibles through a different organ" (*ibid.* 9-11), and that this central faculty of perception "is in itself numerically one and the same, but different in its mode of being (τὸ εἶναι); different, that is to say, in genus as regards some of its objects, in species as regards others. Hence ... one can perceive [numerically different objects] coinstantaneously with a faculty which is numerically one and the same, but not the same in its relationship" (*ibid.* 17-20).

Thus Aristotle conceives the central sense power as a μεσότης τις, an equilibrium poised upon a center point (μέσον), equally disposed to respond to either extreme of all the opposing pairs of sensible qualities which constitute the proper objects of the special senses. As the *special* sense is a μεσότης which unites and distinguishes in a single act opposing qualities *within* the same *genus* (e.g. white and black), so the central sense is a μεσότης which unites and distinguishes in a single act the heterogeneous qualities (white and sweet) presented by the several special senses.[74]

[74] The central sense might be crudely represented as an arrangement of five or six separate balance arms, each corresponding to a different special sense, all centered at the same point so as to register simultaneously the distinctions between as many different pairs of opposing objects. All would be *united* in a single point, the μέσον, which would at the same time serve as the point of *distinction* among them. For with reference to this point each balance arm would measure the difference between its *own proper objects*; while simultaneously the distinction and degree of difference between the various balance arms would be registered with reference to the *same* single point, precisely because it serves as a pivot for all. The essential feature of the arrangement would be the single common μέσον capable of a variety of different relationships. The illustration may not be altogether fanciful. Alexander, in his essay on the problem (Quaestiones III, 9, *Supplementum Aristotelicum* [Berlin, 1892], Vol. II, Pt. II, p. 96, l. 14 ff.), conceives the *point* in question as the center of

Thus far, however, we have considered the central sense power exclusively from the point of view of its form, which can exist, of course, only as embodied in a suitable organ (cf. 424a24-26). Aristotle begins to discuss the material aspect of the central sense in the *De Somno* (*Parva Nat.* 455a14 ff.), where he asserts not only that there is one central sense (μία αἴσθησις) but that the "*controlling sense organ*" (τὸ κύριον αἰσθητήριον) is one, though diverse in its relationship to different genera of sensible objects. This is said to be the organ "controlling all other sense organs, to which the others extend" (*ibid.* 455a33). And the identity of this organ is clearly indicated a few paragraphs later with the statement that "all sanguineous animals have a heart, and the source and center (ἀρχή) of movement and of the *controlling sense* (ἡ αἴσθησις ἡ κυρία) is here" (456a4-6). The function of heart and blood in the nutritive-sensitive life of the organism receives increasing attention in the latter parts of the *Parva Naturalia*, where it is firmly established that "all sanguineous animals have the supreme organ of the sense faculties in the heart, for it is here that we must look for the common sensorium belonging to all the sense organs" (469a10-13).[75]

---

a circle which serves as a common boundary for several straight lines drawn from the circumference. And the circle whose center constitutes the point of balance for every diameter seems to have provided the basic form in Peripatetic mechanics; cf. (Pseudo-A.) *Mechanica* 847b14 ff.

[75] Cf. also *De Gen. An.* 743b26-30; *De Part. An.* 647a25-32, 656a28 ff., 665a11-13. Aristotle's failure to mention the heart explicitly as the organ of central sense in the *De Anima*, *De Sensu*, etc. is due, I believe, to his methodology and not to any basic change of doctrine between the *De Anima* and the latter parts of the *Parva Naturalia*. The argument from silence (cf. Ross, *Parva Naturalia*, pp. 7, 8, 12) is proverbially dangerous; a man does not mention what he assumes as obvious or what he holds as true but does not wish to discuss at the moment. In the *De Anima*, which is explicitly concerned with *soul*, the *formal* aspect of the living organism, Aristotle avoids discussion of the *material* aspect as much as possible, since the bodily organs are described in detail elsewhere. Witness his treatment of the organs of *special* sense in the *De Anima*. Conversely, while treating the *material* aspect of the organism, as in the anatomical and physiological portions of the "biological works", Aristotle is correspondingly chary of *psychological* detail. The *De Anima* and the "biological works" represent, it seems to me, two distinct but complementary approaches to the same complex reality, the living organism.

When in the *De Anima* he establishes the existence of a central sense power

From what we have already seen of Aristotle's conception of the heart it is easy to understand why he considered it the one organ best suited to carry on the function of the central or common sense power. Foı its central location makes the heart most easily accessible to sense impulses from any point in the body (cf. *Parva Nat.* 467b28-34; *De Part. An.* 666a11-17).[76] All blood vessels lead to the heart, and these Aristotle believed to be the channels which carry sense stimuli (κινήσεις) from the special organs of sense (*Parva Nat.* 461b12ff.; *De Part. An.* 667b27-31). And the unity of the animal soul demands that both its sensitive and nutritive functions be centered in the same bodily organ. To assign different ἀρχαί in separate parts of the body to each of these functions would be to effect a real division in the soul, a position which Aristotle is at pains to avoid (*Parva Nat.* 467b18-34, 469a17-19).[77]

Moreover, identifying the central organ of sensation with that of nutrition enables Aristotle to explain waking, sleeping, and dreams as psychic phenomena of the central sense connected with the nutritive processes that take place in the same organ (*Parva Nat.* 453b25ff.). The innate heat, centered in the heart (cf. *De Part. An.* 666a3), expands from this center, carrying with it fresh

---

distinct from the special senses, Aristotle must, to be consistent with his own principles (cf. 424a24-28), assume that it informs an appropriate organ. For he insists that the nutritive-sensitive soul can exist and function only through proper bodily organs. And clearly the only bodily organ suitable, from Aristotle's point of view, to carry the functions of the central sense power is one centrally located and communicating with all the organs of special sense, namely, the *heart*. That Aristotle did not consider the theory of *De Somno* (which assumes the heart as seat of sensation) incompatible with the theory of hylemorphism (cf. Ross, *op. cit.*, p. 12) seems clear from his simultaneous acceptance of both in *Metaphysics* VIII (cf. 1043a35-36 and 1044b15-20) and in the *De Generatione Animalium* (cf. 734b20-735a8, 738b25-27, 743b26-30).

[76] In this connection, cf. the interesting passage in the (Pseudo-A.) *Physiognomonica* (813b7ff.) which describes the effect of stature and quality of flesh upon the general sensitivity and intelligence of man.

[77] The passage in *De Partibus Animalium* (672b10ff.) on the diaphragm and its function in shielding the central organ of sense in the heart from the warm evaporation rising from the stomach does not imply that the ἀρχή of the *nutritive* soul is in the region below and therefore separated from the ἀρχή of sense in the heart, as Plato held. All that Aristotle tells us of the heart as ἀρχή of *both* nutritive and sense life in the *Parva Naturalia, De Partibus,* and *De Generatione Animalium* denies this. Cf. Nuyens, p. 161-162, note 44.

nutriment in the form of blood and rendering the whole body alert and erect, receptive of sense impressions which impinge upon the peripheral organs and are carried by the blood vessels to be united in the heart. This constitutes wakefulness. When the tide of innate heat shifts, repulsed by the cold of the extremities and compressed about the region of the heart, the blood retires to the center, the body droops, sense impressions from outside die out, and the sensory function of the heart is inhibited, except for the dreams formed there from the confused vestiges of sensation remaining in the organs. This is sleep.

Furthermore, the three chambered heart which Aristotle discovers in higher animals is evidently appropriate as the organ of central sense. For the central sense power, being essentially a μεσότης, most fittingly informs a three chambered structure in which the center chamber forms a μέσον between the other two. This is to provide a receptacle for the blood which, it will be remembered, is of *intermediate* quantity and heat in the middle chamber, and of greater purity than the hot, abundant blood of the right or the cold, scanty blood of the left chamber (*De Part. An.* 667a1-4). The purity and moderate quantity and warmth of the blood in the center chamber insures its tranquillity, according to Aristotle (*ibid.* 4-7), for the source and center (ἀρχή) of the entire organism should be as tranquil as possible. The psychological importance of this is clear from the treatise *On Dreams* (*Parva Nat.* 461a7-b5), where we are told that the κινήσεις derived from the special sense organs are transmitted through the blood stream to the ἀρχή, and that they are preserved in their integrity "in proportion as the blood becomes calm and its purer separated from its less pure elements" (*Parva Nat.* 461a24-27; and cf. *De Part. An.* 656b4-7). It is obvious, then, that the blood in the central chamber, being pure and most tranquil, must reproduce most faithfully the movements from the special senses, making the heart in this respect also well adapted to be the organ of the controlling sense whose function is to gather the impressions of the several senses and unite them at a single, central point.

The heart, then, in its gross structure is differentiated like the non-homoeomerous parts; but in its elemental structure it is *homoeomerous*. Again, this fits it for its function as organ of the central sense, since Aristotle maintains that the sense organs proper must be of uniform blend throughout, though the blend will differ according to the object each is designed to receive (*De Part. An.* 647a6-35). As in the other sense organs, the hot, cold, moist, and dry in the heart must be exactly proportionate, i.e. blended in proper balance, if it is to react proportionately or moderately to sensible stimuli. Those whose heart is too dry (solid) are dull and insensitive; those in whom the heart is too moist (fluid) are too quick, i.e. hypersensitive (*De Part. An.* 667a14-15; *Parva Nat.* 450b9 ff., 453a15 ff.). The heart is a *healthy* sense organ, i.e. its response is normal, according to nature, when its opposing elements are blended in a middle state of proper balance (ἐν μεσότητι).

The heart, therefore, since it embodies the principle of μεσότης in every respect, — in its position, in its connection with the other organs, in its inner structure and elemental composition — is well adapted to serve as the ultimate organ of sense, being "one thing, a single μεσότης, but with a plurality of relationships" (431a18-19). For it is a single organ in number and position, poised at a mid-point (μέσον), ready to respond proportionately to either of opposite κινήσεις from the organs of special sense. When the sensible object is presented the heart reacts coinstantaneously to the movements arriving from eyes, ears, tongue, etc., uniting them in itself as a single subject, yet in the same act distinguishing them as coming from different sets of organs. In this way, then, the central sense power, informing and operating in the heart, constructs the composite image which reproduces the sensible form of the total object as manifested through all the avenues of sense.

5. *Summary*

Thus through the notion of μεσότης Aristotle explains the nature of both special senses and the common sense power, and in so doing enriches our understanding of what the term implies for him. First, it appears that when applied to the organism, in connection with the nutritive as well as the sensitive life, the term μεσότης suggests to Aristotle no mere mathematical abstraction but a *dynamic equilibrium of opposing physical parts or powers* informed and activated by the soul. It is true that the basic "intelligible form" indicated by μεσότης is that of a proportion, equation, or balance; but (to put it in Aristotelian terms) that form exists objectively in the concrete living organism only as organizing or "informing" matter. Thus the μεσότης constituting the healthy animal or the accurate sense organ will have a formal and a material aspect. Hence any adequate discussion of the living organism and its activities must consider both aspects, as Aristotle himself reminds us (*De Anima* 403b7-8, 18-19).

Moreover, we find that in the context of his physiology and sense psychology the term μεσότης implies for Aristotle more than a static principle of organization, and even more than the simple *state* of an organ or organism in which opposing elements are disposed symmetrically about a center point or μέσον, — though this condition fulfills the essential notion of μεσότης. But beyond this is revealed a more active or *dynamic* aspect of μεσότης, particularily as the concept is used to explain the nature of the sense powers. Here the form μεσότης, — precisely because it is essentially a balance or equilibrium — is found to give the organ, and the entire organism, a distinctive *power*, the *capacity to react or respond proportionately to stimulus* from the environment.

Something of this has been noted with regard to the μεσότης of health, which not only orders the internal constituents of the organism, but enables it to respond proportionately to changes of the hot, cold, moist, and dry in the environment. But the very essence of sense perception as a proportionate response to the qualities of objects in the environment is explained through the *dynamics* of the balance or equilibrium, i.e. the dynamics of the μεσότης, which constitutes the sense organ. Recall the principle noted pieviously, that proportionate response of the organ or organism depends upon a proper balance or equilibrium of its constituents. In sense perception, the μεσότης of the organ is in fact *ordered to* the activity of proportionate response, to the activity of perception, as any faculty (δύναμις) is ordered to its act. Aristotle's discussion of sense perception, then, underscores the dynamic aspect of a μεσότης as imparting a capacity for proportionate response. This aspect should be kept clearly in mind in considering his discussion of the μεσότης of moral excellence.

Finally, since Aristotle uses μεσότης and related notions to explain the formation and activity of the living organism with regard to health, nutrition, and sense perception, and since he regards the *soul* as the formal-final-efficient cause of the organism, we may be justified in concluding that, in Aristotle's view, the soul normally operates according to the principle of μεσότης where it informs and activates the body. The living organism, in fact, seems to be conceived basically as an embodied μεσότης in which the parts, at least those concerned with nutrition and sense perception, are grouped symmetrically about a single organ, the source and center of the vital processes, the heart. We should expect, then, to find some consistency with this general pattern in Aristotle's account of other vital activities involving the compound of body and soul, among them those responses of emotion and action which constitute the moral life of man.

D. MORAL EXCELLENCE AND THE MEAN

Aristotle's analysis of the living organism, so far as we have seen, leads to the conclusion that its natural, and *best*, condition results when the opposing constituents of the body, under the determining influence of the soul, join in a dynamic equilibrium, — ἐν μεσότητι. On the bodily or physiological level, when the elements, and the organs they compose, are so ordered by the "nutritive part" of the soul, the composite functions properly, growing and maintaining its existence by proportionate response to the environment. It then enjoys its fundamental excellence, health. Again, on the psychological level, the "sensitive soul" informs and constitutes each sense organ as a μεσότης, a natural balance of opposing constituents, which enables it to perform its characteristic function, distinguishing the objects in its environment by responding proportionately to their sensible qualities. In so far as the form, the μεσότης, is realized in its matter, the sense organ enjoys its proper excellence, keenness of perception. On each level it is the μεσότης, the symmetrical disposition or ἕξις of opposing constituents, that enables the organism to fulfill its function by proportionate response. In the light of its physiological and psychological application, then, we may attempt to understand more clearly what Aristotle has in mind when he employs the concept of μεσότης to explain the nature of moral excellence (ἠθικὴ ἀρετή), the ἕξις which governs the responses of man as man.

1. *The Method of the Ethics*

The classical discussion of moral excellence is to be found, of course, in the *Nicomachean Ethics.* Aristotle's treatment of it there, however, must be understood as confined within certain limits imposed by the nature of that treatise. The aim of the *Nicomachean Ethics*, and of the *Politics*, is primarily practical, outlining the principles of the πολιτικὴ τέχνη, the master-craft of all the arts (1094a26-27), of which "the end is not knowledge but action" (πρᾶξις; 1095a5).[78] Consequently Aristotle's discussion in these works is directed toward the practical guidance of human action rather than toward a profound scientific analysis of it, i.e. he is speaking πρακτικῶς rather than φυσικῶς. "For we are inquiring not in order to know what virtue is, but in order to become good ... " (1103b27-28). He presents the physiological and psychological aspect of moral activity, therefore, only in so far as that is necessary to the practical moralist.[79] His principle is that

[78] In the present section citations by Bekker number *alone* will be from the *Nicomachean Ethics*; the *Eudemian Ethics* will be indicated by the letters *E.E.* The text of the *N.E.* will be that of J. Burnet, *The Ethics of Aristotle* (London, 1900). The reader should be warned that the translation of the ethical treatises will often represent a departure from, or a very free adaptation of, the Oxford translation.

[79] For the difference between *practical* and *theoretical* science, cf. *Met.* 993b21-24. Theoretical science has three branches, — mathematics, *physics*, and theology (*ibid.* 1026a18-20). *Physics* deals with substance that is subject to change or motion, i.e. *material* substance, form as united to matter (*ibid.* 1025b19-29). It includes, therefore, *physiology* and *physiological psychology* (*ibid.* 1025b34-1026a7; *De Part. An.* 641a18-33). In the *Nicomachean Ethics*, a *practical* treatise, therefore, Aristotle apologizes at times for speaking φυσι-

each type of inquiry (μέθοδος) should go into detail only in so far as its subject matter demands (1098a26-34; and cf. *Physics* 194a12-b15), and though the statesman must know some psychology (and physiology), Aristotle specifies that he intends to treat this aspect of man "just to the extent which is sufficient for the questions we are discussing; for further precision is perhaps something more laborious than our purposes require" (1102a23-26).[80] The *Nico-*

κώτερον (cf. 1170a13; 1147a24; 1167b28-1168a10), or refrains from φυσικὸς λόγος as foreign to his purpose (e.g. 1102a30-34; 1155b2-11; 1147b9). He holds that each discipline should consider the matter and form of its objects only up to a certain point, i.e. in so far as necessary for its purpose. Cf. *Physics* 194a12-b15. Aristotle compares ethics and politics to medicine (cf. 1103b26ff.; 1105b12ff.; *E.E.* 1216b11ff.); as medicine is interested in producing health rather than knowing what it is, so the "political art" studies the *sources* of virtue to *produce* it, not to know what it is. This accounts for the emphasis, in the *Ethics*, upon virtuous *acts* rather than upon detailed analysis of the essential nature of virtue, for acts are the source of virtue *in practice*. "Since, then, the present inquiry does not aim at theoretical knowledge ... we must examine the nature of actions ... for these determine also the nature of the habits that are produced ..." (1103b26-32).

[80] In view of this express statement of method we should not be surprised to find Aristotle adopting from the "exoteric discourses" (1102a26-28; and cf. Burnet's note *ad loc.*) the popular division of soul into λόγον ἔχον and ἄλογον, and refusing to discuss φυσικῶς the problem of whether the distinction between them is real or only logical (1102a27-32; 1102b25). With certain qualifications about classifying the sensitive-emotional powers as λόγον ἔχον or ἄλογον (1103a1-3; and cf. *De Anima* 432a31-32), this general grouping of faculties serves his purpose well enough. Nuyens (pp. 191-193) regards this division as Platonic and so cites it as evidence that the *N.E.* must be "notably earlier" than the *De Anima*. However, the *qualifications* with which Aristotle adopts the classification in the *N.E.* reflect his criticism of it in the *De Anima* (432a22-b7) and bring it into conformity with his own psychological principles.

Again, when Aristotle in the *N.E.* distinguishes the "excellences of the body" from those "of the soul" (e.g. 1102a16-17) and refers to the various emotions, — desire, anger, fear, etc. — as πάθη ἐν τῇ ψυχῇ (1105b21-23), Nuyens sees this as indicative of a dichotomy between body and soul which Aristotle *repudiates* in the *De Anima* (403a16-19; Nuyens, *loc. cit.*). However, it should be noted that *in the De Anima* also, even *after* making the precision to which Nuyens refers, Aristotle uses this same loose and popular manner of expression. Cf. 408a1-5; 403b17, 409b15, 411a26-28, 431b6-7; and cf. Hick's note *ad* 408b13, p. 275. We must not demand, therefore, that he speak more precisely in the *Ethics*, particularly when he excuses himself for a certain lack of technical nicety (ἀκριβεία) in that treatise. Furthermore, though he does speak loosely in the *Ethics* of the emotions and moral virtues as experiences (πάθη) and excellences of the *soul*, when he comes to contrast them with the operation and excellence of the *intellect* (νοῦς) Aristotle makes it abundantly clear 1) that *machean Ethics*, therefore, deliberately refrains from discussing the nature of moral excellence in all its physiological and psychological details, leaving these to the more technical scientific treatises (e.g. the *Physics*, the biological works, the *De Anima, Parva Naturalia, De Motu Animalium*).

Aristotle's approach to the problem of morality in the *Ethics* is guided by a pedagogical principle which he never tires of repeating, namely, that we must begin investigating the unknown from what is known to us, the unseen from the seen (1095b2-4; 1104a13-14). Accordingly, the method he adopts in explaining the nature of moral excellence is analogical, and the prime analogue is something familiar and visible, — the body and its excellences. Throughout the *Ethics* the correlation is made, explicitly or by implication. The art and science of moral guidance and statesmanship are compared to medicine and gymnastics, the virtuous man is compared to the healthy, the vicious to the sick. The emotional reactions which motivate the moral act are explained through parallel physiological reactions, and the essence of moral excellence (ἀρετή) is defined and established in a context of examples drawn from medicine and dietetics.[81] All that we have seen, therefore, of

---

they belong to the soul as *united to the body*, i.e. to the compound (σύνθετον: 1178a9-24; cf. *De Anima* 412a16, and Hick's note *ad loc.*, p. 308; also *Met.* 1043a30; 1070a14), and 2) that νοῦς alone is comparatively independent of matter, — a position which conforms exactly to the doctrine of the *De Anima*. It is important to note also that in the *Ethics* the niceties of the body-soul relationship are specifically included among the subjects Aristotle is reluctant to discuss. Cf. 1176b17-18; 1178a22-23. In general, then, it seems to me that the *De Anima*, and the other psychological and physiological treatises, should be understood as *complementing*, rather than contradicting, the less technical discussion of the *Nicomachean Ethics*. For a detailed critique of Nuyens' position on the psychology of the *N.E.*, Book X, see J. Leonard, S.J., *Le Bonheur chez Aristote* (Brussels, 1948), Appendix III, pp. 201-208.

[81] There is no need to labor this point, since the article by Werner Jaeger on "Aristotle's Use of Medicine as Model of Method in His Ethics" (*Journal of Hellenic Studies*, LXXVII [1957], 54-61), provides ample evidence that "the medical example, far from being a casual analogy, is present to the philosopher's mind throughout. It belongs to the very foundation of the ethical science, at least in the form it has taken in the *Nicomachean Ethics*" (pp. 56-57). Though for the present I prefer to suspend judgment on some of the broader interpretations and conclusions presented, I am indebted to Professor Jaeger for relief from the burden of documenting Aristotle's use of physiological and medical

Aristotle's physiology and psychology in health and disease should help us to understand what he has in mind in comparing moral excellence to physical, and in defining it as a ἕξις ἐν μεσότητι.

In this connection it might be asked whether Aristotle's basic notion of health and disease in the *Ethics* and *Politics* is the same as that which appears in his other works. In keeping with his methodology in the *Ethics* and *Politics* Aristotle avoids any technical analysis of the nature of health, but his understanding of the matter may be gathered from the terminology of several incidental references. In the *Nicomachean Ethics* we are told, for example, that health involves a proportion or symmetry (συμμετρία) which is not the same in all healthy persons, or even in the same person at all times (1173a24-28). Again, in discussing friendship between opposite types of character, he uses the hot, cold, moist, and dry as examples in formulating the general principle that opposites are attracted to each other only incidentally, in that what they really desire is the *mean* (μέσον) between them. "For that is what is good, e.g. it is good for the dry, not to become wet, but to come to the *mean*, and similarly with the hot and *in all other cases*" (1159b20-24). The universal form of this principle and its physiological application is brought out clearly in the parallel passage of the *Eudemian Ethics* (1239b30-38):[82]

Yet in a sense the love of the contrary is love of the good; for opposites desire one another because of the mean; they desire one another because ... out of the two arises a single mean. For those who are overchilled return to the mean if they are warmed, and those who are overheated, if they are cooled; and similarly in other cases. Otherwise they are always in a state of desire, not being at the mid-points ...

Finally, the *Nicomachean Ethics* provides an example of just such a diseased or unbalanced condition in the case of the "melancholics", who are said to be "in constant need of a remedy (ἰα-

illustrations *throughout* the *Ethics*, and for calling attention to a number of points which the present study attempts to discuss in greater detail and from a different point of view.

[82] Regarding the authenticity of the *Eudemian Ethics* it must suffice to remark that the current tendency among scholars is to regard it as Aristotle's own. Even though this be misguided, the work has at least the value of being the oldest and most authoritative *commentary* on Aristotle's ethical teaching.

τρέια); for even their body is continually nettled because of its blend (κρᾶσις), and they are ever subject to strong desire" (1154b-12-14). The "melancholic" blend or temperament is characterized by excess of cold and moist, resulting in oversensitivity and instability (cf. *Parva Nat.* 453a18-26; 457a27 ff.; 464a33 ff.); internal lack of balance brings pain and abnormal response.

It will not be difficult to recognize in these incidental references from the *Ethics* the terminology and leading notions of the *Physics* passage (246b3 ff.) which describes the bodily excellences, health and well-being (εὐεξία), as "constituted in a blending of hot and cold elements within the body in due proportion (ἐν κράσει καὶ συμμετρίᾳ) in relation either to one another or to the surrounding atmosphere", a state which, as we have seen, unites the opposing constituents in a μέσον, and so is called by Aristotle a μεσότης, — a middle state or mean condition. It is also interesting to note that the *Ethics*, like the biological works (cf. *De Part. An.* 652b17-19; *De Gen. An.* 767a16-20), raise this physical and physiological principle to the status of a universal law: opposites attain their good by joining in a *mean*.

2. *General Comparison of Bodily and Moral Excellence*

Turning now to the meaning of μεσότης in the *Nicomachean Ethics*, we find the term used in connection with moral or *ethical* ἀρεταί of the soul, the excellences of those faculties and functions governed by what Aristotle calls the "*non-rational* part" (μόριον ἄλογον) of the soul (1102a28 ff.). This is distinguished from the "part" of the soul "possessing reason" (λόγον ἔχον), to which the *intellectual* excellences properly belong (1103a4 ff.; and cf. 1097b33 ff.). The "non-rational part" of the soul in its broadest sense governs both the nutritive-reproductive functions and the sensitive-emotional-motor activities of man. With the first, the functions of the "nutritive soul", Aristotle is not immediately concerned in the *Ethics* and *Politics*, since they are not directly subject to rational control and since man, having them in common with plants and animals, is not considered "good" or "bad" *as man* because of them (1102a33-b12). But the second set of "non-rational" functions can, Aristotle believes, be controlled and modified by reason, so that they might even be considered "rational" in that they are "capable of obeying reason" (1102b13-29; 1103a2-4).[83] It is to this "part" of the soul, then, called in general "appetitive" (ἐπιθυμητικόν, ὀρεκτικόν: 1102b31) as source of the emotional drives in man, that ethical "goodness" and "badness" belong; and in explaining their nature Aristotle introduces the notions of excess, deficiency, and μεσότης.

Underlying the explanation of ethical excellence particularly in the early books of both the *Nicomachean* and *Eudemian Ethics* is the analogy, express or implicit, between body and soul, between health, strength, or well-being and moral excellence, between medicine or gymnastics and the "political art". The *Eudemian Ethics* makes the basic comparison explicit (1220a2-4): "Just as good physical condition (εὐεξία) consists of the separate excellences of the parts of the body, so also excellence of the soul ... " And the *Nicomachean Ethics* reveals the method behind this (1104a13-14): "To gain light on things imperceptible we must use the evidence of sensible things" (*viz.* bodily health and strength). In taking up the investigation of moral excellence we are, the *Eudemian Ethics* remarks (1220a18-21), in the position of those who "know that health is the best disposition of the body" but do not know what health is.

In beginning his investigation, the first point established by Aristotle in the *Nicomachean Ethics* (1103a17ff.) is that ethical excellence (ἠθικὴ ἀρετή), as its name implies, arises from repetition or habituation (ἔθος). The *Eudemian Ethics* (1220a38-b4)

[83] N.B. The existence of an element in the soul somehow distinct from reason, yet capable of resisting or obeying reason, is established by Aristotle through *physiological* analogy (1102b14ff.): as a paralyzed limb moves in one direction though the patient intends the opposite, so in the soul of the uncontrolled (ἀκρατής) the impulse to action moves contrary to the direction of reason; but in the self-controlled (ἐγκρατής) and virtuous the impulse to action, like the healthy limb, cooperates with reason. Aristotle calls attention to the method explicitly (1102b22-25): "But while in the body we see that which moves astray, in the soul we do not. No doubt, however, we must none the less suppose that in the soul too there is something contrary to the rational principle ... ".

links this with moral *character* in general (ἦθος), which grows out of habituation (ἔθος) in that "by our often moving in a certain way what is not innate in us is finally trained to be operative in that way." The ἦθος seems to be conceived as the "moral organism" of man, which Aristotle later compares to the physical organism, the body (1128a11-12); and ἠθικὴ ἀρετή is the proper excellence of the ἦθος (cf. 1138b35 ff.), developed through repetition of individual acts.

The moral excellences, and moral character in general, then, are not something man is born with, nor are they developed by nature alone; "rather, we are adapted by nature to receive them, and are brought to perfection by habituation" (1103a24-26). Unlike the senses, for example, which have their particular excellence by nature before it is exercised, moral excellence must be built up by repetition of individual acts, as in acquiring an art (1103a27-33). "We become just by doing just acts, temperate by doing temperate acts, brave by doing brave acts" (1103b1-3).

On the other hand, moral evil or vice (κακία) developes in the same way, by habituation; and the same powers of emotion and action which produce good acts to build up virtues may also produce evil acts to develop corresponding vices. "It is from the same causes and by the same means that every virtue is both produced and destroyed, and similarly in every art" (1103b8-9). For example, "by doing the acts that we do in our transactions with other men we become just or unjust; and by doing the acts that we do in the presence of danger, and being habituated to feel fear or confidence, we become brave or cowardly" (1103b14-18). Moral character, therefore, is the product of repeated individual emotional responses leading to action in a variety of situations. If the individual acts are good, moral character developes moral ἀρετή, a fixed disposition to respond in a good way; if they are bad, the opposite disposition results. "In a word, from corresponding actualizations (ἐνέργειαι) our moral habits (ἕξεις) are formed" (1103b23).

Finally, the ἕξις developed through the repetition of a particular response in the past disposes man to the same response in the future; the emotions and actions corresponding to the ἕξις become easy and pleasant, developing the ἕξις further and increasing the capacity for feeling and acting in that particular way (1104a27-30).

In outlining this theory of the "moral organism" and its formation Aristotle evidently has in mind the parallel of the physical organism. For speaking of the moral excellences, he says (1104a27-b3):

But not only are the sources and causes of their origination and growth the same as those of their destruction, but also the sphere of their actualization will be the same; for this is also true of the things which are more evident to sense, e.g. of strength; it is produced by taking much food and undergoing much exertion, and it is the strong man that will be most able to do these things. So too is it with the virtues; by abstaining from pleasures we become temperate, and it is when we have become so that we are most able to abstain from them; and similarly too in the case of courage; for by being habituated to despise things that are terrible and to stand our ground against them we become brave, and it is when we have become so that we shall be most able to stand our ground against them.

The *Eudemian Ethics* shows the reasoning from bodily to moral excellences even more clearly (1220a23-33):

First, then, let it be laid down that the best state is produced by the best means, and that with regard to everything the best is done from the excellence of that thing (e.g. the exercises and food are best which produce a good condition of body, and from such a condition men best perform exercises). Further, that every condition is produced and destroyed by some sort of application of the same things, e.g. health from food, exercises, and weather. This is clear from induction. *Virtue too*, then, is that sort of condition which is produced by the best movements of the soul, and from which are produced the soul's best works and feelings; and by the same things, if they happen in one way, it is produced, but if they happen in another, it is destroyed.

The moral virtues and vices, therefore, like those of the body, are abiding dispositions (ἕξεις) established by the repetition of acts which are good or bad for the moral organism. What is it then that *distinguishes* one from the other, a good act from a bad, a virtue from a vice? To answer this Aristotle turns again to physiological analogy (1104a11-27):

First, then, let us consider this, that it is the nature of such things to be destroyed by defect and excess, as we see in the case of strength and of health (for to gain light on things imperceptible we must use the evidence of sensible things); both excessive and defective exercise destroys the strength, and similarly drink or food which is above or below a certain amount destroys the health, while that which is proportionate (σύμμετρα) both produces and increases and preserves it. *So too is it*, then, in the case of temperance and courage and the other virtues. For the man who flies from and fears everything and does not stand his ground against anything becomes a coward, and the man who fears nothing at all but goes to meet every danger becomes rash; and similarly the man who indulges in every pleasure and abstains from none becomes self-indulgent, while the man who shuns every pleasure, as boors do, becomes in a way insensible. Temperance and courage, then, are destroyed by excess and deficiency, and preserved by a *middle state* (μεσότης).

The concept of μεσότης, mentioned here for the first time in the *Nicomachean Ethics*, is drawn explicitly, it should be noted, from the bodily virtues, especially *health*, and applied as the distinguishing feature of moral virtue. Furthermore, the things which develop health and moral virtue are said to be τὰ σύμμετρα, as opposed to the excessive and deficient, implying that the notion of συμμετρία is closely connected with μεσότης as characteristic of both health and virtue. It may help to a better understanding of moral virtue, then, if we accept Aristotle's suggestion and follow out the comparison, examining health and virtue φυσικώτερον.

3. *Bodily and Moral Virtue and the Equilibrium of Opposites*

First of all, from what Aristotle tells us in this passage and elsewhere, we know that the body of man is ultimately constituted of certain δυνάμεις through which it is subject to qualitative change from hot to cold or from moist to dry and *vice versa*, i.e. to certain "affections", — πάθη — under the impact of diet and environment (*De Part. An.* 646a13-18; etc.). Over a period of time, through the food (hot-dry), drink (cold-moist), exercise, climatic changes, etc. to which the body is subjected on so many individual occasions, these δυνάμεις (and the corresponding πάθη, — the heat, cold, moisture, or dryness of the body) are increased and strengthened and develop an abiding relationship to one another, — a ἕξις. If the diet, exercise, climate, etc. are properly balanced or proportioned (σύμμετρα), the hot, cold, moist, and dry maintain a certain equality or συμμετριά, balancing each other in equilibrium. Rather than existing in a raw state then as opposites, they blend to a *middle state* or *mean condition*, a μεσότης, as their abiding and stable disposition. They are then well-disposed (εὖ ἔχοντα), the body is free from internal conflict, responds well (i.e. proportionately) to new influences, maintains its equilibrium, and is in *health*. But if diet, exercise, climate, etc. are *not* σύμμετρα, then the δυνάμεις and πάθη of the body grow disproportionate, equality and balance among them is lost. With their equilibrium destroyed, the opposing constituents set up conflict, the stronger tending increasingly to dominate and absorb the weaker, resulting in pain, an abnormal craving for those elements which can restore the balance, and an instability which responds disproportionately, so increasing the internal conflict and pain. This is *disease*, and if it continues the internal conflict of opposites eventually destroys all proportion between them, the body can no longer function, and dies.

Turning now from the physical to the moral organism, we find man's "appetitive" life, like his "nutritive", explained in terms of δύναμις, πάθη and ἕξεις (1105b19ff.; *E.E.* 1220b5ff.; cf. *Hist. An.* 608a10ff. and the Ox. transl. note *ad loc.*). As in the body, the ἕξεις are abiding dispositions of the ἦθος with regard to the πάθη, and may be either good or bad (1105b26-27; *Met.* 1022b10-14); the δυνάμεις, likewise, are "powers" or "faculties" by which man is capable of experiencing the πάθη (1105b24-26). But the πάθη of the moral organism are the feelings or *emotions*, — "I mean desire, anger, fear, confidence, envy, joy, friendly feeling, hatred, longing, emulation, pity, and, in general, the experiences accompanied by pleasure and pain" (1105b21-24; and cf. *Cat.* 9a28ff., esp. 9b34-10a10).

One very important characteristic of the δυνάμεις and πάθη of the physical organism, as conceived by Aristotle, is that they are opposites. How is this true of the *moral* organism? The πάθη of *pleasure* and *pain* are obviously opposites, which Aristotle correlates with the hot and cold: "Heating is the contrary of cooling, being heated of being cooled, being pleased of being pained, so that they admit of contrariety. They also admit of the greater and less ... " (*Cat.* 11b3-6). Each πάθος, as Aristotle mentions above, is accompanied by pleasure or pain, so that one or the other of these opposites gives the basic feeling-tone to each emotion, the basic drive to *pursue* (διώκειν) or *avoid* (φεύγειν). Pleasure and pain, therefore, are the primary opposites of the moral organism as hot and cold are of the body. The *Eudemian Ethics* comments (1221b33-39):

Necessarily, then, the moral character (ἦθος) must be bad or good by its pursuit or avoidance of certain pleasures and pains. This is clear from our classification of the emotions, powers, and habits (πάθη, δυνάμεις, ἕξεις); for the powers and habits are powers and habits of the emotions, and the emotions are distinguished by pain and pleasure. So that ... it follows that all moral virtue has to do with pleasures and pains.

The *Nicomachean Ethics* marshalls some eight different arguments to establish the proposition that "moral excellence is concerned with pleasures and pains" (1104b8ff.). Among them Aristotle makes the point that the moral virtues are concerned with actions and emotions, and that every emotion and action is accompanied by pleasure and pain. He approves Plato's notion that proper education consists in training the powers to take pleasure in and be pained by what it should. In training the ἦθος to a state of virtue, the moral guide corrects excessive tendencies in either direction by their opposites, applying pleasure or pain (reward or punishment) as the physician cures excesses of heat or cold by applying their opposites. For the habits are formed, and men made morally better or worse, by reason of the pleasures and pains they pursue and avoid. And the internal response flows out into act, so that to feel pleasure or pain rightly or wrongly is of great importance for external conduct. Hence "the whole concern of virtue and of political science is with pleasures and pains; for the man who uses these well will be good, and he who uses them badly, bad" (1105a10-13).

Pleasure and pain, then, the basic drives to pursue or avoid, comprise for Aristotle the two primary and opposite powers through which the appetitive soul functions. One or the other of these is the basis of all the variety of emotions and actions which he distinguishes, for "every emotion and every action is accompanied by pleasure and pain" (1104b14-15). A variety of emotions is possible because of the variety of objects and situations eliciting pleasurable or painful response, and because pleasure and pain can be associated with past memories and future expectations as well as with present activities (1168a13-17; *Rhet.* 1370a27-35). Ultimately, then, the ἦθος will be good or bad, virtuous or vicious according to its abiding dispositions with regard to pleasure and pain, for "moral excellence is such as to be productive of the best with regard to pleasures and pains, and vice the contrary" (1104b-27-28).

Moral virtue, therefore, is a habit with relation to pleasure and pain. As we have seen, this habit is 1) increased by what produces it; 2) is destroyed by the same when realized in a different way; 3) once established, it actualizes itself fully (ἐνεργεῖ) with regard to what produces it (1105a13-16). There is here, as has been suggested, a close parallel between Aristotle's moral psychology and his physiology. Each of these principles is verified with regard both to pleasure and pain and to the hot and the cold in the production, destruction, and actualization of moral and bodily virtue respectively. To recapitulate briefly:

(1). As bodily health and strength are built up by a repetition or routine of proper diet and exercise, so moral virtue is developed by repeated acts of proper *emotional* response and actions, which are, fundamentally, controlled experiences of pleasure and pain as diet and exercise are controlled assimilation of the hot and cold, moist and dry. "By abstaining from pleasures we become temperate ... by being habituated to despise things that are terrible and to stand our ground against them we become brave ... " (1104-a32-b3).

(2). On the other hand, as health is ruined and disease brought on by wrong diet and exercise, moral virtue is destroyed and the habit of vice developes by habituation to *improper* pleasures and pains. "But we say men are bad through pleasures and pains, either by the pursuit and avoidance of improper pleasures or pains, or by their pursuit in an improper way" (*E.E.* 1222a1-3).

(3). As the healthy body responds properly and easily to diet and climate, so the habit of moral virtue, once established, actualizes itself in responding ὡς δεῖ to painful and pleasant objects or situations. Having become temperate we are best able to abstain from pleasures; and having become brave we are best able to stand our ground when threatened with pain (1104a32-b3).

Furthermore, we have seen that Aristotle draws upon physiological analogy when confronted with the problem of explaining precisely what it is that distinguishes right from wrong, proper from improper, good from bad in the nature of emotions and actions, habits and habitual responses. The bodily virtues of health and strength are produced by food and exercise that are σύμμετρα, i.e. neither excessive nor deficient in supplying and controlling the hot, cold, moist, and dry; for this συμμετρία of diet and exercise establishes an abiding μεσότης among the opposing δυνάμεις of the body, according to the λόγος or proportion demanded by the form. Analogously, feelings and actions produce the habits of moral virtue when they are σύμμετρα; e.g. a man builds the habit of courage when on repeated occasions he fears neither too much nor too little (1104a18-24). "Temperance and courage, then, are destroyed by excess and deficiency, and are preserved by a 'middle state'" (μεσότης: 1104a25-26).

If we press the analogy between the physical and moral life of man, therefore, we are led to the conclusion that: 1) those emotions and actions are morally right, proper, and good which strike the *mean* in *pleasure* and *pain*; 2) the repetition of such emotions and actions establishes an abiding *middle state* or *mean disposition* in the faculties of pleasure and pain; 3) this abiding disposition tends to produce emotion and acts of similar character and constitutes moral virtue. This seems to be what Aristotle has in mind (1106b-16-28):

I speak of *moral* virtue, for it is this that is concerned with emotions and actions, and in these there is excess, deficiency, and the *mean*. For instance, both fear and confidence and appetite and anger and pity and, *in general*, *pleasure and pain* may be felt both too much and too little, and in both cases not well; but to feel them at the right times, with reference to the right objects, toward the right people, with the right motive, and in the right way, is what is both *intermediate* (μέσον) and *best*, and this is characteristic of virtue. Similarly with regard to actions ... Therefore virtue is a kind of *middle state* (μεσότης), since it *aims at what is intermediate* (στοχαστική γε οὖσα τοῦ μέσου).

Hence the habit of moral virtue in general is established through repeated actualizations of the powers of pleasure and pain in the best way, i.e. through feeling pleasure and pain neither excessively nor deficiently but μέσως, — *proportionately to the person and the situation*, and this is as right reason (ὁ ὀρθὸς λόγος) directs. The man who feels and acts thus does so in the right way; and repetition of the best feelings and actions establishes the best disposition.

The μεσότης of moral virtue, therefore, must mean an abiding disposition of proportion, balance, or equilibrium between the opposing faculties or powers of pleasure and pain, those elementary powers of man's moral organism, the two basic appetites of pursuit and avoidance which, manifesting themselves in a variety of modes, are distinguished as the different emotions. This balance of the powers is developed through actualizing them μέσως, and once the balanced ἕξις or μεσότης is established, the moral organism tends to respond proportionately to pleasant or painful objects and situations, i.e. to produce feelings and actions μέσως, so that virtue is στοχαστικὴ τοῦ μέσου. This is brought out clearly in the *Eudemian Ethics* (1222a6-13):

But since, as we have assumed, 1) virtue is that sort of habit which renders men productive of the noblest actions, and through which they are disposed as well as possible with regard to what is noblest, while 2) what is noblest and best is what is in accordance with right reason, and 3) this in turn is the *mean* between excess and deficiency, relative to us, — it necessarily follows that moral virtue is a *mean condition* (or *middle state*: μεσότης) relative to the individual, and is concerned with certain *means* (μέσ' ἄττα) *in pleasures and pains, in things pleasant and painful* ... For he who is excessive in his feelings of delight exceeds in the pleasant; and he who is excessive in his feelings of pain, exceeds in the opposite way ... while the good man feels in the proper way (ὡς δεῖ).

Thus, in moral virtue as in health, the proper or proportionate response of the organism depends upon the internal balance or equilibrium of its opposing constitutive powers, upon their being ἐν συμμετρίᾳ, ἐν μεσότητι. When this ἕξις is established the organism responds proportionately to its environment, physical and moral, producing healthy or virtuous activity, and fulfills its function easily and pleasantly; but when the interior συμμετρία or μεσότης is destroyed through excess and defiency, the response of the organism is disproportionate, it fails in its proper function, and the result is the conflict and pain of disease or vice.

4. *Further Characteristics of Bodily and Moral Virtue*

(a). According to the passage just quoted, right reason demands that feeling and action be neither excessive nor deficient, but in a mean πρὸς ἡμᾶς; hence the ἕξις which produces such actualizations must be a μεσότης *relative to the individual* (καθ' αὐτὸν ἕκαστον). This qualification is explained, again through physiological analogy, in a well-known passage of the *Nicomachean Ethics* (1106a27 ff.). There Aristotle distinguishes the mean between excess and deficiency as: 1) *absolute* or objective, and this is one and the same for all men, and 2) *relative to us*, which is neither too much nor too little, yet is *not* one and the same for all. His example of an absolute mean is the number six, the arithmetic mean between ten and two, regarded as extremes of excess and deficiency. Whereas he explains the relative mean thus (1106a36-b8):

> But the mean relative to us is not to be taken so; if ten pounds are too much for anyone to eat and two pounds too little, it does not follow that the trainer will order six pounds; for even this is perhaps too much or too little for the person who is to take it, — too little for Milo, too much for one beginning athletic training. The same is true of exercise, e.g. running and wrestling. Thus a master of any art avoids excess and deficiency, but strives for and cleaves to the mean, — not the *objective* mean but the mean *relative to us*.

The same quantity, and quality, of food, then, is not reasonable or suitable for all men; rather it must be *proportioned* to the nature and function of the individual, being neither excessive nor deficient, but a μέσον *with relation to him*. Milo and the beginning athlete may both be healthy, but their capacities for assimilating food will differ because of the differences in their development, bodily structure, and the corresponding functions each is called upon to perform.

As mentioned earlier, Aristotle holds that the συμμετρία of health is not the same in all men, nor the same in each individual at all times (1173a24-28). Young, old, and middle aged men may all be healthy, though the συμμετρία of the middle aged will be more perfect than that of the young (who incline to an excess of warmth and moisture) and that of the old (who tend to an excess of cold and dry). It is natural that the συμμετρία should differ somewhat in all three, and that the function of each should differ accordingly. The same might be said of the natural differences between man and woman, both of whom may be healthy, yet have different physical structure and function.[84]

Another of the bodily excellences, beauty, will involve different proportions of bodily parts according to the nature and function of the individual (*Rhet.* 1361b7-14):

> Beauty varies with the time of life. In a young man beauty is the possession of a body fit to endure the exertion of running and of contests of strength; which means that he is pleasant to look at ... For a man in his prime, beauty is fitness for the exertion of warfare, together with a pleasant but at the same time formidable appearance. For an old man, it is to be strong enough for such exertion as is necessary, and to be free from all those deformities of old age which cause pain to others.

All three men are handsome, but each according to his nature and function. And the same would hold, obviously, for the difference between the beauty of man and woman.

Just as the disposition of health and beauty differ, then, according to the nature and function of the individual, so too the συμμετρία or μεσότης of emotional power and response will, properly,

[84] The connection between nature, function, and excellence is prominent in Aristotle's thought, and is emphasized in the passage immediately preceding his explanation of the absolute and relative mean (1106a15-23). The effect of the peculiar excellence of any organism, he maintains, is to make it good and enable it to perform its function well. He illustrates with the example of the eye and of the horse, and concludes: "Therefore, if this is true of every case, the excellence of man also will be the state of character which makes a man good and which makes him do his work well" (1106a22-24).

be relative to the individual according to nature and function. The virtue of liberality, for example, must be exercised with reference to the endowment of the individual. "For liberality resides not in the multitude of the gifts but in the ἕξις of the giver, and this is relative to the giver's substance" (1120b7-9), and "he is liberal who spends according to his substance and on the right objects, while he who exceeds is prodigal" (*ibid.* 24-25). Likewise, magnificence is a virtue relative "to the agent ... who he is and what means he has; for the expenditure should be worthy of his means, and suit not only the result but also the producer" (1122b-23-26). Again, the virtue of friendship (φιλία) does not demand that the same kind and intensity of affection and service be offered to everyone, but this will depend upon whether the individual be father or son, older or younger, husband or wife, ruler or ruled, etc. (1158b12 ff.). "For the *virtue* and *functions* of each of these is different, and so are the reasons for which they love; the love and the friendship are therefore different also" (1158b18-20). More generally, "the functions of man and woman are different" and, though both are good, each has a proper virtue related to function (1162a22 ff.). Likewise, since different citizens, according to their natural endowments, have different functions in the state, their virtue also will be relative to that function. Hence courage, temperance, justice, etc. will differ, for example, in ruler and ruled as they do in man and woman (*Politics* 1277b17-25; and cf. 1259b-23 ff.). "All must partake of them, but not in the same way, but in such measure as is proper to each in relation to his own function" (*ibid.* 1260a18-20). As in health, strength, and beauty, then, the μεσότης which is to produce *proportionate* or appropriate emotions and actions will not be the same for all, but relative to the nature and function of the individual, — πρὸς ἡμᾶς.

(b). Moral virtue, moreover, has other important properties analogous to those of the bodily virtues. In the healthy body, as we have seen, the opposing powers are joined together in the particular proportion (λόγος) demanded by the nature and function of each part, and blend to a middle state in which none of the extremes exist as such, each being moderated by its opposite. This accounts for the *unity and stability* of each part, and of the healthy organism as a whole. For the proportionate blend of opposites does away with all conflict between the powers within the body, and at the same time makes the body less liable to dissolution under the impact of the environment.

Now Aristotle seems to envisage something analogous to this in the constitution of the virtuous or "healthy" moral organism. We have seen that the virtuous ἦθος is developed through ἔθος, — habituation through repetition, and, once established it becomes a sort of "second nature", a *form* imposed upon the undetermined matter of the opposing emotional powers, and a principle of action which defines the character of the emotional response. When the powers of the soul are subjected to training, "frequency tends to produce nature" (*Parva Nat.* 452a30). "For what has been habituated becomes, as it were, natural" (*Rhet.* 1370a6-7). When the μεσότης of moral virtue is firmly established by ἔθος, then, it reproduces the unity, stability, and regularity of nature at its best.

Virtue brings unity, for the basic appetites, being proportionately developed, balance each other so that one does not tend to dominate the emotional life, there is no conflict between them, and they respond to the direction of reason readily and easily. Among the powers in the soul of the temperate and brave man "all things are in harmony (ὁμοφωνεῖ) with rational principle" (1102b28-29). Appetite should be ruled by reason, and so it is in the soul of the temperate man, "for the noble (καλόν) is the mark at which both aim, and the temperate man craves for the things he ought, as he ought, and when he ought; and this is what rational principle directs" (1119b15-18). The virtuous man is in fact distinguished from the merely self-controlled (ἐγκρατής) by this complete inner harmony of emotions and reason. For the merely self-controlled has excessive or evil desires but resists them in acting according to reason; whereas the truly virtuous man *has no* evil desires to resist, because of the balance of his appetites and their ready response to reason (1151b33 ff.).

Furthermore, this unity of emotions and reason gives the virtuous man an inner stability; all his faculties are united when he feels pain and pleasure, and the *same* things are at all times pleasant or painful to him, "for he is, so to speak, unfluctuating" (ἀμεταμέλητος: 1166a27-30). Uncontrolled and vicious characters, on the other hand, are torn, like the diseased, by internal strife. Their faculties are not united in liking and disliking the same thing, but "their soul is rent by faction, and one element in it by reason of its wickedness grieves when it abstains from certain acts, while the other part is pleased, and one draws them this way and the other that, as if they were pulling them in pieces ... " (1166b18-24). Thus the μεσότης of virtue, like that of health, brings internal unity, stability, and peace; without it, as in disease, there is only inner conflict, instability, and pain.

Again, as the opposing elements of the healthy body tend to maintain their proportionate blend against the assaults of the changing environment, so the balanced character of the virtuous man proves impervious to the pleasures and pains which disturb his moral inferiors. It is the temperate man who is best able to abstain from pleasures, and the brave man who endures terrors best (1104a35-b3). The *Eudemian Ethics* makes the physiological and moral analogy explicit in describing the brave man's endurance of what is truly fearful (1228b30-38):

This is the reason ... why the habit of the brave man is praised; his condition is analogous to that of the strong or healthy. For these are what they are, *not* because, in the case of the one, no toil, and in the case of the other no extreme (ὑπερβολή) crushes them, but because they are either unaffected (ἀπαθεῖς) absolutely, or affected only to a slight extent, by the things that affect the many, i.e. the *majority*. Those, moreover, who are sick and weak and cowardly are also affected by the affections (disturbances: παθήματα) common to the majority, but more quickly and violently than the many; and other things that affect the many leave them wholly unaffected, or affected but slightly.[85]

[85] There seems to be no reason to assume corruption in the text here. The healthy and sick, virtuous and vicious, are *both* compared to the generality of men, though as opposite extremes. The majority, in their physical life, enjoy only mediocre health, and, in their moral life, waver between ἐγκρατεία and ἀκρασία (1152a 25-27), being neither virtuous nor vicious. The truly virtuous and vicious, however, are both ἀπαθεῖς or slightly affected by some of the common affections, but in opposite ways (*Physics* 246b17-20). For example, the virtuous will be moved in all or most instances by the appeal of the καλόν, never or rarely by the merely ἡδύ; the majority, largely ἀκρατεῖς, will be moved at times by the καλόν, more often by the ἡδύ; the vicious are ἀπαθεῖς to the καλόν, moved habitually by the ἡδύ.

Thus the healthy and virtuous, because of their internal unity and stability, tend to resist the disintegrating influence of environment, both physical and moral, responding only so far as is good and reasonable, and maintaining the equilibrium of their constituents.

(c). Another happy result of developing the habit of moral excellence is that virtuous activity becomes pleasurable, with the purest kind of pleasure. Aristotle's distinction of pleasures is well known (1152b1 ff.; 1172a19 ff.). First, there are those that accompany the fulfillment of a felt need, like the pleasures of eating and drinking (1152b33 ff.). But these always presuppose a previous lack of something, with its accompanying pain, and are limited in that they cease with the fulfillment of the need. Hence Aristotle regards them as imperfect and impure.

In contrast to these are the pure and perfect pleasures that accompany the unimpeded actualization of a natural power or faculty when it is in good condition and exercised upon its proper object (1174b14 ff.). These are the *true* pleasures, and not just accidentally pleasant as relieving previous pain (1153a3 ff.). They perfect the actualization of the faculty, not intrinsically in the manner of a ἕξις from which the activity springs, but rather "as a supervening perfection, like the glow of health (ὥρα) in men at their prime" (1174b33-35). And these true or pure pleasures accompany not only the proper activity of every natural faculty, both sensory and intellectual, but also the actualization of those ἕξεις which, like moral virtue, are acquired by habituation and become "second nature" (cf. *Rhet.* 1370a5 ff.). Hence "virtuous action is pleasant, or without pain; it certainly cannot be painful" (1120a27-28).

In fact, the pleasure or pain which one finds in good acts is an index of whether or not he has acquired the corresponding ἕξις. For acts that are difficult or painful in themselves become pleasant through habituation (*Rhet.* 1370a13; 1371a25). Accordingly, Aristotle holds that (1104b4-8):

> We must take as an indication of states of character the pleasure or pain that ensues on acts; for the man who abstains from bodily pleasures and delights in this very fact is temperate, while the man who is annoyed at it is self-indulgent; and he who stands his ground against things that

are terrible and delights in this, or at least is not pained, is brave, while the man who is pained is a coward.

Obviously, then, the virtuous and the vicious, like the healthy and the sick, will not be pleased and pained by the same things. The sick and the vicious, being in an unnatural or unbalanced state, will delight in those things that can restore the balance and relieve their abnormal craving, even though these are not truly desirable or pleasant in themselves but only *per accidens*. The healthy and the virtuous, on the other hand, whose powers of body and soul are in their normal or natural condition of proportion and balance, enjoy the pure and true pleasure that accompanies the exercise of these powers upon their proper object.

(d). Moreover, since nature is known through her best products (*Politics* 1254a34-b3), the reactions of the healthy and the virtuous man are *normative*. What pleases or pains him is truly and in itself pleasant or painful to human nature, i.e. is truly helpful or harmful to the perfection of man's nature. For when the physical and moral organism are balanced ἐν μεσότητι they react proportionately, being pleased or pained by things that are truly good or bad for the human nature which is, as it were, fully expressed or embodied in them. This is of great importance for Aristotle's axiology, since it provides him with a concrete norm of the absolute good and pleasant for man, as distinguished from what appears so to the individual (1113a23-34):

Perhaps we should say that the object of rational desire in the absolute and true sense is the good, but that each person desires that which is presented (or appears: τὸ φαινόμενον) to him as good; and therefore that which appears so to the *good* man (σπουδαῖος) is *truly* good (and desirable), while what strikes him is such for the wicked (φαῦλος), — just as in the case of our bodies, the things that are truly such are healthful for those that are in good condition (well-disposed), while other things are such for the sick; and similarly things that are bitter, sweet, hot, heavy, and so on; for the good man discerns each thing correctly, i.e. in each instance the truth is presented to *him*. For the good and pleasant are peculiar to each according to his state of character, and perhaps the good man differs most from others in this, that he sees what is true in each particular, being as it were the norm and measure (κανὼν καὶ μέτρον) of them (i.e. of the good and pleasant).

Again, in his discussion of pleasure, Aristotle applies the same analogy to distinguish true from false pleasure (1176a11-22):

The same things delight some people and pain others, and are painful and odious to some but pleasant and attractive to others. This happens too in the case of sweet things; the same things do not seem sweet to a man in a fever and to a healthy man, — nor hot to a feeble man and to one in good condition. The same happens in other cases. But in all such matters it seems that what appears so to the *good* man (σπουδαῖος) *is* so. If this is correct, as it seems to be, and virtue and the good man, as such, are the measure of each thing, those also will be pleasures which appear so to him, and those things pleasant which he enjoys.

The *Eudemian Ethics* provides some interesting examples (1235b31-1236a1):

Of the good some is absolutely good, some good to a particular man, though not absolutely; and the same things are at once absolutely good and absolutely pleasant. For we say that what is advantageous to a body in health is absolutely good for a body, but not what is good for a sick body, such as drugs and the knife. Similarly, things absolutely pleasant to a body are those pleasant to a healthy and unafflicted body, e.g. seeing in light, not in darkness, though the opposite is the case to one with ophthalmia. And the pleasanter wine is not that which is pleasant to one whose tongue has been spoiled by inebriety (for such men add vinegar to it), but that which is pleasant to the uncorrupted taste. So with the soul ...

The objects are not good and pleasant, of course, *because* the good man perceives them as such, but rather he perceives them as such because they *are* so and "the good man discerns each thing correctly". It is because his powers are properly disposed that the truly good and pleasant things are good *for him*, and are perceived as such. The *Eudemian Ethics* again comments (1248b27-34):

A good man, therefore, is one for whom the things good by nature are good. For the goods men fight for and think greatest, — honor, wealth, bodily excellences, good fortune, and power — are good by nature, but may be hurtful to some because of their dispositions. For neither the imprudent nor the unjust nor the intemperate would get any benefit from the use of them, any more than an invalid from the food of a healthy man, or one weak and maimed from the equipment of one in health and sound in all limbs. A man is noble and good because those goods which are noble are possessed by him for themselves, and because he practises the noble for its own sake, the noble being the virtues and the acts that proceed from virtue.

In all of this, then, Aristotle draws constantly upon the analogy of the healthy and the morally virtuous man, the sick and vicious. The healthy and the morally virtuous man are for him the concrete "norm and measure" of what is healthy and morally good for human nature, and this in two ways:

(1). The healthy and virtuous man *embodies* human nature as it should be, i.e. at its best, on the "nutritive" and "appetitive" levels at least, and this consists essentially in a μεσότης, a condition of balance among opposing factors, the hot and cold in the body, the faculty of pleasure and pain in the moral organism.

(2). The internal balance of health and virtue results in proportionate *response* to things as they are, i.e. in healthy and virtuous *activity*. This is motivated by pleasure and pain, which prompt the healthy and virtuous man to pursue what is truly good for human nature at its best and to avoid what is truly harmful. Hence the reactions of the healthy and virtuous, his likes and dislikes, *manifest* the activities and objects that are good or bad for human nature as such.

This implies, finally, that in both the healthy and virtuous man internal balance brings the power to *perceive things as they exist objectively*: The proper balance of hot, cold, moist, and dry enables the healthy man to taste as sweet and sour what are objectively so, to take pleasure in or be pained by what is objectively helpful or harmful to the human body in its natural condition. Likewise, the proper balance of emotional powers allows the virtuous man to "discern (κρίνειν) each thing correctly", i.e. to perceive what is objectively good, noble, fitting or helpful to man as man and to be pleased by it so as to pursue it, as well as to perceive and be pained by the opposite and so to avoid it. This seems to imply a close interconnection between the μεσότης of health, of moral excellence, and of sense perception.

To recapitulate: In the *Ethics* the general nature of moral excellence is presented through the analogy of bodily health, and the moral character or organism (ἦθος), like the physical organism, is said to involve δυνάμεις, πάθη, and ἕξεις. The opposing δυνάμεις of the moral organism (like the hot and cold in the body) are those producing pleasure and pain. These two affections constitute the basis of all human emotions (πάθη) and the actions to which they lead. Through repeated activation in emotion and action these δυνάμεις for pleasure and pain develop in time an abiding disposition (ἕξις) with relation to one another. And if, like properly balanced food and exercise, the successive actions and emotions are proportionate to the person and the situation (i.e. if they strike a "mean", being neither excessive or defective) they then develop a properly proportionate habit or disposition in the basic powers so that neither tends to dominate. The emotional powers then balance one another, existing in a "mean state" (ἐν μεσότητι) as demanded by the nature and function of the individual (πρὸς ἡμᾶς).

Now this symmetrical disposition, once developed, tends to produce further emotions and actions that are properly proportioned to the situation and the person, and the moral organism is then thought to have attained its proper excellence (ἀρετή); it is constituted in a state of equilibrium, ready to respond proportionately to the moral environment, i.e. to produce emotion and action as right reason (ὀρθὸς λόγος) demands. And once its opposing constituents are established in equilibrium, the moral organism, like the physical, is free from inner conflict and the resulting pain; rather it enjoys a unity and stability that resists dissolution, and experiences pure pleasure in its unimpeded operation, responding harmoniously to reason in the pursuit of the best object, the καλόν, the truly good and pleasant.

For the virtuous man, like the healthy man, because of the equilibrium of his inner powers, is capable of discerning things as they truly are, and of responding proportionately to them. Consequently the virtuous man, like the healthy, becomes a visible, concrete norm or standard, in so far as he embodies human nature as it should be and manifests in his response of pleasure and pain, pursuit and avoidance, that which is good or bad for human nature as such.

5. *The Psycho-Physical Aspect of Emotion and Moral Excellence*

Though the ethical treatises make it clear that moral excellence is essentially constituted in a "mean state" of the emotional faculties, the opposing capacities for pleasure and pain, nevertheless they leave a number of questions unanswered regarding the psychology and physiology involved in emotion, action, and the habits constituting virtue and vice. What *is* emotion? What precisely does Aristotle mean by these "powers of pleasure and pain" as they exist in the concrete living organism? How are the feelings of pleasure and pain basic to all emotion? And how then are the various emotions distinguished? How are the emotions connected with sense perception, so that "only the virtuous man discerns correctly"? Again, how is emotion related to external action on the one hand, and to intellection on the other? The answer to such questions involves the psycho-physical theory which Aristotle avoids discussing in his moral and political treatises. Yet some features of this theory must be indicated if we are to understand the μεσότης of moral excellence in relation to the total human organism and its activities.[86]

A point of primary importance to the understanding of Aristotle's psychology of emotion and moral excellence appears toward the end of the *Nicomachean Ethics* (1178a9 ff.) where, in comparing the happiness attainable through the exercise of moral virtue to that of the intellectual virtues, Aristotle finds the life of moral virtue inferior to that of the intellectual precisely because the former is bound up with the body. Consequently, the life of moral virtue is merely "human" (ἀνθρωπική), while the intellect, being "separate" from the body, is more "divine". This is because moral excellence has to do with the emotions, and the emotions are experiences of the composite. "Some moral activities seem even to result from the body, and excellence of moral character seems to be closely associated with the emotions in many respects" (1178a-14-16). The moral virtues, "being connected with the emotions,

[86] There is room here for only the barest sketch of this matter, since a full discussion would involve all of Aristotle's physiology and psychology of emotion and action. Cf. Beare, p. 286ff.; Siwek, *passim*.

concern the composite (of body and soul); but the excellences of the composite are merely human; hence so is the life according to these virtues, and the happiness it brings" (*ibid.* 20-23).

The emotions and moral excellence, then, belong to the composite, and so have both a psychic and a bodily or physical aspect. This is disregarded for the most part in the *Ethics*, but the strictly psychological works make it very clear. The *De Anima*, for example, teaches that the πάθη of the soul, such as spirit (θυμός) and fear (φόβος), are never separate from the proper matter (φυσικὴ ὕλη) of animals (403b17; cf. 403a3, 16; 408b1 ff.). Hence Aristotle calls the emotions technically ἔνυλοι λόγοι (403a24 ff.), and insists that any one of them be defined fully as a "change (κίνησις) of such and such a body, part, or power, by such and such a cause, for such and such an end" (*ibid.*). He exemplifies with the definition of *anger* (ὀργή) as a "desire for giving pain in return" which manifests itself physically as "a boiling of the blood or heat about the heart" (403a31). The soul is the *source* (ἀρχή) of every vital change or movement in the compound organism (*ibid.* 408b7 ff.; 411a26-30; 413b11 ff.; 415b21 ff.), but every such change involves both body and soul. The *De Sensu* generalizes, listing as common to body and soul not only the processes of sense perception and memory, but spirit (θυμός), desire (ἐπιθυμία), appetite in general (ὄρεξις), and pleasure and pain (436a8 ff.). Thus all the emotions, the appetites, the basic powers of pleasure and pain, belong to the *composite* and have both a psychic and a somatic aspect.

Regarding the basic feelings of pleasure and pain in general, we have already seen that Aristotle conceives pleasure as arising when a ἕξις is activated without hindrance and according to its nature (1153a14-16). Pain, then, accompanies conflict of some sort, when activation is impeded or when a faculty is activated contrary to its nature. A ἕξις, an habitual disposition or capacity to function in a certain way, requires the presence of an object for its activation. When the object is *proportionate* to the faculty, the faculty operates according to its nature and pleasure results; when the object is *disproportionate*, there is pain. Thus, for example, if an object of sense perception be proportionate to the sense organ and power, pleasure accompanies perception; if the object be disproportionate or excessive, its perception brings pain (*De Anima* 426a27 ff., 435b7 ff.; *Parva Nat.* 439b27-440a6, 440b20, 442a12 ff.). And pleasure moves the organism to *seek* the proportionate object as helpful or "perfecting"; pain moves it to *avoid* the excessive object as destructive.

More specifically, pleasure and pain are conscious reactions, so that perception, awareness of the proportion or disproportion between faculty and object is essential. Now moral excellence involves pleasure and pain of the composite, and these are related to sense perception (*Physics* 247a8 ff.). Aristotle explains the difference between perception and the pleasure-pain reaction as follows (*De Anima* 431a8-10): "Sense perception, then, is like a simple spoken word or act of understanding. But when the object is pleasant of painful, the soul, as if affirming or denying, *pursues* or *shrinks away* from it." At the same time Aristotle maintains that perception and the pleasure-pain reaction are in reality but different operations of the same faculty (*ibid.* 10-14):

Indeed, to experience pleasure or pain is precisely *to be in act with the sensitive* μεσότης *regarding what is good or evil, as good or evil.* Actual avoidance and appetition (ὄρεξις) are just this. And the faculty of appetition is not different from that of avoidance, nor are either of these different from the faculty of sense perception, though they are different formalities.[87]

[87] This is a departure from the Oxford translation. Professor W. J. Oates, in the article cited earlier, calls attention to this passage as connecting the μεσότης of sense perception with that of moral excellence, and comments (pp. 395-396): "In the first place, we should note that the mean is the integrating principle of the sense organ itself. Secondly, this passage emphasizes the judging power of the mean as a prime attribute of the sense organ. The mean in the sense-organ then in Aristotle's view acts as a judge and determines the pursuit or avoidance of certain things, and we must remember that here he conceives of the mean as an entity. It determines mechanically certain actions which occur as a result of sense perception." Again, in an interesting and closely reasoned treatment of "The 'Moral Sense' Aspect of Aristotle's Ethical Theory" in *American Journal of Philology*, LXIX (1948), 42-61, E. H. Ohmsted presents this passage as connecting the μεσότης of perception and moral virtue (p. 47), stressing the connection to the point of maintaining that "what Aristotle meant by saying that virtue was in its essence and according to its definition a *Mean* was to point out that it is of the essence of virtue to be able to deal with the particular moral situation clear-sightedly" (p. 61). My reasons for

The sense power, then, reacting to the environment, not only distinguishes one object from another, but distinguishes some as helpful and others as harmful, giving rise to feelings of pleasure or pain with regard to the object. These feelings 1) constitute the impulse to seek or avoid the object (φυγή, ὄρεξις) and 2) are the beginning of all external action. Thus Aristotle holds that wherever there is sense perception (i.e. in all animals) there is pleasure and pain; and wherever these, there must be desire, appetite (ἐπιθυμία, ὄρεξις) to pursue the pleasant and avoid the painful (*De Anima* 413b22-24; 414b4-5). For the faculty of sense perception and the power of pleasure and pain are ontologically the same.[88]

This identification of the faculty of emotion with the faculty of sense perception, then, explains for Aristotle the feelings of pleasure or pain arising when an object is actually present to the senses. But what of those emotions, the pleasures and pains, which arise in man with memories of the past or anticipations of the future (*Rhet.* 1370a27-35), i.e. when the object is *not* present to the sen-

---

considering this interpretation somewhat too restricted should be clear from the text.

[88] For Aristotle the power of sense perception and that of pleasure and pain are identified not only ontologically but *teleologically* in that both are ordered to the *same end*. The nature of sense perception demands that it be joined to pleasure and pain, for its *purpose* is to preserve the existence of the animal by guiding it to what is good and away from what is bad (*De Anima* 434b10ff.). But this cannot be effected unless the animal, in addition to perceiving the differences of color, taste, touch, etc., also perceives some objects as *helpful*, others as *harmful*, and be moved toward or away from them by the feeling of pleasure or pain. Hence the lowest form of animal must have at least touch and taste "to distinguish in food the pleasant from the unpleasant, so as to flee from the latter and pursue the former" (*Parva Nat.* 436b16-17). Man and other animals possessing the power of locomotion have the higher senses also as "a means of preservation, their final cause being that such creatures may, guided by *antecedent* perception, both pursue their food and shun things bad or destructive" (*ibid.* 20-22). However, it may be remarked here that in *man* the senses serve not only his physical needs but enable him, through the data they present the *intellect*, both theoretical and practical, to *live well* (*ibid.* 437a1-3). The pleasures and pains of man's sensitive or appetitive faculties must be directed, therefore, not only toward what helps or harms his physical existence, but especially toward what promotes or destroys that higher life, the life of the intellect, by which he "lives well", as we shall see.

ses? Aristotle explains these through the faculty of *presentation*, — φαντασία.

Sense perception, originating when a present object affects the organs of sensation, is, as we have seen, a psycho-physical process, a "movement of the soul through the body", involving the special and common sense powers and their organs. Now Aristotle maintains that even when the sensible object is no longer present to the senses, the psycho-somatic modification caused by it may remain in the perceiving subject (*De Anima* 425b24) as 1) a conscious after-image or 2) a "vestige" of the original perception which persists in the material organ and only later affects the central sense power, emerging into consciousness as a dream or as a waking recollection (*ibid.* 408b15-18; *Parva Nat.* 459b5 ff., 460b20). This sense power which "*re-presents*" an absent object to consciousness through a secondary image is called φαντασία. The "re-presentation" or phantasm is produced in the central sense (*Parva Nat.* 450a10). The faculty of φαντασία, then, is really identical with the central sense power, though they differ formally (*ibid.* 459a15-17).[89]

Now each phantasm, like the actual sense perception from which it comes, not only reproduces the sensible form of the object but may also represent the object as helpful or harmful, and so carry with it the power to evoke feelings of pleasure and pain (*De Anima* 431b2 ff.). Hence man is stimulated to emotion and action not only by the perception of a present object but also by the phantasm, which represents an object remembered as helpful or harmful in the past, or anticipated as such in the future (*ibid.*; *Rhet.* 1370a-28 ff.). The phantasm is, in fact, conceived as a controlling factor in the whole mechanism of emotion and action, with which moral excellence is concerned. For Aristotle maintains that "a living creature is capable of moving itself only in so far as it has appetite; but it has no appetite without φαντασία" (*De Anima* 433b28-29; and cf. 433a20 ff.).

[89] That is, φαντασία is the central sense power *qua* reproducing a composite image in the absence of its object. As we shall see, this power of the central sense to reproduce images of *past* sense perceptions is for Aristotle the basis of memory, recollection, and the accumulation of experience.

The central sense power, then, — the power which produces the composite image or phantasm, and corresponding reactions of pleasure or pain — is identical with the faculty of emotion. For every emotion moving man to action is just this: a feeling of pleasure or pain stimulated by the image or phantasm of a helpful or harmful object. Anger or wrath (ὀργή, θυμός), for example, begins with a feeling of pain arising when reason or imagination (φαντασία) discloses a slight or insult to oneself or friends (1149a-33-34; *Rhet.* 1378a32-34). This is succeeded by the thought of vengeance, and that expectation evokes a feeling of pleasure, "because the thoughts dwell upon the act of vengeance, and the imagination (φαντασία) then arising produces pleasure ..." (*Rhet.* 1378b8-9). Thereupon "anger by reason of the warmth and hastiness of its nature ... rushes to vengeance" (1149a31-32; and cf. 1117a6-9). Fear, on the other hand, is "a pain or disturbance (ταραχή) due to a mental picture (φαντασία) of some destructive or painful evil in the future" (*Rhet.* 1382a19-22; cf. *E.E.* 1229a33ff.). Confidence (θάρσος), again, "is the opposite of fear, and what causes it is the opposite of what causes fear; it is, therefore, the expectation (ἔλπις) associated with the mental picture of the nearness of what keeps us safe and the absence or remoteness of what is terrible" (*Rhet.* 1383a15-18). Finally, allied to fear is the feeling of shame (αἰδώς, αἰσχύνη), which is "a pain or disturbance in regard to bad things, present, past, or future, which seem likely to involve us in discredit" (*ibid.* 14-16; cf. *N.E.* 1128b12), that is, "a mental picture of disgrace, in which we shrink from the disgrace itself and not from the consequences" (*Rhet.* 1384a23-25). Thus the *phantasm*, representing an object as helpful or harmful and stimulating the reaction of pleasure or pain, constitutes the λόγος, the reason, the exciting cause and objective, which gives each emotion its particular form and dialectical definition (cf. *De Anima* 403a29 ff.) and distinguishes it from other emotions. This is the *formal* aspect of emotion, the *psychic* aspect produced by the central sense power of the soul.

But as a λόγος ἔνυλος each emotion has its material aspect also. The pleasure-pain reactions to the phantasm are experiences of the composite, and so are united with characteristic changes in the body which set off the whole psycho-physical mechanism of pursuit and avoidance and issue in external action. Specifically, the positive pleasurable emotions bring an increase of heat in the body, while the negative painful emotions are manifest physically in an increase of cold. Aristotle holds, for example, that "anger is productive of heat" (*De Part. An.* 650b35), while "fear is, in fact, a cooling down" (*Rhet.* 1389b32). The increase of heat associated with pleasant emotions moves the body, in general, to greater activity, to aggressive action of pursuit; while the cooling down that comes with painful emotion inhibits activity, causing the organism to move slowly, to shrink from or avoid the object (cf. *Rhet.* 1389a2 ff.). Thus the pleasure-pain reaction to the phantasm and the corresponding bodily affection of heat and cold constitute the formal and material, the psychic and somatic, aspects of a single emotional experience in the composite organism.

In fact Aristotle conceives these two aspects of emotion as being so closely united that not only does the psychic reaction of pleasure or pain bring a corresponding change in the body, but a condition of excessive heat or cold in the body tends to produce the corresponding feeling in consciousness. In describing the opposite emotional characteristics of youth and age, for example, Aristotle explains them as flowing from an excess of heat or cold in the bodily disposition of the young and the old (*Rhet.* 1389a2 ff.).[90] And in the same individual at different times an unbalanced state

[90] *Young* men, Aristotle says, "like those full of wine, are *pervaded by heat* (διάθερμοι) because of their physical structure" (*Rhet.* 1389a18-19; cf. *N.E.* 1154b10). Consequently "their hot tempers and hopeful dispositions make them free from fear and confident; we cannot feel fear so long as we are feeling angry, and the expectation of good makes us confident" (*ibid.* 26-28). They are also pleasure-loving (ἐπιθυμητικοί), desiring particularily the pleasures of sex; fierce, loving honor and victory; they enjoy companionship and laughter; they tend to do all things *excessively* and violently. The *old*, on the other hand, are the opposite. Physically "their disposition is contrary to that of the young; for they are *cooled down* (καταψύγμενοι), while the young are hot" (*ibid.* 1389b 31). Consequently the old are cowardly and live in fearful anticipation, since "old age has opened the way to the cowardice; for *fear is*, in fact, a *cooling* down" (*ibid.* 31-32). The anger of old men is correspondingly feeble, their physical desire extinct or weak; they are querulous, small-minded, and apathetic; and in general they tend to *under-do* everything. Thus the natural excess of heat or cold in the body brings a corresponding tendency toward the positive or negative, the pleasurable or painful, emotions.

of body results in disproportionate emotional response, so that "sometimes on the occasion of violent and striking occurrences there is no excitement or fear felt, on others faint and feeble stimulations produce these emotions, viz. when the body is already in a state of tension resembling its condition when angry" (*De Anima* 403a19-22). The bodily condition may indeed be so influential that "in the absence of any external cause of terror we find ourselves experiencing the feelings of a man in terror" (*ibid.* 23-24).

These two complementary aspects of emotion, then, are produced by faculties which we have already studied. The central sense faculty, poised to respond proportionately to opposite sensible qualities, is also the faculty which produces the *formal* or psychic aspect of emotion, — the composite image or phantasm and the opposite responses of pleasure and pain. The *material* or physical aspect of emotion, on the other hand, is produced by that faculty which controls the opposing powers of heat and cold in the body, enabling it to respond proportionately to changes in the physical environment. Thus together the central sense faculty and the faculty controlling the vital heat comprise the *faculty of emotional response* in the concrete living organism.

Moreover, as we have seen, *both* the central faculty of sense perception and that which controls the vital heat attain their natural excellence and are disposed to respond most perfectly when their opposing constituents are ordered in equilibrium, — ἐν μεσότητι. Now the excellence of the emotional faculty must coincide with that of the faculties which comprise it. The obvious implication seems to be that in the concrete living organism the μεσότης of *moral excellence*, the disposition of the emotional faculty which enables it to respond proportionately to the truly helpful and harmful, being essentially a condition of balance between the opposing powers of pleasure and pain, *is in its formal aspect ontologically identical with the* μεσότης *of sense perception, and in its material aspect with the* μεσότης *of vital heat and cold.*

It follows as a corollary that the primary bodily *organ* of the emotional faculty, the organ in which moral excellence may be most properly said to exist, must be identical with the organ of the central sense power and the controlling center of vital heat. It is the heart, therefore, the embodied central sense power, that produces the composite image and the phantasm; it is the heart first of all that responds with pleasure or pain, shifting its normal balance of hot and cold and thereby inaugurating the movement of pursuit or avoidance throughout the entire organism.[91] Accordingly, Aristotle not only declares in general that anger and fear increase the heat and cold of the body (*De Part. An.* 650b35; *Rhet.* 1389b32), but describes the physiological aspect of anger precisely as "a boiling of the blood or heat about the heart" (*De Anima* 403a31-32) and assumes that "to be angry or to fear is for the heart to be changed in this way or that" (*ibid.* 408b7-8; cf. 432b31-32). For the unique position of the heart ἐν μέσῳ qualifies it pre-eminently as the source and center of emotion and movement, as well as of the other vital activities, in the organism:

For it is in the front and center (μέσον) of the body that the heart is situated, in which we say is the principle and source (ἀρχή) of life and of *all motion and sensation* (*De Part. An.* 665a11-13) ... Moreover, the *motions of pain and pleasure*, and generally of all sensation, plainly *have their source in the heart*, and find in it their ultimate termination. This indeed reason would lead us to suspect. For the source must, wherever possible, be one; and of all places, the best suited for a source is the center. For the center is one, and is equally or almost equally within reach of every part (*ibid.* 666a12-16).

Thus in the living organism the faculty of emotion is identical with the nutritive and perceptive faculties in that the principal organ of all three is the heart. Hence the three faculties, and their specific operations, are so closely identified that deviation from the normal μεσότης in one means deviation and disturbance in the others. For example, any deviation in the structure of heart and blood vessels, or in the composition of the blood, disposes the organism to abnormal emotional response, as well as to defective perception

[91] The rest of the bodily organism is capable of obeying the *faculty of pleasure and pain* because its organ, the *heart*, is also the center of vital heat and of the vascular and respiratory systems. A full discussion of the mechanics of emotion and action is beyond the scope of this study. But Appendix III will attempt to indicate briefly how the bodily changes of emotions and actions are initiated and controlled by the heart, the principle and source at the *center* of the physical organism.

(*supra*, p. 220). Thus the size of the heart affects character (ἦθος). For an excessively large heart predisposes to fear and cowardice, since "the affection normally produced by fear is present beforehand. For the vital heat is not proportionate to the size of the heart and, being small, has little effect in the large space, so that the blood is comparatively cold" (*De Part. An.* 667a17-20). Hence "those with large hearts are cowardly, while those with smaller, i.e. *middle-sized* (μέσαι) hearts are more courageous" (*ibid.* 15-17). And "it is much the same with the blood vessels and chambers of the heart; if large, they are too cold" (*ibid.* 23-25). Likewise, if the heart be too moist or too dry perception and emotional response are disproportionate. Thus melancholia, an excess of black bile (cold-moist), is associated with excessive moisture in the cardiac region, resulting in overactivity and instability. The result is that memory is poor, emotional response to phantasms violent, and control difficult (*Parva Nat.* 450b8-11; 453a15-23; 457a27-34; *N.E.* 1150b25-29; 1154b12-16). Obviously, then, emotional reactions and character are strongly conditioned by the size and blend of the heart.

So, too, with the blood. Animals whose blood is too liquid or watery are inclined to fear. "For fear chills the body; so that in animals whose heart contains so watery a mixture the way is opened for the operation of this emotion ... " (*De Part. An.* 650b27-30). "Such animals, on the other hand, as have thick and abundant fibres in their blood are of a more earthy nature, and of choleric temperament, and liable to bursts of passion. For anger is productive of heat; and solids, when they have been made hot, give off more heat than fluids" (*ibid.* 33-35). Hence Aristotle concludes, "the nature of the blood affects both the character (ἦθος) and the sense perception of animals in many ways" (*ibid.* 651a13-14). There is, then, a very close correlation between emotional balance and the proper blend of material constituents in the body. Health, therefore, becomes closely related to moral excellence.

There is also a very close correlation between emotional balance and the proper functioning of the central sense power. For accurate sense perception and clear-cut phantasms require a certain stable equilibrium in the heart and blood stream. Hence deviations of heat and cold, such as accompany antecedent pleasure and pain, affect the material aspect of images and phantasms arising in the heart, distorting them and thus hindering the central sense power from attaining objective truth (Cf. *Parva Nat.* 461a8 ff.; *De Part. An.* 656b5-6). Moreover, correct judgment is difficult when the appetitive faculty is actually operative in pursuit or avoidance, for the mind tends to identify a mere resemblance with the object, and so is easily deceived (*Parva Nat.* 460b3-10):

To return to our original question, one thing may be laid down as clear from what has been said, namely, that even when the external object of perception has departed the impressions it has made persist and are themselves objects of perception, and moreover that *we are easily deceived about our perceptions when we are excited by emotions*, different persons according to their different emotions; for example, the coward excited by fear, the lover by amorous desire; so that, with but little resemblance to go on, the former thinks he sees the enemy approaching, the latter that he sees his beloved. And the more deeply one is under the influence of emotion the less similarity is required to cause these illusions. In the same way in fits of anger and *in all states of desire* (ἐν πάσαις ἐπιθυμίαις) *all men become easily deceived*, and more so the more their emotions are excited.

The deceptions resulting from antecedent emotional activity are, in fact, basically the same as the illusions resulting from the disturbed condition of the body in disease. Aristotle continues (*ibid.* 11-16):

This is the reason also why persons in the delirium of *fever* sometimes think they see animals on the walls, an illusion arising from a slight resemblance afforded by the patterns traced upon the wall; and this sometimes increases proportionately with the afflictions of the patients in such a way that if they are not very ill they know well enough that it is an illusion, but if the illness is more severe they actually move according to the appearances.

Thus the feelings of pleasure and pain, like disease, disturb the equilibrium of hot and cold in the heart and interfere with the formation of clear images and phantasms. Truly accurate perception and imagination, therefore, require not only health but the emotional stability of moral excellence.

Recapitulation and conclusion: To review the psycho-physical aspect of the sensitive organism, then, we may recall that the faculty of perception and presentation attains its characteristic excellence when the sense organs, and particularily the central sense organ, the heart, are properly formed and habitually balanced in equilibrium, their opposing constituents blended to a mean. Thus disposed, the sense power is psychologically neutral, at peace, poised to respond impartially toward either extreme of sensible quality. The formation of an accurate, integral image or phantasm requires that this balance be swayed only by the sensible object perceived, and that the impulse be conveyed and presented to the central sense through the medium of blood that is pure in blend and tranquil.

If these conditions are to exist, however, the faculty which governs the vital heat must maintain the opposing powers of hot and cold in constant balance, so that neither dominates. Again, it is in the heart primarily that this stable equilibrium of health is achieved, since the heart determines the balance of hot and cold in the rest of the body. When the hot and cold blend evenly in the heart, then, movement ceases, the body enjoys the repose of dynamic equilibrium, and the sense faculty is most free to perceive accurately and to form clear, integral phantasms.

This equilibrium of hot and cold, however, is disturbed not only by disease but also by *emotion*, the active experience of pleasure or pain, which inaugurates external activity. For the faculty of emotion is identical, in its formal and material aspects, with the faculties of sense perception and of vital heat. Hence the reaction of pleasure oı pain brings an excess of heat or cold in the heart, resulting in bodily changes which distort sense perception and imagination and move the mind to false judgments. Just as the opposing constituents of the body, then, must be brought to a condition of balance and blended to the stable equilibrium of health by proper diet and exercise, so through proper moral training the opposing capacities for pleasure and pain must be brought to a condition where neither feeling (and its bodily affection) dominates the organism, but the two habitually balance each other in the stable equilibrium of *moral excellence*.

For if the emotional powers are habitually disposed in an unbalanced condition of excess and deficiency, the conscious life is dominated by either pleasure or pain, which means that the organism is in a constant state of active craving for pleasure or for relief from pain. The satisfaction of this craving then becomes the end or objective of all its activities, and moves it to seek what appeals as pleasant to its disordered nature (the apparent good) rather than what is truly good and pleasant (τὸ καλόν). "Living by emotion they pursue their own pleasures and the means to them, and avoid the opposite pains, and have not even a conception of what is noble and truly pleasant, since they have never tasted it" (1179b-13-16; and cf. 1113a24-b2; 1114a31 ff.; 1140b12 ff.). At the same time the body is in a state of habitual disturbance which interferes with sense perception, imagination, and judgment (cf. 1147a10 ff.). Moreover, its reactions to what is presented as helpful or harmful are disproportionate, since it is antecedentally disposed to excess in the direction of either the painful or the pleasant emotions, according to its dominant tendency (1150b19-29; 1151b23-32). Thus an habitually unbalanced condition of the opposing emotional powers leaves the organism unstable, restless, in conflict with itself and its environment (cf. 1166b7-22).

On the other hand, the habitual condition of moral excellence is characterized by order, tranquillity, freedom from the disturbance of pleasure and pain. For when the emotional powers are properly ordered in the stable equilibrium of moral excellence, the organism is habitually free from strong antecedent emotion or appetite. That is, when the opposing powers of the compound organism are evenly balanced (ἰσάζῃ), "their activity appears neither pleasant nor painful" (1154b23-24).[92] Conscious life is then tranquil, steady, emotionally neutral until stimulated by a proportionate object. This constitutes that state of tranquillity which the brave man enjoys before action (1116a9; cf. also 1117a31),

[92] Aspasius explains Aristotle's meaning in this difficult passage as follows (*Commentaria in Aristotelem Graeca*, XIX, part I, page 157): "He means by this that we do not have a body composed of a *single* material substance but rather of *several*, so that if one of the active principles within us act upon another of a different nature, a case of 'contrary to nature' obtains; for example, when the hot acts violently upon the cold, this is a case of opposition. But when all are equally balanced, then it seems that what takes place is neither pleasant nor painful ..."

the painless state which the prudent man cultivates (1153a31), the neutral state which men of unbalanced nature cannot endure (1154b7). Thus the *Eudemian Ethics* (1218a21-24) instances the controlling moral virtue of temperance (σωφροσύνη) along with bodily health and strength as goods consisting essentially in a certain order and tranquillity (τάξις καὶ ἠρεμία); while the brave man and the healthy are characterized as "unaffected (ἀπαθεῖς) absolutely or only to a slight extent (ἠρέμα) by things that affect the many" (*ibid.* 1228b34-35). Accordingly, Aristotle seems willing to admit the popular definition of moral virtue as a certain "absence of feeling or tranquillity (ἀπάθεια καὶ ἠρεμία) with regard to pleasures and pains" (*ibid.* 1222a3-5), though he insists that this be understood with the proper qualifications (cf. *N.E.* 1104b24-26).

This habitual disposition of equilibrium between the opposing powers of pleasure and pain constitutes the *best* state of the "appetitive" or "affective" soul, then, because it brings order, unity, stability, tranquillity into the emotional life of man. But what is perhaps more important, in freeing consciousness of the disturbances of pleasure and pain, leaving the sense power unimpeded to perceive accurately and the emotional powers poised to respond proportionately to the truly good and evil, the μεσότης of moral excellence disposes the "affective" soul to perform its natural function, that is, to serve and obey the highest faculties in man, his practical and theoretical reason.

As the child should live according to the direction of his tutor, so the appetitive element should live according to rational principle (λόγος). Hence the appetitive element in a temperate man should harmonize with the rational principle; for the noble (τὸ καλόν) is the mark at which both aim, and the temperate man craves for the things he ought, as he ought, and when he ought; and this is what rational principle directs (*N.E.* 1119b13-18).

But since man is by nature composed of a ruling and a subject part, each of us should live according to the governing element within himself ... and thus it is with regard to the faculty of contemplation ... Therefore whatever choice or acquisition of things good by nature best produces contemplation of the divinity ... this is most excellent ... but any that through deficiency or excess hinders the worship of the divinity in contemplation, this is bad. Thus it is with the soul, and this is the best norm for the soul, — to be as far as possible *unconscious* of the *irrational* part of the soul, as such (*E.E.* 1249b12-23).

Action "according to right reason" takes place when the irrational part of the soul does not hinder the rational from carrying on its own proper activity. For then only will action be "according to right reason". Let me explain: We have within the soul an inferior part and a superior part. But the inferior part always exists for the superior. Thus in the case of body and soul, the body is for the sake of the soul. Now we will declare the body to be in good condition when it is so disposed that, far from hindering the soul from accomplishing its proper function, it rather joins in and strives along with the soul in the performance of its function (for the inferior is for the sake of the superior in that it is to work along with the superior). Similarly, then, when the *emotions* do not hinder the *intellect* (νοῦς) from carrying on its proper function this will be a case of something done "according to right reason" (*Magna Moralia* 1208a9-21).

6. *Moral Excellence and the Intellect*

Moral excellence, then, involving as it does both the μεσότης of health and of sense perception, is the best state of that part of the human soul which is itself without reason (ἄλογον) but whose function is to assist the operation of reason (λόγος) and to translate its direction into human action. How is this accomplished? How does moral excellence contribute to intellectual activity, and to human happiness as a whole? Before concluding, we must attempt to indicate some answer to these problems, however briefly.

By way of preliminary, we must first recall some points of Aristotle's doctrine on the nature of the intellect (νοῦς) and its connection with the other, non-rational faculties of the soul:

(1). For Aristotle the intellectual faculty is quite distinct from the sensitive, though both are powers of the one soul. The senses, operative through their material organs, have as their object the sensible qualities of the particular, changing thing; while the intellect informs no material organ and has as its object the intelligible form, i.e. the immaterial, unchanging essence which makes each thing what it is (*De Anima* 429a10-b23).

(2). Yet the function of the intellect, and its knowledge, *depend* upon the senses. Why? Because Aristotle holds that the intelligible forms or unchanging essences exist only in particular, concrete, sensible things. Hence they are always presented to the intellect under a particular sensible form. "For though the act of sense-perception is of the particular, its content is universal, — is man, for example, not the man Callias" (*Post. Anal.* 100a17-b1). Consequently the intellect is extrinsically dependent upon the senses for its operation, since they "bring in the tidings of many distinctive qualities of things, from which the knowledge of truth, speculative and practical, is generated in the soul" (*Parva Nat.* 437a1-3).

Since according to common agreement there is nothing outside and separate in existence from sensible extended substances, the objects of thought are in the sensible forms ... Hence no one can learn or understand anything without sense perception (*De Anima* 432a3-7).

(3). The *process* by which the intellect forms the universal, the ἀρχή of all intellectual operations, is detailed in the *Posterior Analytics* (99b36-100a8). The importance of sense perception and its persistence as a phantasm, memory, and experience should be noted:

Though *sense perception* is innate in all animals, in some the sense impression comes to *persist* ... and when such persistence is frequently repeated a further distinction arises between those animals which out of the persistence of such sense impressions develop a power of systematizing them and those which do not. So out of sense perception comes to be what we call *memory*, and out of frequently repeated memories of the same thing develops *experience*. From experience again, i.e. from the *universal* now stabilized in its entirety within the soul, the one beside the many which is a single identity within them all, originate the *skill* of the craftsman and the *knowledge* of the man of science, skill in the sphere of coming to be and science in the sphere of being.

Thus the primary universal is associated with, and dependent upon, the *phantasm* built up through successive sense perceptions (*De Anima* 429a1-2); it is not identical with the phantasm, but it never exists without a phantasm (*ibid.* 432a13-14).

(4). Since the intellect produces science and art from sense experience, and experience is produced from memory (*Met.* 980b-27 ff.), memory is obviously important in all intellectual activity. Yet for Aristotle the memory is entirely a function of the *sense* power, so that intelligible forms or universals are not directly remembered as such, but only the phantasm with which each is associated (*Parva Nat.* 449b30-450a25). Now since the central sense power in the heart produces the phantasm and is identical with the faculty of memory (*ibid.*), it is clear that the disproportion and disturbance of undisciplined emotion or disease will interfere with memory (*ibid.* 450b1-12) and with the intellectual operations depending upon it. Moreover, the power of deliberate *recollection*, found in men alone and closely associated with intellect (*ibid.* 453a9-10), operates through the chain of phantasms linked in memory, and so is also impeded by the disordered movements of emotion, disease, growth and decay, etc. (*ibid.* 453a15-b11). Thus the intellect depends upon the phantasm, both in the initial formation of universals and in their retention and recall, — processes which are basic to all intellectual activity.

(5). Aristotle distinguishes intellectual activity (1139a26-29) as *speculative* (θεωρητικός), which aims simply at knowing, and *practical* (πρακτικός), aiming at action (πρᾶξις) or production (ποίησις). Correspondingly, he divides the intellect into two faculties according to the objects with which each is concerned in the pursuit of these activities (1139a7-13):

> And let it be assumed that there are two rational parts of soul, — one by which we contemplate the kind of realities whose first principles do not admit variation, and one by which we contemplate variable things; for where objects differ in kind the part of the soul answering to each is different in kind ... Let one of these parts be called the *scientific* (ἐπιστημονικόν) and the other the *calculative* (λογιστικόν).

Thus the activity of the *scientific* or *theoretical* faculty is to know, and to contemplate, the eternal, unchanging realities; the activity of the *calculative* or *practical* faculty is to deliberate on the concrete and changing, with a view to acting or producing, pursuing or avoiding. And in both types of activity the intellect is dependent upon the senses and the phantasm, which presents the object of action or of contemplation; for "there is no intellectual activity without phantasms" (*Parva Nat.* 449b31).

> To the thinking soul phantasms serve as present perceptions, and when it asserts or denies them as good or bad, it flees or pursues. For this reason the soul never thinks without a phantasm (*De Anima* 431a14-17). And whenever one contemplates, it is necessary at the same time to

contemplate some phantasm; for phantasms are like present perceptions, but without the material object (*ibid.* 432a8-10).

From these considerations at least one general conclusion will be obvious: Whatever hinders the central sense faculty from producing accurate perceptions and phantasms, from remembering and recollecting, will *ipso facto* interfere with the operations of the intellect. But we have seen that the disturbances of emotion and disease interfere with the central sense power in all these functions. Hence the stability and tranquillity of moral excellence and health are required for the *unimpeded* operation of sense faculty *and* intellect, both practical and speculative.

We may now consider each aspect of intellect in relation to moral excellence.

a. *Moral Excellence and the Practical Intellect*

Moral excellence, the best condition of the non-rational faculties producing emotion and action, is naturally most closely involved with the *practical* intellect, whose function is to control action. It is precisely the capacity to control his emotions and actions through practical reason, i.e. to exercise deliberate choice, that makes man self-determining and responsible, distinguishing him from the other animals (1111b7-9, 1112a1-17, 1139a32ff.). And since the repetition of individual actions, for which he is responsible, builds and maintains a certain habit of emotional response and action, man is also held responsible for his moral habits, and is praised or blamed for his virtue oı vice (1113b3ff., 1114b6ff.).

A number of questions arise at once: How does Aristotle find it psychologically possible foı the practical reason to control the non-rational faculties of emotion so that man can be said to determine his own act? How does a series of such acts form a habit? How does reason participate in the acts of moral excellence once the habit is formed? How does the habitual μεσότης of moral excellence help the practical intellect to function properly? We must be satisfied here to indicate briefly a possible solution to these problems.

Aristotle has difficulty, we may recall, in classifying the "affective part" of the soul as either *non-rational* (ἄλογον) or *rational* (λόγον ἔχον), for without itself possessing reason it yet participates in reason in so far as it can "pay heed to and obey" reason as a child obeys its father (1102b29-1103a3). In other words, practical intellect has the power somehow to direct emotion and action. How does Aristotle explain this control psychologically?

In examining moral excellence from the psycho-physical aspect, we found that Aristotle regards the phantasm as the controlling factor in the whole mechanism of emotion and action (*supra*, p. 251). Moreover, we have seen that every operation of either the practical or speculative intellect must be accompanied by a phantasm. Hence when the practical intellect conceives an object or course of action as good or bad this is accompanied by a phantasm which can elicit the response of pursuit or avoidance (*De Anima* 431b2-9):

The faculty of thinking then thinks the intelligible forms in the phantasms and, as in the case of present sensation what is to be pursued or avoided is marked out for it, so where there is no present sensation and it is engaged with the phantasms it is moved to pursuit or avoidance ... But sometimes by means of the phantasms or thoughts which are within the soul it calculates as if it were looking at the objects, and deliberates upon the future with relation to what is present. And when it pronounces that an object is pleasant or painful it immediately takes to flight or pursues. And so generally in human action.

Human action, then, is determined not simply by present sensation, nor by the phantasm alone, but by the phantasm associated with an intellectual judgment, a phantasm produced by what Aristotle calls *calculative* or *deliberative* imagination (*ibid.* 433b27-30):

To sum up, then, ... inasmuch as an animal is capable of appetite it is capable of self-movement; it is not capable of appetite without possessing imagination; and all imagination is either (1) calculative or (2) sensitive. In the latter all animals, and not only man, partake.

The general psycho-physical process is described by the *De Motu Animalium* as follows (702a16-19):

For this reason we might say that as soon as a man thinks (νόει) he should go he goes, unless something else stands in his way. For the organic parts are suitably readied for action by the emotions, the emotions by the appetite, the appetite by the phantasm or image; and this arises either through intellection (νόησις) or sense perception.

Now because in man appetite, emotion, and action may be stimulated by both the perception of a present object or the phantasm of a future object suggested by reason (*De Anima* 431b6-10), there can arise a conflict of appetites "when a principle of reason and a desire are contrary" (*ibid.* 433b6), as is the case when "intellect (νοῦς) bids us hold back because of what is future, while desire prompts to what is present" (*ibid.*; and cf. *Rhet.* 1372b12 ff.). This simultaneous awareness of a present good or evil sensibly perceived and a future good or evil suggested by reason and presented in a phantasm establishes the conditions for *deliberation* and *choice*. For in comparing the future course of action presented through the phantasm to that prompted by present sense perception, the mind is faced with alternatives; hence it can *deliberate* (βουλεύεσθαι), i.e. weigh one against the other (λογίζεσθαι) by a single standard (*De Anima* 434a5-10):

Sensitive imagination (φαντασία) ... is found in all animals, deliberative imagination only in those that are calculative (λογιστικοί). For whether this or that shall be enacted is now the function of calculation. And the measuring must be done by a single standard; for that is pursued which is greater [cf. *Rhet.* 1369b19-27]. It follows that what acts in this way must be able to make a unity from several phantasms.

When man has compared the two objects or courses of action, he may decide (κρίνειν) to follow the alternative suggested by reason, i.e. he may *take* the future good *in preference* to the object of present desire (προ-αιρεῖσθαι: cf. 1112a15-18). Thereupon the phantasm, representing the future object as good or helpful, will stimulate the proper emotional response and action. Accordingly Aristotle describes choice (προαίρεσις) as "a deliberate desire for things in our power; for we first deliberate, then, having decided, we experience desire toward the result of our deliberation" (1113a9-12).

It is in this ability to *choose*, to follow the object suggested by reason rather than the prompting of present sensation, that man's self-determination and moral responsibility lie. Aristotle classifies the moral character of men accordingly: (1) The *uncontrolled* person (ὁ ἀκρατής) is described as one in whom "when the intellect commands and thought orders to avoid or pursue something, there is no movement, but he acts according to *desire*" (*De Anima* 433a1-3; cf. 1095a4-12, 1151a20ff., 1156a32-34). (2) The *self-controlled* (ὁ ἐγκρατής), on the other hand, are those who, "while experiencing appetite and desire refuse to enact that for which they have appetite, but follow their intellect" (*ibid.* 433a7-8; cf. 1151a25-27). In them the non-rational part of the soul obeys the rational (1102b-27-28), however reluctantly, and they are thought to *choose* deliberately, while the uncontrolled disregard the suggestion of reason and yield to present desire. Thus Aristotle says "the uncontrolled acts under the influence of desire and not through rational choice, while the self-controlled, on the contrary, acts from rational choice, not out of desire" (1111b14-16).

Now the disposition of the emotions has great influence in the moment of choice. For the uncontrolled individual, we are told (1147a ff.), follows the present impulse because *antecedent emotion* keeps him from an effective realization of his true good as directed by reason. He knows habitually what is truly good or evil, so that "he does not consider his action right *before* he comes under the influence of emotion" (1145b30-32); yet in the state of emotion stimulated by present perception this knowledge is useless (1095a8-10), like the knowledge of one who is asleep, mad, or drunk (1147a10ff.).

> It is plain, then, that uncontrolled people must be said to be in a similar condition to men asleep, mad, or drunk. The fact that they use the language that flows from knowledge proves nothing, for even men in these states utter scientific proofs and verses of Empedocles ... so that we must consider the uncontrolled to speak like actors on the stage (1147a18-23).

Thus undisciplined emotion, stimulated by the present object, must disturb the central sense power, as we have seen, so that the phantasms are disordered and the operation of the intellect impeded. This either prevents deliberation and rational choice entirely, or causes man to disregard the result of deliberation (1150b19-22), leaving him to be ruled by sense impressions and phantasms, like the animals (*De Anima* 429a4-8, 433a10-12).

Now it is precisely the object of moral training to bring these wayward emotions to order and tranquillity in the stable equilibrium of moral excellence, that state in which the faculty of pleasure and pain remains unmoved by the sensible objects that move most men by their presence, yet responds readily to the phantasm associated with reason. And the "appetitive soul", as we have seen, is brought to this state by habituation, i.e. by the *repetition of acts of choice*, acts in which the individual disregards present pleasures or pains and follows the direction of reason.

In the beginning such acts may not be easy, even for the man who is generally self-controlled; for he is beset with strong and evil desires (1146a10ff.) which he can overcome, perhaps, only through careful forethought and preparation. "For some men ... are not overcome by their emotion, whether pleasant or painful, because they perceive it beforehand and, foreseeing it, arouse themselves, i.e. their reasoning faculty, in advance" (1150b22-25; and cf. 1117a20-22). However, though at first the act of choice may mean a painful struggle, the memory of having done the good or noble thing makes it pleasant (*Rhet.* 1370b1-4). And when the act of choice is frequently repeated a habit is formed, a "second nature", as we have seen (*supra*, p. 240), which, when activated in further acts of choice, brings pleasure, like the unimpeded operation of any natural faculty.

Habit, whether acquired by mere familiarity or by effort, belongs to the class of pleasant things; for there are many actions not naturally pleasant which men perform with pleasure once they have become used to them (*Rhet.* 1369b16-20) ... For as soon as a thing becomes habitual it is virtually natural; habit is a thing not unlike nature ... (*ibid.* 1370a6-7; and cf. *N.E.* 1180a1).

Eventually, therefore, through repeated acts of choice, it becomes "natural" for the appetite, the faculty of pleasure and pain, to remain poised until the practical intellect decides the course of action, and then to respond to the phantasm suggested by reason. At that point man has acquired *moral excellence.*

For this poise of the emotional powers in the presence of pleasant or painful objects distinguishes the truly virtuous man, as we have seen. Unlike the merely self-controlled, who feel, and must resist, strong and evil desires, the virtuous man *has* no excessive desires

(1146a10-13): "He is so constituted that he feels no pleasure contrary to reason, while the self-controlled feels such pleasure but is not led by it" (1152a2-3). This does not mean that the man of moral excellence is unfeeling (cf. 1119a5 ff.); rather, his feelings are perfectly subordinate to practical reason, tending neither to pleasure or pain as such, but directed to that which reason indicates, — the truly good and pleasant (τὸ καλόν).

Hence appetite must be in harmony with reason in the temperate man. For the aim of both is τὸ καλόν. That is, the temperate man desires the right thing in the right way at the right time, while this is also what reason enjoins (1119b15-18).
And in the temperate and brave man the non-rational element is perhaps even more tractable; for in him everything is in harmony with reason (1102b28-29).

Thus through habituation the emotional powers are brought into stable equilibrium, and the virtuous man finds pleasure in responding to the phantasm under the direction of reason. His virtuous habit is a ἕξις προαιρετική in that it has been built up through acts of choice (1112a2-5); and virtue, once established, disposes him to choose in the future what reason directs, resisting easily and "naturally" the prompting of desire (1104a27 ff., 1114b27-29).

Aristotle recognizes that in so far as moral excellence implies a balance of the emotional powers, identified with the nutritive-sensitive part of man, it may be shared by non-rational beings, i.e. by children or animals, and may exist through good natural endowment or training. But in its proper sense as an emotional balance enabling the adult to feel and act according to his own reason, moral excellence is strictly *human*, a perfection of man alone. Accordingly, Aristotle distinguishes "*natural* virtue" (φυσικὴ ἀρετή) from *moral* virtue in the true sense (ἀρετὴ κυρία); the former is a good natural disposition which "even children and brute beasts possess", while the latter requires intelligence (1144b-1 ff.). Properly trained children may act *reasonably* (κατὰ λόγον) through the intelligent direction of another; and "if a man of good natural habits acquire intelligence (νοῦς), then he excels in human action, and the habit which merely resembled virtue will then be true virtue" (1144b13-14). But only the adult can act *rationally* (μετὰ λόγου), determining his actions according to his own intelligence. Hence "moral excellence is not merely a disposition *conforming to* right reason, but one *accompanied* by right reason. And right reason concerning matters of conduct is *prudence*" (φρόνησις: 1144b26-28).

Prudence is the virtue of the practical intellect which necessarily accompanies and governs the stable equilibrium of the emotions in true moral excellence. Aristotle thus maintains that "in the ethical faculty (τὸ ἠθικόν) there are two dispositions, natural virtue and true virtue, and of these true virtue does not exist without prudence" (1144b16-17). He defines prudence as "an abiding disposition for attaining truth in action through reason where good or evil for man is concerned" (1140b6-7), and holds that *both* prudence and moral excellence are necessary to make man capable of the proper choice in every situation (1139a23-26):

> In as much as moral virtue is a disposition concerned with *choice*, and choice is *deliberate* appetition, it follows that if choice is to be good both the *reason* (λόγος) must be true and the appetite right, and appetite must pursue what reason affirms.

Moral virtue makes the appetite right, directing it to the καλόν rather than to the merely pleasurable, thus fixing the end of action; and prudence directs the faculty of pursuit or avoidance to the right *means* (1144a15 ff.). If moral excellence is missing, man uses his intellect to seek mere pleasure, and prudence degenerates into *cleverness* (1144a23-36); if prudence is missing, emotional balance becomes a blind and inhuman capacity for action (1144b1-12). Hence "it is by prudence *and* moral excellence that man fulfills his function; for moral excellence makes our objective right, and prudence prescribes the right means" (1144a8-9).

Conclusion: Moral excellence, then, is the condition of the non-rational soul which enables it best to serve the practical reason in the production of correct human action. As the excellence of the *non*-rational part of man, moral virtue is subordinate to prudence, the characteristic excellence of practical reason. And in every act of *human* virtue the determining factor must be the *intellect*; to be virtuous in the full sense, an act must not only spring from a firm and unchanging habit, but must be done with full awareness and knowledge by deliberate choice, the act being consciously chosen for its own intrinsic nobility (1105a29-b1; cf. 1115b18-24).

Yet the practical intellect cannot exert its proper determining influence upon man's activity unless the non-rational part of man is disposed to cooperate. Hence the μεσότης, — the stable equilibrium — of moral excellence is essential to correct human action in that it establishes the necessary conditions for the unimpeded operation of practical intellect and enables man to follow, habitually and easily, the direction of reason. Thus:

Moral excellence controls *antecedent* appetite, feelings of pleasure and pain, keeping the emotional faculty poised in balance until reason indicates the truly good and pleasurable (τὸ καλόν).

This emotional poise allows the sense power to function freely in (a) perceiving the present situation *as it is*, and (b) recalling with ease and accuracy the phantasms associated with rational principles of action, — thus enabling the practical intellect to *deliberate* and *decide* objectively.

Through habituation the balanced emotional powers *respond* readily and with pleasure to the phantasm connected with reasoned opinion, reacting proportionately to the particular situation, without excess or deficiency, moving the human organism to pleasure or pain, pursuit or avoidance, as reason directs.

This interdependence of moral excellence and prudence appears in Aristotle's final comprehensive definition of moral excellence as "an abiding disposition related to choice, consisting specifically in a *middle state* or *mean condition* (ἐν μεσότητι: cf. *Physics* 210a-18-20; *Met.* 1023b23-25), — a middle state relative to each — determined by *rational principle* (λόγος), that is, as (ὡς) the *prudent* man would determine it. It is a mean state between two that are evil, one by way of excess, the other by deficiency. It is a mean state, moreover, in that the vices fall short of, or go beyond, what is proper in emotion and action, while moral virtue detects and cleaves to the *mean*. Consequently, with regard to its substance and the definition of its essence moral virtue is a *middle* state or *mean* condition. But in point of what is best and well-disposed it is an *extreme*" (1106b36-1107a8).

b. *Moral Excellence and the Speculative Intellect*

Moral excellence, then, which disposes the non-rational part of man to follow reason in feeling and action, is naturally more intimately involved with the *practical* intellect than with the *speculative*, whose aim is knowledge rather than action (1178a17-23). Yet in discussing the speculative intellect we must always bear in mind, first, that even the highest intellectual activity is carried on by the *same* soul which informs the body, since Aristotle maintains the essential unity of the human soul; it is not merely the *intellect* that knows and thinks but the *man*, the human composite of body and soul (*De Anima* 408b1-30). Secondly, though the intellect informs no bodily organ it is nevertheless extrinsically dependent upon the sense power, as we have seen, in the conception of universals, in their retention and recall, and in *all* its operations. Consequently whatever disturbs the composite organism and the sense powei will interfere with the operation of the speculative intellect. Hence moral excellence and health contribute to its activity, at least negatively, by controlling the disturbances of emotion and disease, establishing within the human organism a stable equilibrium, the internal condition of the compound in which the νοῦς is best able to operate. This internal connection between moral excellence and the operation of the speculative intellect may become somewhat less obscure if we consider the nature of the highest intellectual virtue and activity.

As prudence is, for Aristotle, the characteristic excellence of the practical intellect whose object is the particular and the changing, so *wisdom* (σοφία) is the excellence of the speculative intellect, which knows the universal and the eternal (1143b15-17, 1144a2-3). And just as the object of the speculative intellect, being universal and eternal, is higher and "more divine" than that of the practical intellect, so wisdom, the excellence of the speculative intellect, is a higher excellence than prudence; indeed, prudence is not only inferior but *ordered to* wisdom, as medicine is to health (1145a4-10):

> And it is clear that deliberate choice will not be correct without prudence any more than without moral virtue ... Yet prudence is not in authority over wisdom nor over the higher part of the intellect, any more than medicine is over health; for the one (prudence, medicine) does not *use*

the other (wisdom, health) but looks to *producing* the other, and gives orders *for the sake of* the other, but not *to* the other.

Wisdom, then, is the highest excellence of man's highest faculty and the objective of all human action produced through prudence and moral virtue. Wisdom is described by Aıistotle as follows (1141a17-20):

It is clear, therefore, that wisdom must be the most perfect of the forms of knowledge. Hence the wise man must not only know what follows from the primary principles (ἀρχαί), but must attain truth in the primary principles themselves. Therefore wisdom must be a combination of *intelligence* (νοῦς) and *scientific knowledge* (ἐπιστήμη), — a completely developed scientific knowledge of the highest realities.

By *intelligence*[93] Aristotle means that virtue by which the intellect grasps the unchanging first principles of knowledge (1140b31 ff., 1143b2-3), the undemonstrable universals which are the basis of science and art and correspond to the unchanging essences of things (*Post. Anal.* 72a25 ff., 99b20 ff.). Now we have already seen how these are formed through a kind of intellectual intuition of the intelligible form presented by the phantasm, originating in sense perception, memory, and experience (*ibid.* 99b36 ff.), all functions of the sense power.

The second constituent of wisdom is *scientific knowledge*, which is a comprehension of the universal, necessary, and eternal, developed by deduction from the primary principles, and possessed with certitude (1139b18-36, 1140b31-33). Again, the development and use of scientific knowledge in the individual depends upon the proper functioning of his sense power, since 1) "science and art come to men through *experience*" (*Met.* 981a3-4), i.e. through sense perception and memory, and 2) since all intellectual operations, including scientific demonstration, must be carried on with the help of the phantasm (*Parva Nat.* 449b31 ff.). It is true that the scientific knowledge constituting wisdom has as its object "reali-

[93] The word νοῦς is used here in a restricted sense. We have seen a number of instances where νοῦς bears the general meaning of "*intellect*", signifying the rational part of the soul as a whole. Here it means rather a *particular virtue* of the intellect, "*intelligence*" or "rational intuition", by which the intellect grasps the undemonstrable first principles of knowledge. Combined with a fully developed scientific knowledge it constitutes the virtue of wisdom.

ties of the most exalted nature" (1141b3), particularly that reality which is most divine, — the eternal, immaterial, and most universal cause, the Prime Mover and Supreme Good — along with the lesser, though more visible, divinities which compose the celestial system (*Met.* 982a20-983a11, 1026a15 ff., 1074a31; and cf. *N.E.* 1141b1-4). Yet even the non-extended and eternal must be intellectually conceived and contemplated through phantasms involving extension and time (*Parva Nat.* 450a4-13). Hence both constituents of wisdom, — intelligence and scientific knowledge of the highest realities — depend upon the functions of the central sense power.

Now the very nature of thought (νόησις) and knowledge (ἐπιστήμη) demands that the phantasm upon which they depend be undisturbed and tranquil. For, psychologically, thought and knowledge are not movement but the *termination* of movement, a *settling down* or repose of the mind in the possession of its object, which depends upon a corresponding tranquillity in the sense power serving it. Thus Aristotle says that "thinking (νόησις) seems more like a coming to rest or a halting (ἠρέμησις καὶ ἐπίστασις) than a motion; and the same is true of a deduction" (*De Anima* 407a32-34). Again, in his analysis of the practical syllogism (*ibid.* 434a16-21) he states that "the faculty of knowing (τὸ ἐπιστημονικόν) is never moved, but *remains at rest*"; the universal premise, identified with the intellect, influences action while itself "remaining in a state more like rest" (ἠρεμοῦσα μᾶλλον). Likewise, in describing the formation of the primary universals, as we have seen (*Post. Anal.* 100a3 ff.), Aristotle says that art and scientific knowledge take their origin "from experience, or from the complete universal *coming to rest* (ἠρεμήσαν) in the soul" (*ibid.* 6-7). Throughout the passage the point is made emphatically that the universal is formed and intellectual comprehension takes place when certain elements in the stream of sensation "come to a halt" or "take a stand" (στάντος, ἔστη, στάντος, ἵσταται, στῇ; *ibid.* 100a12-b2; and cf. *De Interpretatione* 16b19-21). Finally, this notion of thought and knowledge as constituted in a state of repose or stability is presented vividly in the sequel to that *Physics* passage with which we began the discussion of Aristotle (247b1 ff.). There, defending the proposition that generation and change take place only in extended substances, Aristotle says (247b1-13):

Again, the states (ἕξεις) of the intellectual part of the soul are not alterations, nor is there any generation (γένεσις) of them ... since that which *potentially* has knowledge comes to the state of *actual* knowledge not by being set in motion itself but by *something else coming to be present to it*; for whenever the *particular* occurs the mind knows the *universal* by means of the particular. Again, there is no generation of the actual use, i.e. of the *actualization*, of these states, unless one thinks that there is a generation of seeing and of touching also; for the actual use or actualization is similar to these (cf. *De Anima* 417b18-23). And the original acquisition of knowledge is not a generation or an alteration. For we say that by *coming to rest* or coming to a halt (τῷ ἠρεμῆσαι καὶ στῆναι) the mind *knows* (ἐπί-στασθαι) or understands (φρονεῖν); and there is no generation of a state of rest (τὸ ἠρεμεῖν) ...

Thus the original acquisition and subsequent actualization of intellectual virtues like scientific knowledge or wisdom are brought about not by any movement in the intellect itself, but by "*something else* coming to be present" so that the intellect "rests" or "halts" upon it. From what we have seen of the intellect and its operation we may infer that Aristotle has in mind here the *phantasm*, which presents the *universal*, — the intelligible form — as embodied in *particular* sensible qualities. This is confirmed by the rest of the passage, which brings out vividly the dependence of intellect upon the sense power, explaining the acquisition and actualization of knowledge as occurring not through any change in the intellectual faculty but through the *settling down* or tranquilizing of the body (*ibid.* 247b13-248a9):

Furthermore, when someone has changed from a state of intoxication, sleep, or disease to the contrary state we do not say that knowledge has been generated in him a second time, even though he was incapable of exercizing his knowledge in the previous state. So likewise we do not say that knowledge has been generated in him when he acquires it in the first place. For one becomes rational (φρόνιμος) or "knowing" (ἐπιστήμων) by the soul's *calming down* (τῷ καθίστασθαι) *from physical disturbance* (ταραχή). This is the reason why children are not able to learn or discriminate through sense perception as well as their elders; for the disturbance or motion within them is great. The *calming down* or *tranquilizing* is effected in some respects by nature itself, and in others by other agencies; but in either case it is through the alteration of certain elements in the *body*, as when the use or activation of knowledge is recovered upon the return to sobriety or wakefulness. It is evident, then, from what has been said, that being altered, or alteration, takes place in perceptible objects and in the *sensitive* part of the soul, and in no other except *per accidens.*

Thus the virtue of knowledge and its activation, being dependent upon the phantasm produced by the sense power, are impeded or rendered impossible by the physical disturbances accompanying drunkenness, sleep, disease, and growth, as well as the dissolution of old age (cf. *De Anima* 408b23-25). But we have seen that undisciplined emotion creates exactly the same type of disturbance, so that Aristotle frequently groups the state of emotion along with these others as interfering with the normal operation of sense and intellect (cf. *De Anima* 429a7-8; *Parva Nat.* 450b2-9, 453a15-b8, 456b18 ff., 460b27-461b12; *N.E.* 1154b3-16, etc.). Accordingly he compares the emotionally uncontrolled (ἀκρατής) to the young, the intoxicated, and to those who are asleep, in that the physical disturbance caused by emotion, wine, and the processes of nutrition and growth prevent the acquisition or actualization of knowledge in all four (1095a4-10, 1147a10 ff.). And "if one asks how this state of *un-knowing* (ἄγνοια) is dissipated so that the emotionally uncontrolled returns to a state of knowing, the explanation is the *same* as in the case of intoxication and sleep, and is not peculiar to this condition; we must go for it to physiology" (1147b6-9).

The physiological explanation, as we have seen, is that knowledge is acquired and activated only when the body, and the sensory system in particular, *calms down*, being freed from disturbance and brought to a state of stable equilibrium in all respects, i.e. to a state of maturity, health, sobriety, and moral excellence. Some of these may be produced by natural processes alone; others, like health and moral excellence, may require assistance from the physician and trainer, the moral guide and statesman. All are necessary to establish the compound organism in that state of order and tranquillity which allows the speculative intellect to operate unimpeded in the acquisition and activation of wisdom through intelligence and scientific knowledge.

7. *Moral Excellence and Happiness*

For Aristotle, man's nature and activities can only be explained by reference to an ultimate end or objective (τέλος), a supreme good toward which his natural powers and functions are ordered, and in the attainment of which they are perfected or fulfilled (cf. 1097b22 ff.; *E.E.* 1219a1 ff.). Being the fulfillment of man's nature and function, this supreme good constitutes the ultimate standard (ὅρος) of his dispositions and actions; for these must be considered *excellent* in so far as they contribute to the attainment of the good toward which human nature is ordered, and *defective* in so far as they deviate from or hinder the attainment of that good. Our understanding of moral excellence as a μεσότης of man's emotional powers will hardly be complete, then, until we have considered its relation to his ultimate good.

In the early chapters of the *Nicomachean Ethics* Aristotle concludes that the ultimate good for which all men act is happiness (εὐδαιμονία), for that alone is thought to be absolutely ultimate (i.e. always chosen for its own sake), self-sufficient or perfect in itself, and most desirable (1097b1 ff.). In seeking to determine the specific nature of happiness, Aristotle relates it to man's faculties and functions, deciding in general that it must consist in the actualization or active exercise (ἐνέργεια) of man's powers, particularly those powers of the soul not shared with brute animals; and these must be actualized according to the *proper excellence* of each (κατ' ἀρετήν) for a whole lifetime (1097b22-1098a20; and cf. *Parva Nat.* 455b17). And if there are several such powers and excellences, man's fullest and most characteristic happiness will be in the actualization of the *best and most perfect* among them (1098a18-19).

Accordingly, the proper functioning of the "nutritive soul", manifesting the bodily excellences of health, strength, and beauty, is demanded for *perfect* happiness (cf. 1099b3-4, 1153b16-18, 1178b33-35) but is not characteristic of *human* happiness. The same may be said of the sense power and its various functions, which are shared with the animals (1097b34-1098a3). The emotional powers leading to action, however, though in themselves non-rational, are in man capable of "obeying" or being determined by reason when they have been brought to the stability and equilibrium of moral excellence, as we have seen. Hence moral excellence, in so far as it cooperates with reason, is a human excellence, enabling man to fulfill his function *as man*; its actualization in emotion and action, therefoie, constitutes an important part of specifically human happiness. The power which sets man above all the other animals, however, is his intellect (νοῦς). Cooperating with moral excellence, the piactical intellect directs him toward what is truly good and pleasant through its characteristic excellence, prudence. But we have seen that this faculty and function of intellect is ordered to another, as medicine is to health. For the faculty and excellence which Aristotle considers most sublime are the speculative intellect and its highest excellence, wisdom. Man's fullest and most characteristic happiness, then, is found in the activity of his speculative intellect through the actualization of wisdom.

As each virtue or power produces its proper act, so wisdom produces the activity which Aristotle calls *contemplation* (θεωρεῖν). In contemplation the speculative intellect is actively engaged with the objects of wisdom, — the universal, necessary, and eternal, the first principles and causes of the universe, among them the Supreme Good and Prime Mover (*Met.* 981b2-983a11). And contemplation is not so much the pursuit and acquiring of wisdom as the active possession and enjoyment of it (*ibid.* 1072b22-25; and cf. *N.E.* 1177a25-28):

For that which is capable of receiving the intelligible object, i.e. essence (οὐσία), is *intellect*. But intellect is activated in the *possession* of the object. Therefore the possession rather than the receptivity is the divine element which the intellect seems to contain, and the *act of contemplation* (θεωρία) is what is most pleasant and best.

As the activity of man's highest faculty upon its highest object, contemplation comprises the essence and culmination of human happiness. It fulfills all that happiness requires: It perfects man's true self, his most dominant part; it is accompanied by the purest and most lasting pleasure; it is in itself most noble, self-sufficient, and desirable (1177a12 ff.).

But if happiness consists in activity in accordance with excellence, it is reasonable that it be activity in accordance with the *highest excellence*, and this will be the excellence of the *best* part of us. Whether this be intellect or something else that is thought to rule and direct us according to nature and to hold cognizance of the noble and the divine ... it is the activity of this part of us in accordance with its proper excellence that will constitute *perfect happiness*; and ... this is the activity of *contemplation* (1177a13-18).

But the activity of the intellect seems to excel in importance, being of the nature of *contemplation*, and to aim at no objective beyond itself, and to hold its own peculiar pleasure ... Moreover, self-sufficiency, leisureliness, such freedom from weariness as is humanly possible, and the other attributes of the happy man appear to be connected with this activity. This, then, must be *perfect human happiness*, provided it cover a full span of life (1177b19-25).

As the fulfillment of man's nature and his supreme good, then, the activity of intellect in contemplation is the end toward which all other human faculties and functions are ordered and subordinate. The body is organized by the "nutritive" soul to serve the higher faculties of sensation, emotion, and reason; the non-rational faculties of sensation and emotion are ordered to serve the intellect and to follow practical reason in choice and action; and practical reason is to direct action to the attainment of wisdom and to its exercise in contemplation (*Politics* 1333a17 ff., 1334b13-29). For "in the world of nature as well as of art the lower always exists for the sake of the higher" (*ibid.* 1333a22-24).[94] This natural subordination of man's faculties and operations establishes for Aristotle a corresponding scale of values in human activities according as they perfect man in themselves or serve as means to his perfection, and particularly as they contribute to his complete happiness in the leisurely contemplation of the universal and eternal (*ibid.* 1333a31-b2):

Life as a whole is also divided into its different parts, — action and leisure, war and peace; and in the sphere of action we may further distinguish acts which are merely necessary, or merely and simply useful, from acts which are good in themselves. The preferences which we give

[94] The translation of the *Politics* here is substantially that of Ernest Barker (*The Politics of Aristotle* [Oxford, 1946]). This work will be cited henceforth as "Barker".

to the parts of life and their different activities will inevitably follow the same general line as those which we give to the parts of the soul and their different activities. War must therefore be regarded as only a means to peace; action as a means to leisure; and acts which are merely necessary, or merely and simply useful, as means to acts which are good in themselves ... Men must be able to lead a life of action and war; but they must be even more able to lead a life of leisure and peace ... they must be able to do necessary or useful acts; but they must be even more able to do good acts.

We are now in a position to consider how moral excellence and the active exercise of the moral virtues fit into this scheme of things. First, it must be noted that, as the perfection of man's emotional powers, moral excellence is good and desirable *in itself*, like health and prudence (1097a15-20, b2-4, 1145a3-4). Yet, as we have seen, moral excellence is also desirable as a means to a higher good, being ordered to serve the intellect, cooperating with prudence in choice and action to secure the acquisition and activation of wisdom (cf. 1097b1-7). Accordingly, acts of bravery, justice, magnanimity, etc. are good and noble in themselves as springing from virtue and appropriate to human nature; but the active life of war and politics, in which moral virtue and prudence are chiefly exercised, is inferior and ordered to the higher life of contemplation, enjoyed in leisure, in which man's perfect happiness consists (1177b1-25). It is true, Aristotle admits, that the exercise of moral excellence and prudence brings a certain happiness; but this is a secondary kind, a purely "*human*" happiness, involved with emotion and action, belonging to the composite of body and soul (1178a8-22). The activity of moral excellence and prudence is desirable in itself, but more desirable in that it contributes to the life of contemplation. For the life of contemplation "is above the merely human; for it is not in so far as he is man that he will live so, but in so far as something divine is present in him; and by so much as this is superior to our composite nature is its activity superior to that which is the exercise of the other kind of virtue" (1177b26-29).

We may ask, in conclusion, why precisely a μεσότης, — a stable equilibrium — of the emotional powers should constitute moral excellence, the *best* disposition of the "affective soul"? Why is this "mean condition" desirable in itself, and how does it contribute to the realization of man's perfect happiness in the activity of contemplation? To answer this is simply to recapitulate what we have already discovered in examining moral excellence:

A μεσότης or "mean condition" is desirable in itself as perfecting the emotional powers; for in being developed to a state of balance or equilibrium, these opposing powers are integrated and stabilized, i.e. brought to order and tranquillity. This leaves the "affective" life of man free from conflict, free from habitual feelings of pain or craving for pleasure.

This state of the emotions is also best for the bodily organism, leaving it undisturbed by the physical changes of habitual or violent emotion, i.e. by deviations in the balance of hot and cold resembling those of disease, drunkenness, etc.

When the emotional powers are poised in a state of equilibrium, the faculty of sense perception (which is really identified with that of pleasure and pain) operates undisturbed, responding proportionately to the sensible object, reporting the sensible qualities without distortion, integrating the composite image and recording it as a memory in the central sense, recalling it easily and accurately as a phantasm, serving the intellect in the perception of the particular and in the formation or recall of the universal.

Again, an habitual state of equilibrium enables the emotional powers to serve the practical intellect best in choice and action. For (a) it keeps the feelings of pleasure and pain in check while the sense power, without disturbance or prejudice, reports the particular situation accurately and cooperates with the intellect in deliberation and judgment. Then (b), since the "affective soul" is equally balanced between pleasure and pain, it responds exactly, i.e. proportionately, to the intellectual judgment as presented in the phantasm, moving the body to pursue or avoid as reason directs. Internal proportion or equilibrium is, as we have often seen, the basis of proportionate response.

Finally, the condition of balance ἐν μεσότητι constitutes the best disposition of the "affective soul" with relation to man's ultimate good, for in a number of ways it helps him achieve his highest activity and happiness in contemplation. First, when poised in a state of balance the emotional powers are not directed toward the mere satisfaction of appetite, i.e. toward sense pleasure or release from pain, but toward the *good* (καλόν), the object indicated by reason as truly desirable, perfecting human nature as such. And this means, above all, the activity of contemplation. Hence the equilibrium of his emotional powers not only frees man from the prompting of irrational appetite but disposes him positively to seek his true happiness in contemplation, the noblest and best of human activities.

Secondly, in the state of emotional equilibrium man is best able to perceive things as they are, and to exercise prudence in selecting the *best means* to his ultimate good. Moreover, his emotional powers are then best disposed to cooperate with prudence in the external action necessary to secure these means. In order to enjoy the activity of contemplation man may have to engage in war to secure peace, engage in business to provide for his leisure. Even though devoted to contemplation he must live and deal with other men, take his part in government, share the help of friends (1178a-24 ff., b32 ff.). And emotional equilibrium responding to the direction of prudence produces the virtuous action by which these external means and conditions of contemplation will be best achieved.

Thirdly, there is, as we have seen, an important *internal* connection between emotional stability and the activation of the speculative intellect. For contemplation, the act of knowing, is psychologically a kind of intellectual *repose* in the possession of intelligible form presented by the phantasm. It requires that the senses be undistracted by sense appetition, that the body be undisturbed by active emotion. And this tranquillity of appetite and emotion is secured when the powers of pleasure and pain are poised in stable equilibrium. With all disturbing motions of the body set to rest, the sense power can serve the intellect unimpeded, recalling phantasms easily, producing them steadily and accurately in the central sense, as the intellect requires. This is the internal disposition of order and peace which allows man to exercise his highest faculty freely upon its highest object in the contemplation, the intellectual possession, of the universal, necessary, and eternal.

When man's powers and activities are thus subordinated to the ultimate good and brought to perfection in the performance of their proper function on each level, he enjoys happiness that is essentially complete and attains the state of goodness or nobility (καλοκἀγαθία: cf. *E.E.* 1248b8 ff.). His physical and emotional powers are balanced in the equilibrium and tranquillity of health and moral excellence, his sense power is free to operate unimpeded, and his true self, the intellect, unconscious of his non-rational or merely human functions, is absorbed in the contemplation of the eternal and unchanging realities. In this unimpeded activity of his highest faculty man attains his supreme pleasure, approximating that of the Divinity, who "enjoys a single, simple pleasure for all eternity; for there is not only an activity of motion but also an activity of immobility, and pleasure is to be found less in motion than in repose" (1154b26-29). In this state man achieves his supreme good, since his activity is then most noble, self-sufficient, continuous, effortless, unchanging, — in Aristotle's word, "*most divine*" (1177b27 ff.). For man then approaches as closely as his compound nature will allow the simplicity and activity of the self-thinking Thought whose unchanging goodness imparts order and goodness to the whole changing universe.

Therefore the activity of God, which surpasses all others in blessedness, must be contemplation; and of human activities, therefore, that which is most akin to this must be most of the nature of happiness (1178b22-23).

If then, God is always in that good state in which we sometimes are, this compels our wonder; and if in a better this compels it yet more. And God *is* in a better state. And life also belongs to God; for the actuality of thought is life, and God is that actuality; and God's self-dependent actuality is life most good and eternal (*Met.* 1072b23-27).

E. THE POLITICAL ORGANISM AND THE MEAN

Aristotle conducted his study of "things human" in the fields of politics and ethics ... side by side with a study of "things natural" (physics, medicine, and general biology); and the methods and findings of his study of "things natural" were naturally linked with the methods and findings of his study of "things human". The man of science appears again and again behind the student of politics ...

Ernest Barker, *The Politics of Aristotle*, xviii

Nature is for Aristotle a dynamic principle working within all things that change, imparting to each the organization and impulse to fulfill its specific function (*Physics* 192b9 ff.; 198b34 ff.). Thus the nature of man, — his soul or essence operating as an inner principle of development and motion — governs the generation and formation of the human composite, forming material organs to serve the vital functions on various levels, — vegetable, sensitive-emotional, and intellectual (*De Gen. An.* 740b26-741a3). However, as we have seen, the innate faculties provided by nature are not alone sufficient to bring man to his proper emotional and intellectual development (*N.E.* 1103a14-26). This requires his cooperation under the direction of a teacher and moral guide (*Politics* 1337a2-3; 19-22). Still, even in the self-determined activities through which he perfects his emotional and intellectual faculties, man must *imitate* nature by conforming to the principles of proportion and subordination everywhere operative in nature. Man is self-determined and responsible for his individual actions, but nature sets the objective and the norms by which these actions must be directed if he is to attain his proper perfection in the happiness of contemplation. "In general art partly completes what nature cannot bring to a finish, and partly imitates her" (*Physics* 199a15-16).

Nature in man, however, does not for Aristotle reach its complete fulfillment and exhaust its creative power in the production of the perfect *individual*. Rather, the individual is essentially a part of a larger unit, that organized society which Aristotle calls the πόλις. This is the self-sufficient whole which nature operating in man strives to produce, a whole of which each individual and every basic human partnership are natural but subordinate parts. But as nature arrives at perfection in the individual through the subject's cooperation under the guidance of the moral philosopher, so nature requires the cooperation of free and virtuous citizens under the guidance of the statesman or political philosopher to produce the best possible political organism. Hence the direction which Aristotle provides the statesman in his *Politics* is not only a necessary complement to that of the *Ethics*, but, in a sense, represents the culmination and climax of his philosophy of human nature (*N.E.* 1181b15-16). The political science is, as he remarks, the most authoritative of all, since it orders all others to their ultimate end (*N.E.* 1094a27-b12; *Politics* 1282b14-17). To complete the consideration of Aristotle, then, we must glance through the *Politics*, in the hope that what we have learned of his philosophy of nature may illuminate certain aspects of his political thought.[95]

In the discussion we shall assume the following preliminary notions, treated at length by other commentators: 1) The polis is a natural organism, of which the individual is a constitutive element. 2) The polis has the same ultimate objective as the individual, namely, the attainment of happiness through virtuous activity. 3) Different forms of polis, like different animal organisms, are characterized by different arrangement and distribution of functional parts. 4) Different constitutions (or forms of polis) are better *relatively* for different populations (i.e. proportionate to different matter. 5) Some constitutions are better than others *absolutely*, and the best is possible only in the best population. 6) The best form is monarchy, while aristocracy and polity are inferior but still "according to nature". 7) Democracy, oligarchy, and tyranny are all "contrary to nature", and one is worse than the other, in that order.

[95] In the present and subsequent section all passages cited by Bekker number alone are from the *Politics*. In cases of significant variations in the text, I have followed the reading of Ross in his edition of the *Politics* (Oxford, 1957). Translations from the *Politics* are taken, or adapted, from that of Barker. However, the adaptation is frequently so free that it bears but slight resemblance to the original and must be accounted the writer's responsibility. Rackham's translation (Loeb Classical Library Ed., London, 1950) has also been consulted, and utilized, throughout.

With regard to (a) the order of the books in the *Politics* and (b) the date of their composition, I shall follow the position maintained by Barker (Introduction xxxvii-xlvi) and Ross ("The Development of Aristotle's Thought", *Proceedings of the British Academy*, XLVI [1957] 68-76). Both hold to the traditional order of the books, regarding them as conforming to a more or less unified scheme. And they assign the treatise as a *whole* to the last period of Aristotle's life. In this they depart from the position of Jaeger and Nuyens, who divide the *Politics* into earlier "idealistic" books (II, III, VII, VIII) and later "realistic" books (I, IV-VI). This is not the place to summarize the reasons presented by Barker and Ross for their position. However, some additional evidence for the support of that view will be suggested in passing. See also J. Aubonnet, *Aristote: Politique*, Livres I et II (Paris, 1960), Introduction cvii-cxx.

That Aristotle held these notions may be no very difficult matter to document from his *Ethics* and *Politics*; but his reasons for adopting some of them are not quite so evident. *Why* does Aristotle regard monarchy, aristocracy, and polity as "according to nature", while considering the others as deviations "contrary to nature"? Why is one of the three "natural" constitutions considered *best*, and the others "better" absolutely as they approach this? What order does he find in nature which leads him to distinguish this way?

In seeking to answer these problems we may perhaps detect in Aristotle's theory of the political organism the application of two leading principles which he sees at work in all organic life, and which represent for him nature's constant mode of operation.

1. *The Principle of Natural Subordination*

In what we have already seen of Aristotle's theory of man and nature it is possible to discover, implied at least, what might be called the *principle of natural subordination*. This has three aspects, which may be described briefly as follows:

First, there exist structural, qualitative, and functional differences among the various independent organisms, and among the parts or "faculties" within each organism, which Aristotle classifies as "higher" or "lower" according as they approach the pure activity of form or the inert determinability of matter. Thus *among* the various organisms man is the "highest" because he is capable of thought and self-determination, with a faculty most independent of matter. Plants, on the other hand, are the "lowest", capable only of assimilating matter in growth and of reproducing themselves, being of all the most earthy in composition. Similarly, *within* a single complex organism like man Aristotle distinguishes various components, parts, and functions. The soul or form is distinguished from the body or matter, and the functions of man cover a range from the "highest" (the operations of pure form or soul only extrinsically dependent upon matter, — the activity of the intellect) to the "lowest" (the functions of assimilation, growth, and reproduction, shared with plants).

Second, in all organic nature the "lower" or less perfect is *subordinated to*, ordered to assist or serve, the "higher" and more perfect. "In the world of nature as well as of art the lower always exists for the higher" (1333a22-23). The lower fulfills its function, attains its end, achieves its characteristic excellence in so far as it contributes to, and cooperates with, the higher. Thus "plants exist for the sake of animals, and the other animals for the sake of man" (1256b16-17). And in each single organism 1) all faculties, parts, and their functions, being incomplete and less perfect than the whole, are ordered to the good of the whole, while 2) all the lower of these are by nature ordered or subordinate to the higher. Thus within man all the lower faculties and their functions are ordered to his highest faculty and its function, particularily in the exercise of contemplation, which is the essence of his happiness. And it is the intellect, the highest faculty by nature, which governs the lower faculties and their activities, directing them to the good of the whole and to the culminating activity of contemplation (cf. 1333a17-37, 1334b10-28).

Third, in the *most perfect* organisms the natural structure and quality of each interrelated part is perfectly adapted to a specific function, which it performs permanently and exclusively. This imparts greater efficiency, stability, and harmony to the activities of higher and lower parts operating in the whole (*De Part. An.* 683a20-26):

Now it is better, when possible, that one and the same organ or instrument shall not be made to serve several dissimilar uses, but that there shall be one organ to serve as a weapon (which can then be very sharp) and a distinct one to serve as a tongue (which can then be of spongy texture and fit to absorb nutriment). Whenever, therefore, nature is able to provide two separate organs for two separate uses, without one hampering the other, she does so, instead of acting like a coppersmith who for cheapness makes a spit and lampholder in one.[96] It is only when this is impossible that nature uses one organ for several functions.

[96] It is perhaps significant that mention of this strange instrument occurs again in *Politics* 1299b10 with the same principle applied to the polis; and another combination instrument is mentioned with the same application in *Politics* 1252b2. Cf. Ogle's note *ad loc.* in the Oxford translation of *De Partibus Animalium.*

This principle of subordination Aristotle finds operative in all of nature, and particularily in the human male at his natural best (1254a22-39):

Ruling and being ruled not only belong to the category of things necessary but also of things advantageous; and in some instances things are marked out from birth as proper for being ruled or for ruling. And there are many forms of ruler and ruled ... In every compound where a number of elements are combined to form a single common entity, — whether the constituents be continuous or discrete — there always appears a ruling factor and one that is ruled. And this characteristic is present in living things from the whole of nature ... Living beings first are composed of soul and body, of which the one is by nature the ruler and the other the ruled. But it is necessary to study what is "by nature" in subjects that are in a natural condition, not in those that are corrupt. Hence we must examine man in the best possible disposition of body and soul ...

Within the perfect individual Aristotle discerns the prototype of various kinds of rule according to natural capacity (1254b3-16):

As we were saying, then, it is possible to see first of all in the living organism the rule both of slave-master and of statesman. For the soul rules the body with the rule of master over slave, and the intellect rules the appetite with the rule of statesman, or rather of king [cf. *N.E.* 1103a1-3 and *Politics* 1260a4-14, 1259a38-b18]. In these cases it is clearly according to nature and advantageous that rule be exercised over the body by the soul, and over the affective part of the soul by the intellect, the part possessing reason ... And the same principle holds true in relation to man and the other animals ... all these should be ruled by man ... Again the relation of male to female is by nature that of superior to inferior, of ruler to ruled. And in the same way this principle must hold good among all human beings generally.

Thus the principle of subordination, drawn from organic nature, is applied rigorously to human society, composed as it is of individuals who differ in their natural endowment and capacity for excellence. The principle demands that *within any well-organized society*, — whether household, village, or polis — *those less perfect be subordinate to and governed by those more perfect* in the pursuit of the common good, — τὸ ζῆν or τὸ καλῶς ζῆν.

Now in the family unit nature has settled the basis of subordination by making the male superior to the female, the adult superior to the child. But what is the natural basis of superiority in the polis, which Aristotle presumes is primarily a free association of adult males? Some claim superiority and the right to rule on the basis of free birth or the possession of wealth, since both are necessary for the existence of a polis. However, Aristotle maintains that the natural end of the polis is the *noble life*, and that those who contribute most in the common pursuit of this objective are the *most virtuous*, the citizens possessing the highest intellectual and moral excellence. But "those who contribute most to an association of this character have a greater share in the polis than those who are equal to them in free birth and descent, but unequal in civic excellence, or than those who surpass them in wealth but are surpassed by them in excellence" (1281a4-8). Therefore the natural end of the polis, by distributive justice, demands that intellectual and moral excellence constitute the basis of superiority and rule in the polis at its natural best.

Claims to rule must be based on those things which combine to constitute a polis. Therefore with good reason the well-born, free, and wealthy claim honor and office. For there must be free men and taxpayers; no polis could be composed entirely of poor men, any more than of slaves. But granting the need of these, there is obvious need also of *justice and civic excellence*, for without these men cannot live together in a polis. The difference is that the first two are needed for the *existence* of a polis, the last two for its *living nobly*. Therefore in relation to the existence of a polis, all or some, at least, of these might rightly lay claim to honor and office; but in relation to its living well, *education and moral excellence* would make the claim most justly (1283a-14-16).

Accordingly, in another passage Aristotle concludes that "the one conjuncture in which it is noble to follow another, and just to obey him, is when a person emerges who is superior to others in virtue, and (we may add) in practical ability to *do* what is best" (1325b10-13). The best or most natural form of polis, then, is that in which the inferior components willingly accept a subordinate role and follow the direction of those superior in virtue. "In relation to the best constitution a citizen is one who is able and willing to be ruled and to rule with a view to life in accordance with virtue" (1284a1-3). Such a constitution, like the individual at his best, embodies nature's principle of subordination in that the most excellent part orders and governs the whole, while each inferior component cooperates harmoniously according to its natural capacity and virtue, bringing order, tranquillity, and stability to the whole.

Now among his correct constitutions Aristotle designates two which, in addition to governing for the advantage of all, are distinguished by this, that moral and intellectual excellence form the basis of rule in each. These are *kingship* and *aristocracy*; and "to consider the best constitution is the same as to discuss the constitutions so named, for each implies a constitution based on virtue equipped with means for its exercise" (1289a32-34).

We have laid it down that there are three correct constitutions, and of these the best must be that administered by the best men. Such is one in which there has come to be some one individual who surpasses all others together in excellence, or else a whole family or a considerable number who do so, where men are capable of being ruled and of ruling with a view to the way of life most desirable of all (1288a33-37).

These two constitutions are both in a broad sense "aristocracies", since both assign predominance to the *best* citizens (1293b41-42; 1294a9-11; 1310b33-34), i.e. to one individual or to a group surpassing all others in excellence.

If there is one person (or a number of persons, yet not enough to form the full measure of a polis) so preeminently superior in excellence that there can be no comparison between the excellence and political capacity which he shows (or several show, when there is more than one) and what is shown by the rest, such a person, or persons, can no longer be treated as part of the polis. Being so greatly superior in excellence and political capacity, they will suffer injustice if treated as worthy only of an equal share; for a person of this order may very well be like a god among men (1284a3-12).

When a person of this type appears in the community, Aristotle concludes, it would be unjust either to banish him or to make him share the rule with others (1284b23-33). "The only alternative, — and this would appear to be demanded by nature — is for all others to obey such a man gladly, so that such men will be the permanent kings in their states" (*ibid.* 33-34).

This last notion, — that rule should be *permanently* in the hands of some individual preeminently qualified by nature and virtue — may perhaps suggest the reason why Aristotle considers kingship, rather than aristocracy in its specialized sense, the "first and most divine" form of government (1289a40). For, as we have seen, nature at her best assigns one function to one organ or faculty preeminently suited for it (1252b1-5; and cf. 1273b10-12, 1299a35-36):

Nature makes nothing in a spirit of stint, as smiths do when they make the Delphic knife; she makes each separate thing for a separate end; and she does so because each instrument or organ is brought to its finest perfection when it serves a single purpose and not a variety of purposes.

Hence, since the king must be superior by nature (cf. 1259b15-16) and excellence to all others, he will be suited to govern permanently, and this Aristotle considers most desirable. "But since such permanence of function is better also in the case of the political community, it is clearly better for the same men always to be rulers whenever this is possible" (1261a38-40; and cf. 1332b12-23).

Under such a king, then, the source of authority in the government has a natural unity, stability, permanence, and efficiency which Aristotle apparently thought would be lost to some extent in an aristocracy. It is true that a group of men superior in nature and virtue would be suited to rule permanently over the rest of the population. But, according to Aristotle's principles, they would have to share the supreme command by taking turns. For among themselves they would be equal; and "in a group whose members are equals and peers it is neither expedient nor just that one man should be sovereign over all others" (1288a1-3). Hence the members of the ruling group in an aristocracy could only *imitate* natural subordination among themselves by ruling and being ruled in turn (1261b1-7), alternating their functions.

For among equals it is right and just that office should go on the principle of rotation, which is demanded by the ideas of equality and parity. But that equals should be given unequal shares, and men on a footing of parity treated on a basis of disparity, is a thing contrary to nature; and nothing contrary to nature is right (1325b7-10).

The one man of surpassing excellence, on the other hand, will "have sovereign power without any limitations, — not turn by turn with others" (1288a28-29). Hence the principle of subordination, prevailing throughout all nature, will be worked out most perfectly in the polis under kingship.[97]

However, since every form requires a proportionate matter, kingship, the absolutely best constitution, can only be realized in "a populace by nature capable of producing a family of surpassing excellence for political leadership" (1288a8-10), from whom will come a line of men as superior to the rest as a father to his children (cf. 1259b10-18), differing from others "as widely as we believe the gods or heroes to differ from human beings" (1332b17-18). Aristotle admits that "this condition is not easy to secure; and we have nothing like the gulf between kings and subjects which Scylax describes as existing in India" (*ibid.* 23-25). "There are men of equal quality in great numbers, and no one outstanding enough to measure up to the grandeur and dignity of the office of king" (1313a7-8).

Where there is no population suitable for kingship, then, the next best form which nature strives to attain is *aristocracy* in the specialized sense. The material necessary for this form is a population which "by nature produces a mass of citizens capable of being governed in the manner of free men under those qualified by excellence for political leadership" (1288a10-13). As we have seen, the primary condition demanded here is that the population produce, not an individual, but a *group* whose collective excellence surpasses that of all the rest, so that the rest follow its direction willingly and for all time (cf. 1284a3-10; 1332b12-23).

This form, again, is close to the order of nature, conforming to the principle of natural subordination (though not as perfectly as kingship), and setting up nature's absolute norm of goodness,

[97] Other characteristics of the ideal kingship are mentioned from time to time by Aristotle: It is *absolute* monarchy (1285b30-34; 1288a15-29); it guards the interests of all classes equally (1310b40-1311a2); it aims at what is noble and not the merely pleasant, seeks honor rather than wealth, rules over willing subjects (1311a5-8; 1285a25-30); it is most rarely, of all forms, destroyed by external factors, hence most stable and long-lived (1312b38-40).

i.e. intellectual and moral excellence, as the basis of rule (1293b41-42; 1294a9-11).

> The only constitution which can with strict justice be called an "aristocracy" is one composed of men who are not merely "good" in relation to some standard or other but the best (ἄριστοι) *absolutely* in point of excellence. Only under such a constitution is the absolutely good man the same as the good citizen; under all others they are "good" relative to their own particular constitution (1293b3-8; and cf. 1288a38-b3).

However, Aristotle seems to think that the material suitable for aristocracy is also comparatively rare, since a population producing a group of men outstanding in virtue is not often found. "For good breeding and virtue (cf. 1283a34-41) are found in few persons, but the qualifications of wealth and freedom in a much larger number. Nowhere are there a hundred well-born and virtuous; but in many places there are rich and poor in great number" (1301b40-1302a2, reading with Ross).

Aristotle seems to find the generality of men lacking in solid virtue, fluctuating between mere self-control and lack of restraint (*N.E.* 1152a25-27), governed for the most part by their feelings, particularily by pleasure and fear (*ibid.* 1179b10-16; and cf. *Politics* 1319b32-33). He does recognize that most are capable of achieving a certain military valor, enabling them to fight for their polis as citizens in time of war (1279a40-b5; 1288a13); yet even this apparently is not true courage but an inferior kind ("citizen courage") motivated by penalties and rewards (cf. *N.E.* 1116a17-b2). Among the majority of people as they are, then, no natural order of subordination on the basis of excellence is possible. Where all are equally undistinguished it is "neither expedient nor just for one to be sovereign over all" (1288a1-2). "But that equals should be given unequal shares, and those alike be given unlike shares, is a thing contrary to nature; and nothing contrary to nature is right" (1325b8-10).

Is there any natural form of government possible, then, for most men *as they are*? Though Aristotle holds that in the best or most natural form of government preferment is based upon the possession of superior excellence, nevertheless, we may recall, he does admit that the possessors of *freedom* and *wealth* may also with some justice lay claim to a share in government, since both free men and taxpayers are equally necessary for the existence of the polis (1283a14-19; and cf. 1309b7-9):

Claims to rule must be based on those things which combine to constitute a polis. Therefore with good reason the well-born, free, and wealthy claim honor and office. For there must be free men and taxpayers; no polis could be composed entirely of poor men, any more than of slaves.

Now, though Aristotle in his study of actually existing populations finds few men qualified by virtue, he does see a great many qualified by the possession of freedom, and a good number qualified also by the possession of wealth (1301b40-1302a2). In any particular polis, moreover, he finds the wealthy generally few in comparison with those who are free but poor (1280a2-7; 1290b3-4). Populations as they are, therefore, tend to resolve themselves into two opposing parts, — a larger group whose poverty leaves them only the claim of freedom and numbers, and a smaller group demanding preferment on the basis of freedom and wealth. And each of these, the numerous poor citizens and the few wealthy, strive to set up constitutions favorable to themselves, the one democracy on the basis of freedom *only*, the other oligarchy on the basis of freedom and wealth (cf. 1279b37-1280a7). In fact Aristotle regards these two groups, the rich and the poor with their rival claims to preference, as *the* two mutually opposed and essentially disparate parts of the population (1291b2-13; and cf. 1315a31-33, 1318a31-32):

The different capacities (faculties, powers: δυνάμεις) belonging to the other parts, it seems, may possibly be possessed simultaneously by one and the same group of persons. The same persons, for example, may serve as soldiers, farmers, and craftsmen; the same persons, again, may act both as a deliberative council and a judicial court ... But one thing is impossible: The same persons cannot be both rich and poor. This will explain why these two classes, — the rich and poor — are regarded as parts of the polis in a special and peculiar sense. Nor is this all. One of these classes being small, and the other large, they also appear to be *opposite* parts. This is why they both form constitutions to suit their own interest. It is also the reason why men think that there are only two constitutions, — democracy and oligarchy.

In the city-states of his time Aristotle finds the total population sharply divided between these two groups, which explains for him the actual prevalence of oligarchy and democracy over other forms of constitution (1296a22-34):

What has been said also serves to explain why most constitutions are either democratic or oligarchic. In the first place, since the middle class is in most states generally small, whichever of the two main classes (the property owners or the masses, i.e. those who deviate from the middle one way or the other), — whichever of these gains the upper hand draws the constitution in its own direction, so that it becomes either a democracy or an oligarchy. In the second place, since factions and quarrels arise between the masses and the wealthy, whichever happens to prevail over the opposite group, that one refuses to establish a constitution based on common interest and equality, but takes over an excessive share of the government as the spoils of victory, and institutes a democracy in one case or an oligarchy in the other.

Thus Aristotle sees the majority of his contemporaries split into opposing groups and living under democracy or oligarchy, — both unnatural or deviate forms of government. The problem then arises: Is there any *natural* form of constitution attainable by these populations as they are? Aristotle's solution applies a second leading principle of nature to the political organism.

2. *The Principle of Order among Equal and Opposite Parts*: Μεσότης

Returning to nature's mode of operation on the most elementary level of the living organism (*supra*, pp. 163-178), we may recall that the opposing elemental bodies, of which the homoeomerous parts are composed, tend of themselves to conflict, and one will destroy the other if they are brought into contact in disproportionate mass and "power". However, if they are brought together in equal or nearly equal power, i.e. in *proportion* (συμμετρία), and then mixed or blended (κρᾶσις) to a "middle state" (μεσότης), they form a new compound in which no one dominates but all contribute their "powers", each extreme being moderated by its opposite and forming with it a stable whole. This dynamic equilibrium of opposites then possesses internal unity, strength, and stability in proportion as the opposing constituents are more nearly equal and more perfectly blended. While the parts of the body retain their stable blend of components they are healthy and function properly, responding proportionately and retaining their integrity. But when one or other of the opposing constituents grows excessively strong the equilibrium or blend is disturbed, conflict and pain arise, malfunction ensues, and the compound is *diseased*, becoming painfully unstable and tending to break up into its conflicting elements in death.

Again, on a somewhat higher level, we have found (*supra*, pp. 230-260) that Aristotle conceives man's emotional organism (ἦθος) as being in a condition of health or excellence (ἀρετή) when his opposing "powers" of pleasure and pain are trained to an habitual "middle state" in which neither dominates but one counterbalances and modifies the other. Man is then emotionally integrated, stable, at peace, and ready to respond proportionately in feeling and action as reason directs.

Nature's constant way, therefore, of bringing order and stability to a compound of equal but opposite constituents is to keep the opposing extremes *proportionate* and to *blend* them together in a "middle state" or "mean condition" (ἐν μεσότητι). This internal state of balance is also the basis of proportionate response, enabling the organism to function properly.

Now Aristotle invokes this principle specifically in discussing which constitution and way of life is best, not absolutely, but the best attainable by *most men as they are* (1295a25 ff.).

The issues we have just raised can all be decided in the light of one body of fundamental principles. For if it has been well-said in the *Ethics* that the happy life is one led without impediment in accordance with virtue, and that virtue is a "middle state" (μεσότης), then a *middle* way of life (μέσος βίος) must be best for a polis, the "middle state" being that which is possible for each to attain. But the norms that determine the excellence or the evil of a polis must apply also to its *constitution*. For the constitution is, as it were, the mode of life of the polis (1295a35-b2).

Divided as they are, then, between the multitude of poor who desire democracy and the few rich who want oligarchy, most populations will be brought to the best way of life attainable by them through an application of the principle of μεσότης. The best form or constitution for them, therefore, will be one which reconciles the two extremes of rich and poor, bringing them to unity and comparative stability; it will order the population in a "middle state", one midway between those in which either the rich or the poor dominate and tend to destroy the other. Such a constitution will be natural, good, preservative of the whole, while a form favoring either extreme will be unnatural, bad, and destructive.

This middle form of constitution is what Aristotle calls "constitutional government" or "polity". In general it is characterized as the form in which "a multitude govern the polis for the common interest" (1279a38-39) or "the form in which membership of the governing body depends on property assessment" (*N.E.* 1160a34). Specifically, Aristotle explains that "polity is, in its absolute sense, a mixture (μίξις) of oligarchy and democracy" (1293b33-35). Despite popular misnomers, the term is properly applied to that form of constitution which "aims at mixing the rich and the poor, wealth and freedom" (1294a15-17, 23-24).

Aristotle suggests three different principles to be followed in framing such a "synthesis or mixture" of oligarchy and democracy (1294a30 ff.). The first is to adopt two separate measures characteristic of each extreme form, keeping both intact. The resulting combination will then be something "common and intermediate (μέσον), and for that reason characteristic of polity, which is a mixture of the two" (1294a41-b1). The second is to adopt a measure which is the *mean* (τὸ μέσον) between two others, one democratic and the other oligarchic. The third is to combine elements from both, and to mix part of an oligarchical law with part of a democratic. "This, then, is the general method of mixture" (1294b-13-14; and cf. 1297a38-b1).

Aristotle then presents the criterion of a good mixture, obviously applying the concept of the "intermediate blend" basic to his physiological theory (1294b14-40):

It is the criterion of a good mixture of democracy and oligarchy when it is possible to say that the same constitution is both democratic and oligarchic; for it is clear that those who pronounce it so respond in this way because the mixing has been done well; and this is also true of the *mean* (τὸ μέσον), for in it each one of the extremes appears ... In a well-mixed polity both extremes should seem to be present, — and neither; and it should be preserved intact through its own intrinsic stability, and not through external influences; and through an intrinsic stability based not upon the will of a mere majority (for that would be present even in a bad constitution) but on the fact that not a single part of the entire polis would want any other form of constitution.

Thus two opposing extremes, oligarchy and democracy, are mixed to a blend, forming a new intermediate constitution in which the extremes temper each other, elements from each being balanced one against the other. The new form then favors both rich and poor equally, allowing neither to dominate the other, but drawing them into the unity and stability of a "middle state" or "mean condition". This is nature's constant way of forming the equal but opposing elements of an organism into an ordered whole. Hence Aristotle includes polity among the three constitutions that are natural and good.

However, when Aristotle maintains that this is the best form *possible* of attainment by men as they are, he does not necessarily mean that it is immediately applicable to any contemporary population in its present state. For since every form requires a *proportionate* matter, polity too can be realized only in a population properly disposed to receive it. Now the form or constitution of a polis is embodied primarily in a ruling citizen or group of citizens (1278b9-12). In a polity, obviously, these cannot be the extremely rich or extremely poor, who naturally favor oligarchy or democracy. A population suitable for a polity, then, must include a class mid-way between the extremes, men of moderate wealth, the *middle-class* (οἱ μέσοι). And these must be numerous enough to merit the greatest proportion of political control, according to distributive justice.

But where the multitude of the middle class outweighs that of both extremes together, or even one of them, there a polity can exist in permanent stability. For there is no fear that the rich will unite with the poor against these, since neither will ever be willing to be subject to the other; and they will never find, even if they try, another constitution more to their common interest than polity. Neither class would tolerate a system under which either ruled in turn, for they do not trust each other. But everywhere the neutral arbitrator is trusted most, and the "man in the middle" (ὁ μέσος) is such an arbitrator. The more perfectly the constitution is mixed, then, the more permanently stable will it be (1296b39-1297a8).[98]

Hence a population containing a large middle class constitutes the proper matter for the mixed constitution or polity. Invested with political control, the middle class will unite both extremes in itself and temper the influence of each, holding them in dynamic equilibrium. And even if the middle class be only larger than one or other extreme, "by adding its weight to either it counterbalances

[98] In this description of the population suitable for a polity Aristotle prescinds from other aspects which he mentions elsewhere. In the present context he emphasizes the need of a large class of moderately wealthy citizens. These, however, are understood to be citizens wealthy enough at least to provide themselves with armor, for this is what distinguishes them from the poor (1289b31-32) and gives them the right to be included among the ruling class (1279b4-5). In a polity this property qualification must be adjusted so that the middle class will predominate: "It is true that the constitution of a 'polity' or mixed state must be based on a citizen body composed only of those who have arms. But it is not possible to define this qualification absolutely, or to say that it must consist in a fixed amount in all cases. We must seek to discover in each case, and to fix for each, the highest amount compatible with the principle that those who share in the constitution must be more numerous than those who do not" (1297b1-8). Aristotle considers it characteristic of polity that it be "composed of those who bear arms" (1265b29), that its citizens attain a degree of military valor, and that in it "the class that carries on war is most sovereign" (1279b1-5). He believes that polity developed historically as methods of warfare changed, coming into existence with the rise of large populations and citizen armies (1297b16-28; 1289b33-40). This background makes it easier, then, to understand Aristotle's description elsewhere of a population suitable for polity as "one where there naturally comes to be a large number of persons possessing military capacity, who can be ruled and rule under a law which, according to the standard of qualitative proportion (κατ' ἀξίαν: cf. 1301b29-37, 1296b17-24, and *N.E.* 1160b13-14), distributes the offices among those who possess wealth" (1288a13-15; and cf. Barker's note GG, p. 151).

the other and prevents the opposing excesses from coming into existence. Hence it is the greatest blessing for a polis that its citizens possess a moderate but sufficient property" (οὐσίαν μέσην καὶ ἱκανήν: 1295b38-40).

A polis governed thus by a large middle class holding power under a polity will, Aristotle believes, be controlled by the best men in the population, judged by the standard of the ordinary citizen (1295a25-32, b35-40). For those who dominate and control the extremes of rich and poor will be themselves in a "mean condition" with regard to property; and "since it is admitted that what is moderate (μετρίον) or in the middle (μέσον) is best, obviously in the ownership of all goods of fortune a middle amount will be best" (1295b4-6). Aristotle cites a number of ways in which men of moderate wealth are superior in political life (1295b6 ff.). 1) They find it easiest to obey reason, being free from the insolence and enormities born of excessive good fortune, and from the malice and pettiness that come with misfortune. 2) They are properly disposed toward rule, neither desiring office too much nor avoiding it. Unlike the over-prosperous, they know how to obey and are willing to be ruled, yet, unlike the unfortunate, they are not cringing or servile. 3) Again, they are not contemptuous or envious of one another, as are the rich and the poor, but live in friendship, which holds together the political community. 4) They also live in the greatest security, neither envying the rich, like the poor, nor being envied by the poor, like the rich; hence neither plotting nor plotted against. 5) Finally, the middle form of constitution is best because "it alone is free from faction. For where the middle class is large there is least likelihood of faction and dissension among the citizens" (1296a7-9; and cf. 1302a14-15).

The best polis attainable by men *as they are*, therefore, requires a population in which the constituents are properly proportioned, so that the middle class, being large enough to dominate and control the opposing extremes of rich and poor, can unite or blend both extremes in itself and rule for the good of the whole under the mixed constitution or polity. Thus a proportionate blend of opposing constituents ἐν μεσότητι will constitute for Aristotle the basis of health or well-being both in the physical body and in the body politic. For this is nature's way of ordering equally necessary constituents in a compound.

3. *Unnatural or Deviation Forms*; *the Relative Mean*

After examining the natural forms of constitution and the principles they follow, we may perhaps understand more clearly why Aristotle considers democracy, oligarchy, and tyranny to be in themselves *contrary* to nature (1287b40-42). First, the distinctive mark of all three is that a *part* (the poor, the rich, or an individual) rules for its own advantage and not for that of the *whole*. This is obviously contrary to the principle of subordination and to the whole teleology of nature, which postulate that the part is subordinate to the whole and ordered to the good of the whole. Secondly, all three forms violate this same principle in that one part presumes to rule the rest not on the basis of intrinsic excellence qualifying it for rule, but on the basis of some extrinsic circumstance, — free birth, wealth, the force of arms. Thirdly, the μεσότης of equally necessary but opposing constituents demanded by nature is specifically excluded by those who set up oligarchy and democracy. For they insist that one extreme (the rich or the poor) must dominate rather than share equally with its opposite in the compound. And the tyrant goes even beyond this, forcing his rule not only upon equals but also upon his natural superiors (1295a19-21). Clearly, then, all three constitutions are contrary to nature and nature's principles.

Again, Aristotle regards democracy, oligarchy, and tyranny as deviations from, unnatural perversions of, the three natural constitutions. For the deviation forms are founded on distortions of the principles of natural or absolute justice basic to the correct forms. Thus the proponents of democracy pervert the principle of equality upon which polity is founded, while the advocates of oligarchy, and the tyrant, parody the principle of subordination upon which aristocracy of either kind rests (1280a8-15, 23-25):[99]

[99] Aristotle not only holds that each deviation form is a perversion of a particular correct form from the *theoretical* point of view, but he also teaches

And first we must ascertain what standards are proposed in oligarchy and democracy, and what principle of justice underlies oligarchy and democracy. For all hold to some kind of justice, but do not carry it far enough, and fail to express the principle of absolute justice in its entirety. For example, justice is thought to be *equality*; and it *is*, — but equality for those who are *equal*, not for everybody. Again, *inequality* is thought to be just; and indeed it *is*, — but for those who are *unequal*, not for all. But the partisans repudiate this factor of the persons *to whom* the principles apply, and go wrong in their judgments ... and further, they are misled into thinking that, because they profess some principle of justice which is true up to a point, they are professing what is *absolutely* just. Thus one group think that if they are unequal in one respect, — for example, in wealth — they are unequal in all; others think that if they are equal in one respect, — for example, in freedom — they are equal in all.

Thus the advocates of democracy demand an equal share in government for all who are equally free; this puts control of the polis in the hands of the poor, who are everywhere more numerous than the rich (1317a40-b10). The advocates of oligarchy, on the other hand, demand an unequal or dominant share for those who are unequal or superior in property, and so give control to the wealthy. While tyranny, the enforced domination of a single inferior man over his equals and natural superiors, completely perverts the natural rule of the king surpassing in excellence and so willingly accepted by all. Being founded upon false or unnatural principles, then, the deviation constitutions and the laws that conform to them are simply bad or unjust by absolute standards (1282b8-13; 1301a36-37; 1318a19-23).

Moreover, since the deviation constitutions, like the unnatural forms of animal organism, involve an essential deordination, conflict, pain, and instability are intrinsic to them in varying degrees. For whenever one part dominates and exploits the rest for its own advantage, the body politic becomes no longer an association of free men but a sort of despotism, as between master and slave (1279a19-22). Consequently, "in the deviation forms friendship,

---

that in the *practical* order kingship degenerates most easily into tyranny, aristocracy into oligarchy, and polity into democracy. Cf. *N.E.* 1160b10-23. However, the deviation forms can come about in other ways also. Cf., for example, 1286b8ff., 1307a6ff.

like justice, has little place" (*N.E.* 1161a31-32). Rather, the injustice basic to each of these forms generates conflict, since the part of the population unjustly subordinated strives constantly to restore a just balance, or to gain the mastery themselves. "For those who are inferior form a faction to win equality, and those who are equal do so to win greater power" (1302a29-31; and cf. 1301a26-b29).

In oligarchy and democracy the rich and poor both try to enforce their own partial but erroneous norm of justice, and from this class-war ensues (1301a26-39; and cf. 1296a24-33). In democracies the demagogues always split the polis by making war upon the wealthy, while in some oligarchies the rulers even take an oath to "be hostile to the people and plan every evil possible against them" (1310a2-11). Like the excessive element in a diseased body, each group not only dominates but tends to destroy its opposite. And in tyranny the intrinsic conflict is worst of all. For here a single individual forces an entire population unwillingly to submit to his own private advantage and pleasure (1295a19-23; 1311a3-7; 1313a13-16). Consequently, to maintain this unnatural condition the tyrant generally turns to violence, cutting down men of spirit and ability, reducing all to poverty, fostering mutual suspicion and quarrels among his subjects to prevent their uniting against him, etc. (1313a35-1314a29). Therefore "in the deviation forms friendship, like justice, has little place; and least of all in the worst" (*N.E.* 1161a31-33).

Again, like the abnormal animal organism, the polis is unstable, shortlived, in proportion to the intensity of conflict intrinsic to its structure. Stability in any form of polis derives from the common consent of its constituents: "For if a constitution is to be preserved intact, all the parts of the polis must join in willing its existence and its continuance in the same form" (1270b21-23). Stability is threatened, therefore, as soon as this common consent is lost and part of the polis sets up opposition to the constitution. In the deviation forms, however, common consent never exists; from the moment of its inception a deviation government is opposed by that part of the polis which is dominated and treated unjustly. The initial injustice and conflict upon which they are founded, then, make the deviation forms intrinsically unstable. "Constitutions of this sort never endure. The reason is simple. When one begins with an initial error, it is inevitable that one should end badly" (1302a4-7).

There are, however, degrees of inner conflict and instability among the various deviation forms. We have seen that polity is the best and most stable form possible for most men as they are; for when the extremes are evenly blended through a large middle class "not a single part of the entire polis would want any other form of constitution" (1294b39-40). Consequently, "the more perfectly the constitution is mixed the more permanently stable will it be" (1297a7-8). Conversely, the further a constitution deviates from the μεσότης of well-blended polity, the more *unstable* it will be (cf. 1296b3-9; 1302a8-16; 1307a7-20). And within the deviation form of polis, as in the diseased or deformed animal organism, the dominant and excessive part may grow until it destroys the constitution and brings ruin on the whole (1302b33-1303a3; 1309b20-35). Accordingly, 1) *among* the three main deviation forms Aristotle distinguishes one as being less stable in itself than another; and 2) *within* each of the main types he distinguishes several sub-species according to their relative stability.

This aspect of Aristotle's discussion constitutes a sort of political pathology, in which the norm of health is the proportionate blend of the mixed constitution or polity. The three general types of deviation or pathological condition (democracy, oligarchy, tyranny) are first compared to one another, and then each is distinguished into several stages marking the advancing deterioration of the body politic. Aristotle's etiology here seems basically that of the biological writings, where disease is conceived as a disturbance of the equilibrium of opposites and a destruction of their normal blend. If, for example, the opposing parts of the population are present in equal strength but *not blended* in a middle class, there is no stability; for the balance is too precarious, the constitution may be too readily overthrown. Where there is no blend, therefore, it is somewhat less dangerous if the opposites are not equally balanced; for "if either of the parts has a clear preponderance, the other is unwilling to risk a struggle with one obviously stronger" (1304a38-b4). Aristotle considers it axiomatic, therefore, that in the deviation forms where there is no common consent, the constitution must be supported by a majority, by a group stronger than the opposition, if it is to keep any stability at all (1296b-15-17; 1309b16-18). And yet this unnatural condition becomes more and more dangerous as the dominant element grows more extreme, uncontrolled, and destructive of its opposite. All unnatural or deviation forms, then, are bad (1289b9-10; 1302a4); but tyranny, the opposite of the best, is "the worst and farthest removed from polity, oligarchy the second farthest ..., and democracy the most moderate" (1289b2-5).

(a). *Democracy*: For several reasons Aristotle judges that "democracy is the least bad of the deviations" (*N.E.* 1160b20). 1) It has the closest affinity to polity, and "is only a small deviation from polity" (*ibid.* 18-22). 2) It is more secure and long-lived than oligarchy; in it the middle class is generally larger and has more share in the government than under oligarchy (1296a13-16). 3) It enjoys greater freedom from faction than oligarchy; while the people oppose the wealthy, they rarely split into groups themselves (1302a8-13). And there is most room for justice and friendship in democracy, for the people, being equal, have much in common (*N.E.* 1121b8-10). 4) In general, it is more stable because "the greater number forms a stronger support, and the masses are more content when they share equally" (1307a17-18). "Generally speaking, then, the large number of its citizen body keeps democracies intact" (1321a1-2).

However, democracy degenerates in proportion as the mass of citizens, declining in quality, grows more numerous and irresponsible. Aristotle distinguishes four stages, from moderate to extreme democracy:[100]

[100] The following is proposed merely as *illustrative*, since this is hardly the place to attempt an adequate discussion of the various types of deviation forms and the constitutional devices characteristic of each, — matter which occupies a large portion of Books IV-VI in the *Politics*. For the types of *democracy* and their characteristics, cf. 1291b14ff., 1292b23ff., 1297a36ff., 1298a10ff., b13ff., 1317a19ff.; for *oligarchy*, cf. 1292a39ff., 1293a13ff., 1297a14ff., 1298a34ff., b27ff., 1320b18ff.; for *tyranny*, cf. 1295a1ff., 1285b20ff., 1313a34ff. See also Barker, *The Political Thought of Plato and Aristotle* (London, 1906), pp. 448ff.

1. Citizenship qualification: Must be born of citizen parents; and *some* low property qualification. Ruled under law by magistrates and citizen boards. Popular assembly: meets infrequently as possible; to elect on merit or wealth, and to audit magistrates; to legislate and reform. No state subsidy for civic service. (Best; closest to polity).

2. Must be citizen born, but *no* property qualification. Ruled under law by magistrates, selected by vote or lot from all. Assemble for election, legislation, audit, war and peace. No state subsidy.

3. Citizens by birth or naturalization, i.e. all free men. Ruled under law, by magistrates who are elected only where experts needed, otherwise by lot. Assemble for election, audits, war, etc. No state subsidy.

4. Citizens: all free men. *Not* ruled under law; and magistrates make only preliminary decisions. All assemble for all matters; rule by decree. State revenues and subsidy provide leisure for poor to engage in civic service. (Worst; unrestricted rule of the masses; lowest and poorest citizens have most leisure and influence; demagogues arise. Not strictly a constitution, but amorphous mob rule: 1292a31-38).

(b). *Oligarchy*: Aristotle classifies oligarchy with tyranny as being in itself shorter-lived than the other constitutions (1315b12-13), and his reasons are evident from the comparison with democracy. For oligarchy does not have democracy's stability of numbers; and it arouses opposition more quickly in that its leaders are more prone to insolence and avarice (1307a17-20). Again, all citizens are equal in democracy, where the standard is freedom; whereas oligarchy sets up wealth as its standard, in which even the members of the ruling class are unequal. Consequently "oligarchies are prone to *two* sorts of faction, — one within the ranks of the oligarchical party itself, and the other between that party and the people" (1302a8-11). In this rather precarious state where the minority dominates the majority, then, it is clear why, above all else, the rulers must control the armed forces (cf. 1297a29-36; 1321a3-27).

In contrast to democracy, oligarchy degenerates in proportion as the property qualification for full citizenship is set higher, and the ruling group grows smaller and more arrogant. Aristotle distinguishes four stages or types of oligarchy, as follows:

1. Citizenship: (in addition to citizen birth:) Moderate property required (so that poor excluded from rule); all qualifying become members of ruling council; relatively numerous. *Necessary* offices open to

all, but only wealthier eligible for highest offices. New citizens admitted when qualified by property, automatically. Council meets infrequently. Rule under law. (Best form of oligarchy; most moderate, well-blended, closest to polity. Cf. 1298a39-40; 1320b21-23).

2. Higher property qualification required. Rulers chosen from citizens with large property, and *they* choose from the rest those who are to make up ruling body (cooptation); less numerous. Rule under law.

3. Still higher property qualification, and fewer in ruling group. Keep offices and deliberative council in their own hands and make them hereditary. Continue to rule under law, however.

4. Rulers exceedingly wealthy, with strong retinues. Hand on rule by inheritance, and become supreme *instead of* the law, ruling at their own will. Dynasty. (Worst, most extreme form of oligarchy).

Thus the partisans of democracy and oligarchy, pushing their false standards of freedom or wealth to extremes, develop ever more monstrous and unstable forms of constitution, until they end with no constitution at all (1309b22-35).

(c). *Tyranny*: Aristotle distinguishes several kinds of monarchy (1284b35ff.; 1295a1ff.), of which the last is tyranny. We have seen that the tyrant dominates and exploits an unwilling population by force (1313a10-11) entirely against the laws of nature. Therefore this must be the most unstable of all constitutions, the antithesis of the "first and most divine" (1289a40).

Tyranny is a compound of the extreme forms of oligarchy and democracy, and that is why it is more injurious to its subjects than any other form of government; it is composed of two bad forms, and it combines the perversions and errors of both (1310b3-8).

Consequently, tyranny is exposed to all the causes which overthrow the extreme forms of democracy and oligarchy, since "these forms are themselves no more than collective tyrannies" (1312b33-38). Aristotle insists upon this identification of extreme democracy and oligarchy with tyranny (1292a18-24; 1298a33-34); and he regards these three as the most dangerous pathological conditions of the body politic (1320b32-39; and cf. 1319b2-4):

Regarding the dynastic and tyrannical form of oligarchy, which corresponds to the most extreme form of democracy, — in so far as it is the worst form it requires the greater safeguard. For just as our bodies in a good state of health ... admit more mishaps without being destroyed by them, while bodies in a morbid condition ... cannot survive even slight mishaps, so it is with constitutions: the worst require the greatest safeguarding.

So much, then, for the deviation *forms* or constitutions. Now Aristotle explains the production of monstrous or unnatural animal organisms largely through the inadequacy of the *matter* supplied by the parents (*De Gen. An.* 767a36-b18; 769b4 ff.; 770b10-18). For natural or healthy forms can only be realized in proportionate matter (*ibid.* 767a14 ff.). Similarly in the political organism: populations are not always adequate to receive a natural form or constitution. They may be capable only of realizing an unnatural or deviation form, if they can be organized into a polis at all. Even the lowest natural form, the mixed constitution or polity, requires its proper matter, i.e. a population containing a large middle class. Where a middle class is lacking, therefore, an ordinary population can only realize some deviation form, democracy or oligarchy, at best. This is the "natural" or necessary result of the limitations, the inadequacy of the matter (cf. *De Gen. An.* 770b9-18).

What is there in the population, then, that determines *which* of the deviation forms it can realize? In other words, which deviate constitution is proper to what population?

Then, in regard to constitutions generally, we have to inquire which constitution is desirable for which sort of people. For it is possible that democracy rather than oligarchy may be necessary for some, and oligarchy rather than democracy for others (1289b18-20).

This is a problem which occupies Aristotle at some length; we shall attempt only to indicate a few leading principles which he offers to guide the statesman in this matter (1296b13 ff.).

(a). The first principle has already been mentioned: In every polis at least the *majority* or more powerful element must favor the constitution if it is to have any stability, i.e. "the part of a polis which wishes the constitution to continue must be stronger than the part which does not" (1296b15-17). Where a population is split into opposites, then, the constitution must be one which at least the preponderant element will favor.

(b). In reckoning the most powerful or preponderant element

in the population two factors must be taken into account, — *quantity* and *quality*; and these must be weighed one against the other in a proportion (cf. 1301b29-37). Aristotle explains (1296b17-24):

Here we must remember that quality and quantity both go to the making of every polis. By "quality" we mean free birth, wealth, culture, and nobility of descent; by "quantity" we mean superiority in numbers. Now quality may belong to one of the parts which compose a polis, and quantity to another. For example, those who are low-born may be more numerous than the high-born, or the poor than the rich; but the superiority in quantity on the one side may not be sufficient to balance the superiority of quality on the other. Quantity and quality must thus be placed in the balance against one another.

If the opposing parts are close to even balance and blended in a considerable middle class, the population constitutes material suitable for a polity, as we have seen (*ibid.* 38-40). But if one part is excessive and its opposite deficient, the disproportionate nature of the population demands one or other *deviation* form, thus (*ibid.* 25-33):

First, where the great number of the poor preponderates in the aforesaid proportion, there it is natural that a *democracy* exist ... But where the class of rich or notables *over*balances by quality its deficiency in quantity, there it is natural to set up an oligarchy ...

Thus Aristotle determines the *general* deviation form proportionate to a particular population.

(c). The *specific* type of democracy or oligarchy appropriate to a particular population will be determined by the *make-up*, i.e. by the quality and composition, of its predominant element, whether that be the people or the wealthy class. Thus:

(1). Where the predominant poor are for the most part small farmers in an agricultural population, there the first kind of democracy, closest to polity, will be most natural. For, though poor, the majority can meet some property qualification; and as busy farmers living at a distance from town, will be content with infrequent assemblies and the rule of magistrates under law. However, if the majority of poor are not homesteading farmers but shepherds and cattleman, the second type of democracy with no property qualification will be proper. But where the polis grows and the largest group among the poor are craftsmen and traders, or wage-earning laborers, there the third or fourth types of democracy alone are possible. Thus as the predominant poor grow in number and decline in quality, the constitution must deviate ever more widely from the healthy and natural form if it is to hold the polis together at all (cf. 1296b25-31; 1317a23-29; 1318b7ff.).

(2). According to the same principle, where the wealthy preponderate, the oligarchic form demanded by the population must deviate progressively as the ruling class grows wealthier and more powerful, but fewer in number. Where the wealth is distributed in smaller estates among a large number of citizens, the population may sustain the form of oligarchy closest to polity. But as wealth becomes concentrated in fewer hands, the property qualification must be raised and more deviate forms admitted to keep the safest proportion between population and constitution (1296b31-35; 1320b18-33).

Thus the polis can only realize the form for which its population has the potency, deficient as that may be.

(d). Finally, where a population is deficient, the best constitution for that particular polis is the one closest to polity which *it* is capable of attaining. This is the *relative* best (1296b7-13):

> The constitution nearest to the best must be better than all the rest, and the one farther removed from the middle must always be worse, — that is, unless one judge with relation to particular circumstances. I make this qualification because it often happens that, though one constitution be in itself more desirable than another, the other constitution may be more advantageous for some particular people.

Thus, an individual polis may never be capable of realizing the perfect μεσότης of well-blended polity, but only some form of democracy or oligarchy, blended as far as its particular population will permit. This is the middle state *relative to the individual* (μεσότης πρὸς ἡμᾶς), like the best state of health attainable, for example, by a woman, a child, an old man. It may be very imperfect, compared with the disposition of nature at her absolute best (the well-blended polity, the healthy adult male). But it is the best attainable by *this* individual composed of this particular matter at this particular time. And "that is always the most desirable for each individual which is the highest possible for him to attain" (1333a29-30).

To produce the relative best, then, is in one sense the ultimate practical objective of the statesman. For he, like the physician, must always deal in the practical order with *individuals* (*N.E.* 1097a11-14; 1104a4-10; 1180b8-16).

F. CONCLUSION: THE EXPERT AND HIS NORMS OF VALUE

1. *Objectives of Physician, Moral Guide, and Statesman*

In Aristotle, as in Plato, the "political art" is often compared to medicine, and the expert in that art, the moral guide and statesman,[101] to the physician or trainer. Weakness and vice in the moral order, and dangerous or disturbed conditions of the political organism, are constantly compared to pathological states of the body, for which the moral guide and statesman must find a remedy, bringing the individual or the polis to the most natural or "healthy" condition possible.

To accomplish this, however, each needs expert knowledge as well as experience of individual cases, for "a physician or trainer or anyone else can best treat the individual when he has a general knowledge of what is good for all, or for people of such and such a kind ..." (*N.E.* 1180b13-15).

For it is not within the power of just anyone to bring about a proper disposition in all who may present themselves; but if anyone is capable,

[101] Since the *polis* implies for Aristotle not only an organized community but an entire way of life, the science and art of "politics" is the supreme practical science, directing man to his final happiness both as an individual and as a member of the political community. Hence "politics" comprehends the material of both ethics and political philosophy; and the "politikos", the expert in this field, represents for Aristotle a combination of moral guide and statesman. In the following discussion, however, we shall distinguish these two aspects by their separate titles, since we have no composite term suggesting both. Cf. *N.E.* 1094a19-b12; 1099b29-33; and Barker's discussion of the Greek polis in *Greek Political Theory, Plato and His Predecessors*, 3rd ed. (London, 1947), pages 1-29.

it is the man with expert knowledge, just as in medicine and all other professions involving treatment and the exercise of prudence (*ibid.* 25-28).

It appears, in fact, that Aristotle proposes to supply, in his collection of constitutions and his treatise on politics, a type of scientific knowledge analogous to that which the medical writers offered the practicing physician. He observes that, while no one becomes an expert merely by reading, still the medical handbooks "do try to explain not only general methods of treatment, but also how particular groups of patients, classified according to bodily disposition, may be cured, and what manner of treatment should be used for each" (*N.E.* 1181b3-5). Since such writings prove useful to the physician, therefore, Aristotle undertakes in the *Politics* to study, on the basis of his collection of constitutions, "what sort of things preserve states intact and what destroy them, and what preserve or destroy particular types of constitution, and for what reasons some are well governed and others the opposite. For when these have been studied we may perhaps see more comprehensively what sort of constitution is best, and how each kind is best organized, and what laws and customs it should use" (*ibid.* 18-23).

This program, outlined at the end of the *Nicomachean Ethics*, is repeated with greater clarity and detail at the beginning of the central book of the *Politics* (1288b10 ff.) in a long and interesting passage which constitutes a sort of table of contents for the whole work. And the *Politics* passage reveals even more clearly the close parallel that Aristotle sees between the science and art which treat the physical organism and those which treat the political organism. Aristotle outlines first the knowledge required by the *physical* expert in fulfilling his function and solving his problems (1288b10-20):

There is a rule which applies to all the practical arts and sciences, when they have come to cover the whole of a subject, and are no longer engaged in investigating it bit by bit. Each of them severally has to consider the different methods appropriate to the different categories of its subject. For instance, the art of physical training has to consider (1) which type of training is appropriate to which type of physique; (2) which is the ideal type of training, i.e. the training best for the physique of the best endowment and the best equipment (for the ideal type of training must be one suitable for such a physique); (3) which is the type of training that can be generally applied to the majority of physiques, for that too is one of the problems to be solved by the art of physical training. Nor is this all. (4) There may be men who want physical training, but do not want to attain the standard of skill and condition needed for competitions. It is then no less the duty of the trainer and gymnastic master to produce the degree of capacity which such men want. And this is likewise obviously true of medicine, shipbuilding, tailoring, and all the other arts.

The same kind of expert knowledge is required by the analogous function and problems of the *political* expert (1288b22-39):

It follows that the study of politics must be equally comprehensive. First, it has to consider which is the best constitution, and what qualities a constitution must have to be most ideal, supposing no external factors stand in the way. Secondly, politics has to consider what sort of constitution suits which sort of people. For the attainment of the best constitution is likely to be impossible for the majority; and the good lawgiver and true statesman must therefore have their eyes open not only to what is the absolute best, but also to what is the best in relation to actual conditions. Thirdly, politics has also to consider the sort of constitution which depends upon an assumption. In other words, the student of politics must also be able to study a given constitution, just as it stands and simply with a view to explaining how it may have arisen and how it may be made to enjoy the longest possible life. I have in mind the case where a polis has neither the ideally best constitution (or even the conditions needed for it) nor the best constitution possible under the actual conditions, but has only one of an inferior type. Fourthly, in addition to all these functions, politics has also to provide a knowledge of the type of constitution which is best suited to states in general ... We have not only to study the ideally best constitution. We have also to study the type of constitution which is practicable, and with it, and equally, the type which is easiest to work and most suitable to states generally.[102]

[102] These various aspects of the art and science of politics are first mentioned at the end of the *Nicomachean Ethics* in a passage to which Ross has called attention as outlining the program of the *Politics* and proving its unity (cf. note 95 above). The present passage presents essentially the same program or guide to the contents of the *Politics*, and constitutes, I believe, an even stronger argument for the *unity* of that treatise. Aristotle here states in effect that the expert must control *simultaneously both* the ideal and the practical aspects of his art; he must know at the same time both the *absolute* best (Bks. I-III, VII, VIII) *and* the means of bringing existing constitutions to the best condition practical for each particular community (Bks. IV-VI). We shall see, in fact, that this prescription is linked essentially with Aristotle's whole theory of production, which demands that the producer have in mind before all else the *form* of the

On the basis of these passages, then, and with the help of Aristotle's remarks elsewhere regarding the physician, trainer, moral guide, and statesman, we may outline the parallel functions of these experts, — a parallel consistent with Aristotle's assumption that the same basic principles of organic life govern the physical organism of the body (σῶμα), the moral organism of the individual (ἦθος), and the political organism of the state (πόλις).

The common objective of physician, moral guide, and statesman is to imitate nature and make up for her deficiencies, assisting her efforts to produce the perfect organism on all three levels, — the physical, moral and political (cf. *Physics* 199a15-16; *Politics* 1337a2-3; *N.E.* 1099b21-23). To do this, however, each expert must first of all possess correct knowledge of the *form* which nature at her best assumes on each level, — the form of the absolutely best physical, moral, or political organism. This, being identical with the final cause, must be present before all others in the intention of the expert, though as formal cause of the product it is the last to be fully realized in the order of execution or production, if indeed it is ever realized in its perfection. Aristotle is fond of illustrating this by the example of the physician (cf., for example, *Met.* 1032b1-32; *De Part. An.* 639b12-19, and cf. 646a25-b4; *Physics* 194b32-195a3): The physician must know clearly the nature of perfect health, the qualities or disposition of a healthy body, before he can begin to restore health where it does not exist, or bring it to greater perfection where it exist in some degree.

But from art proceed the things of which the form is in the soul. (By form I mean the essence of each thing and its primary substance). ... And

---

product at its best, and then that he produce it as best he can within the limits of the materials available. This, therefore, seems to be an additional reason against assuming, with Jaeger, that in the *Politics* Aristotle's sketch of the ideal (Bks. II, III, VII, VIII) must belong to an *earlier* interest which was to some extent superseded by a more "scientific" or practical approach (Bks. IV-VI) in *later* years. Rather, this passage (appearing, incidentally, at the beginning of Jaeger's "later" books) shows that Aristotle regarded *both* aspects as being at the same time necessary and important to the statesman in diagnosing the condition and treating the ills of the body politic. His theory of production demanded that the *Politics* contain both the ideal and the practical aspects, at whatever time it was written.

health is a formulated essence (λόγος) in the soul, that is, the knowledge of health. Now the healthy subject is produced through the following process of reason: Since *this* is health, if the subject is to be healthy, *this* must be present (e.g. a uniform disposition of body); and if this is to be present, there must be heat. And the physician goes on reasoning thus until he arrives at something final which he can produce. Then the process from this point onward, i.e. the process towards health, is called "production" (ποίησις). Therefore it follows that in a sense health comes from health (*Met.* 1032b1-12).

Now as the physician knows health, the moral guide and statesman must know first of all the true nature or essence of human happiness, — the unimpeded and enduring exercise of perfect virtue, with all that entails. The moral guide must understand the requirements of happiness in terms of the best dispositions of body and soul in the *individual*; and the statesman must employ the same norm to determine the constitution and laws which will best help to attain happiness for the *polis*.

The object we have in view is to discern the best constitution; the best constitution is that under which the polis is best governed; the best governed polis is that which has the greatest capacity for happiness; it is clear then that we must know the nature of happiness ... happiness is the completely perfect activity or exercise of virtue, and this in a mode that is absolute, not qualified (1332a4-11).

Thus physician, moral guide, and statesman must begin with knowledge of the essence of health and happiness as they are realized in the absolutely best, i.e. the *most natural*, dispositions of body, soul, and polis.

However, the experts in any art, particularily in the productive arts, must know not only the form but the *matter*, and the operations necessary to realize the form in its proper matter (*Physics* 194a22-b8). In other words, they must not only know the end clearly, but also the means and material necessary to achieve the end. "In the arts and sciences both these things must be mastered, — the end and the practical means to the end" (1331b38-39). And if the knowledge of the expert is not to remain sterile, he must have on hand, or be able to supply, matter proportionate to the form. For the actual process of production cannot begin until suitable matter is at hand (*De Part. An.* 639b27-31; and cf. 646a25-b2):

For if a house or other end is to be realized it is necessary that such and such material be at hand; and it is necessary that first this and then that be produced, first this and then that be changed, and so on until the end-object is produced, the final cause for which each prior thing is produced and exists. As with the productions of art, so also it is with the productions of nature.

Thus, if they are to assist nature in producing the absolutely healthy or virtuous individual and polis, the physician, moral guide, and statesman must have, or be able to develop, matter proportionate to the best form on each level (1325b33 ff.; 1332a28-b11).

However, it will be rare that natural endowment alone can provide the physical, emotional, and social constituents needed for perfect health, perfect moral virtue, or the best constitution. At least, as we have seen, Aristotle seems to find among men a sad lack of suitable material for the highest forms of moral and political life. If the expert is to attempt to make men better at all, then, he must be content to work for the most part with the very imperfect material which he finds about him.

This necessity, of course, complicates the expert's problem and proves the real test of his skill. For, as we have seen, it is largely due to the disproportionate or imperfect matter of which they are formed that the variety of inferior individuals and states come into existence (cf. 1328a37-b2; 1331b39-1332a2). It is not enough, then, that the expert know the absolutely best form on each level, and the material proper to it alone. He must also know the intermediate and deviation forms, and be able to diagnose the actual condition of the organism *in each particular case*, assessing the causes of unnatural conditions and prescribing means by which they may be altered to bring the subject into a better and more stable condition.

Thus the physician must know the various possible dispositions of body with their characteristic excesses and deficiencies, and the cure or treatment proper to each (*N.E.* 1181b4-5); the trainer must be able to recognize the various types of physique and know the regimen suited to each (1288b12-13); the statesman must know the main types of constitution and their varieties, as well as the things which hurt or help each, if he is "to be able to bring aid to *any* existing constitution" (1289a6-8; and cf. 1288b28-33, *N.E.* 1181b18-19). And in his attempts to improve the individual or the polis, the expert may do more harm than good if he does not know the *limitations* of his material, — for example, what exercise or what constitution each inferior type of physique or population is capable of sustaining (1289b18-23), and what is the best attainable by most physical types or populations as they actually exist (1288b15-16, 34-35).

Equipped, therefore, with this complex knowledge not only of the best possible individual or polis but also of all various inferior forms and the means of their improvement, the physician, moral guide, or statesman must approach his subject realistically as a *particular* individual or group, endowed with certain natural qualities and limitations, and disposed in a certain condition. His task then is to diagnose and evaluate the particular condition of this subject and, if possible, bring it step by step to a *better* one. Here, however, he is faced with the problem of the *absolute* and *relative* good. Things absolutely good (healthy, pleasant, helpful, just) are such for individuals and groups *at their natural best*, but not for all (*N.E.* 1176a8-29). Food that is pleasant and nourishing for the healthy may be unpleasant and harmful to the sick; wealth and power may help the virtuous but make the vicious worse. Hence in dealing with all inferior types, the physician, moral guide, and statesman must begin with the *individual*, prescribing at first the activities or things that are *relatively* good, until he can bring the patient or subject to the condition in which those that are absolutely good are also good *for that individual.* Thus, for example, Aristotle sees the process of both moral development and learning as a transition from what is good and intelligible for the individual to what is also good and intelligible absolutely or by nature (*Met.* 1029b2-13):

For it is of advantage to advance from the less to the more intelligible. Learning proceeds for all in this way, — through what is less intelligible by nature to what is more. And just as in human conduct our task is to start from what is good for each individual and to make what is *absolutely* good good for the individual, so it is our task to start from what is more intelligible to oneself and to make the things which are by nature intelligible intelligible to oneself. Now what is intelligible and primary for particular types of people is often intelligible to a very small extent, and has little or no reality. And still, starting from what is intelligible in a very low degree, but is intelligible to ourselves, we must try to understand the absolutely intelligible realities, advancing, as we said, by way of those very things which are intelligible to us.

Accordingly, Aristotle advises the generality of men not to pursue riches and power, but to desire and choose the things that are good *for them*, and to pray that what is good absolutely may also become good for them (*N.E.* 1129b4-6). For they shall then have attained the perfection of natural development.

The objective of the expert, then, like that of nature, is to bring the individual subject to the highest perfection *of which it is capable*. Where the subject is richly endowed by nature the expert may be able, through careful guidance, to bring it to the absolutely best form of physical, moral, or political life. But in most cases, perhaps, the matter or potentiality of the subject will be so imperfect that it cannot be made proportionate to the absolutely best form by *any* effort of the expert over any period of time. In such cases the objective of the expert can only be to bring his subject to approximate the absolute best in the degree which its matter will allow. This will be the relative best, the best attainable by this subject with its particular nature and circumstances. "For that is always the most desirable for each which is the highest possible for him to attain" (1333a29-30).

a. *The Physician-Trainer*

Thus, since health admits of degrees and does not involve the same equilibrium of elements in all persons (*N.E.* 1173a26-28), the physician's task is not only to produce the most perfect health, but his help is required perhaps even more to produce a relative degree in persons of inferior constitution. If he is called in to care for a woman or an old man, for example, the physician may relieve the pain and disturbance of disease by restoring a certain balance among the elements of the body. But because the patient is by nature imperfectly constituted, the physician can never hope in this case to achieve the perfect μεσότης of the absolutely best condition. His objective, then, must be the *relative* μεσότης possible to a woman or old man, and to this particular patient with her, or his, individual limitations. And, as we have seen, the more precarious the state of the patient, the greater will be the skill and care required of the physician (1320b34-39).

This matter of the absolute and relative best in health, with the degree of effort needed to attain each, is in fact given a cosmological application by Aristotle in the *De Caelo* (292a22-b20) to explain the relatively complex activity of inferior beings in the universe. Significantly, he uses perfect health as the analogue to the perfect condition of the highest being, and we may perhaps be permitted to quote the passage at length for the light it casts not only upon the immediate matter of absolute and relative health and their attainment, but upon the analogous conditions of the moral and political organism and, indeed, upon Aristotle's whole scheme of values:[103]

It is reasonable for that which is in the best state to possess the good without taking action, for that which is nearest to the best to obtain it by means of little, or a single, action, and for those things which are farther from it to need more; in the same way as one body may be healthy without any exercise at all, another by means of a little walking, a third may need running and wrestling and violent exertion, and again a fourth despite tremendous efforts cannot preserve this particular good, but only something else. To succeed in many things, or many times, is difficult ... We must consider that whenever A must be done as a means to B, B as a means to C, and C as a means to some further end, then if the intermediate steps are one or two, it is easier to attain the end, but the more they are the more difficult it becomes ... That which is in the best possible state, on the other hand, has no need of action. It is its own end, whereas action is always concerned with two factors, occurring when there is on the one hand an end proposed, and on the other the means toward that end ... To sum up, there is one thing which possesses, or shares in, the best, a second which reaches it immediately by a few

[103] The text and translation is that of W. K. C. Guthrie in the Loeb Classical Library edition of *De Caelo* (London, 1945), pp. 207-211. The complicated matter of Aristotle's axiology is treated in detail by Professor Whitney J. Oates, *Aristotle and the Problem of Value* (Princeton, 1963).

stages, a third which does not even attempt to reach it, but is content merely to approach near to the highest. For example, if health is the end, then one creature is always healthy, another by reducing, a third by running in order to reduce, a fourth by doing something else to prepare itself for running, and so going through a larger number of motions; another creature cannot attain to health, but only to running or reducing. To such creatures one of these latter is the end. To attain the ultimate end would be in the truest sense best for all; but if this is impossible, a thing gets better and better the nearer it is to the best.

b. *The Moral Guide*

Analogously, we have seen that between the perfection of virtue or moral excellence and its utter perversion in vice there are various states of self-control and lack of control, so that "not every man is good or bad, just or unjust, but there is also an intermediate state" (*Met.* 1023a6-7). Consequently, unless a man is completely bad or "incurably vicious" he is capable of reform and can be helped to improve (cf. *N.E.* 1165b18-20).

For the bad man, when once introduced to better pursuits and ideas, may advance, be it ever so little, toward being better. And if he shall once make even a little progress, it is clear that he can either change completely or at least make great progress. For he becomes ever more easily moved toward moral excellence, even though he made very slight progress at the beginning, so that he is likely to make greater and greater advance. And as this continues he is established finally in the opposite disposition, — that is, provided his time does not run out first (*Cat.* 13a24-31).

And apparently there are degrees of perfection within the state of virtue itself, as there are of health, depending perhaps upon how strongly the μεσότης is established and how invariably it responds to the direction of reason. For Aristotle observes that as pleasure admits of degrees, "the same is true of justice and the other virtues, in respect of which we plainly say that people of a certain character are so more or less, and act more or less in accordance with these virtues; for people may be more just or brave, and it is possible also to act justly or temperately more or less" (*N.E.* 1173a18-23).

The function of the moral guide, then, like that of the physician, will be to bring his subject to the most perfect state of which he is capable. And, obviously, the expert's task will be more difficult, and his services more necessary, at the *beginning* of the process of improvement, becoming progressively easier and less necessary as the subject approaches the perfection of virtue. But in the process, as Aristotle mentions, the efforts of the moral guide to bring the subject to perfection may be thwarted by the limitations of time. The subject may, for example, be initially so timid that the moral guide can only bring him to a state of self-control during his lifetime, death intervening before he ever attains the virtue of courage. In such a case the moral guide must be content with the *relative* best. As again he must if the subject, for example, actually attains the μεσότης of *moral* excellence but spends his life exercising it in *external* activity, and never arrives at the perfection of human activity in contemplation.

Again, Aristotle apparently holds that a great proportion of human beings cannot hope to attain the absolute best because they are imperfectly endowed at birth and, regardless of how long they live, will never possess the necessary means, i.e. the matter or faculties needed for the highest human perfection. Thus women, inferior men, natural slaves are deviations, lacking in some degree at least the physical constitution and intellectual capacity of human nature at its best; consequently they will have different, and subordinate, functions in society; and their characteristic excellence, related to their nature and function, will be of an inferior order, as are those of the lower faculties of the soul when compared to the highest (1260a5-24; and cf. 1277a6-13, 1277b13-30):

For in the soul there is by nature a ruling part and a part ruled; and each has a different excellence, one the excellence of the rational part, the other the excellence of the non-rational. It is clear therefore that the same holds also in the case of other things, so that there exist by nature various types of ruler and ruled. One is the rule of free over slave, another that of male over female, a third that of father over child. The different parts of the soul are somehow present in all these persons, but present in different ways. The slave is entirely without the faculty of deliberation; the female has it, though in an ineffective way; and the child has it, but undeveloped. Now it must be understood that corresponding differences prevail with regard to their *moral* virtues also; that is, all these persons must share in moral excellence, but not all in the same manner. Rather, each must share in so far as required by his or her *function.* The ruler, accordingly, must possess moral excellence in its perfect form (for his function is, to put it simply, that of executive director, and his *reason* is the executive director); but each of the others in so far as it falls to their lot. It is clear, therefore, that there is a moral excellence of all the persons mentioned; but the temperance of a man and of a woman are not the same, nor are their courage and justice ... But one is the courage of a ruler, the other of a subordinate, and similarly with regard to the other virtues.

In the case of women, natural slaves, children, and others of inferior natural endowment, then, the moral guide sets as his objective the *relative* best, — the μεσότης of the emotional powers adapted to their inferior nature and function. This will not be the perfect μεσότης of the adult male at his best but only an approximation proportioned to the subject, — the relative μεσότης. In every case the moral guide aims at bringing the individual to the highest excellence of which the person is capable; but he can only hope to achieve the absolute best when nature provides the best material. Like the physician with health, the moral guide may bring a few to perfect happiness, but the rest only to some share in it according to their capacity (cf. 1328a38-40).

c. *The Statesman and Lawgiver*

Finally, the polis, like the individual, is an organism whose "end or object is the best life that is possible for it" (1328a37-38). Objectively the best or happiest life is found in the activity of perfect excellence; yet while this may be possible for some populations, others can share in it only to a small extent, or not at all (*ibid.* 38-40). All must have forms or constitutions, then, adapted to *their* particular nature and circumstances; for generally people will live best under the best constitution possible *to them* (1323a18-19).

The primary function of the true statesman, therefore, is to study a polis, a race, or any social group and to determine "how they will share in the good life and in the happiness possible to them" (1325a8-11). And he must then help each to attain this, if not under the absolutely best constitution, at least under the best constitution possible for them (1288b24-28). Like the physician, he must be prepared to assist any particular population, existing as it may be under an unnatural form of government; he must be able to diagnose the cause of its difficulties, to stabilize it, and, if possible, bring it to a better and more suitable constitution (*ibid.* 28-33; 1289a2-8).

But just as most individuals never fully develop the highest intellectual faculties and virtues, so most populations do not produce men of surpassing excellence and can never attain the absolute best constitution. And as the moral guide can at best develop in most individuals a certain emotional equilibrium of moral virtue, so the statesman can only hope to guide most populations to relative stability under the balanced or mixed constitution, which resolves the conflict between rich and poor, bringing them to a condition in which they respond readily to direction under law. To accomplish this, however, the statesman must develop in the population a perfect μεσότης, — a middle condition or equilibrium in which the opposing constituents, rich and poor, are blended to a mean in a large middle class. He must also stabilize and maintain this balance through a constitution which mixes and blends the features of democracy and oligarchy, and so appeals to all parts of the population. This then will be the most lasting and natural form possible for most. If such governments do not survive, Aristotle maintains, it is generally because the constitution was not a good blend of these two opposing forms (1307a6-8). Hence the better the constitution is mixed the more permanent will it be (1297a6-7).

However, Aristotle seems to regard most populations actually existing at his time as comparable to those individuals of defective natural endowment who cannot attain even ordinary health, but only a condition less evil than some other; for he finds them almost entirely without a middle class, torn between extremes of rich and poor, and laboring under some unnatural type of constitution (1296a23-33). The statesman, then, must be ready to take each of these as it is, study its components and the cause of its conflicts, and stabilize its condition as far as possible. To achieve this he must guide the particular population to that type of constitution which will put the strongest element in control, and at the same time will *balance* or *moderate* their power so that the weaker element will not be destroyed or driven to revolt. In other words, the statesman must aim to develop in each defective population the *relative* μεσότης possible to it. And the closer it can be brought to the balanced blend of the mixed constitution the more stable will it be.

Perhaps, then, because the statesman will be called upon to expend his greatest effort and skill upon the prevailing deviation forms, Aristotle instructs him in unusual detail regarding the various types of democracy and oligarchy, the causes that destroy them, and the means of improving them. And in his counsels for improving them, for bringing them closer to the balanced blend of polity, he insists upon the same practical principle which appears frequently on the moral and physical levels, — τοῦ μέσου στοχαστέον. To stabilize and improve each constitution as it exists the statesman must "aim at the mean", whether that be the absolute mean of the mixed constitution or the relative mean possible to this particular imperfect population.

An adaptation of the phrase actually occurs in Aristotle's general admonition that the statesman "should always include the middle class in any constitution; if the laws he frames are oligarchical, he should aim at the members of the middle class (στοχάζεσθαι τῶν μέσων), while if democratic, he should embrace them in the laws" (1296b35-38). Thus, while in a defective population the appropriate constitution must always favor the stronger element, that element and its power must not be allowed to grow excessively out of proportion (1309b20-35; 1319b13-17); nor, on the other hand, can the constitution remain stable if some other group, or individual, disturbs the proportion (1302b15-18, 33 ff.; 1308b11-20). Therefore the statesman is advised not to forget "what is at present forgotten in the deviation forms of government, namely, the *mean*" (τὸ μέσον: 1309b18-20). Moreover, every effort must be made to moderate and reconcile the opposing elements in the polis, protecting and cultivating the wealthy under democracy and the poor under oligarchy (1298b14-38; 1309a15-27; 1310a2-13), and bringing them to a condition of equality as far as the constitution will permit (1309a28-33). Finally, the statesman must try to intermingle and blend the members of opposing majority and minority groups who threaten the constitution with faction because of their disparity (1308b25-31; and cf. 1319b25-28):

> The remedy for this difficulty is always to entrust the conduct of business and the authority of office to the opposite sections (I mean the people of quality as opposed to the masses, the poor as opposed to the rich) and to try either to mingle the multitude of the poor with that of the wealthy, or to increase the middle class (for this dissolves the factions caused by inequality).

Thus the statesman may bring democracies and oligarchies to a condition in which they are relatively "well blended" and stable (1320b19-22; and cf. 1318b7-38). And even a tyrant may best be advised to retain his power by adopting a policy of moderation, as far as that is possible (1314a30-1315b17).

Finally, in advising the statesman regarding details of the best possible material conditions for a polis, Aristotle sets up solutions which may guide him in dealing with any particular population; and the solutions are most obviously based upon the principle of "aiming at the mean". By applying this in all details, the statesman will best be able to balance opposing factors and give the greatest stability possible to the constitution of each particular population, with its peculiar character, size, location, neighbors, economy, etc. Thus the size of the population must be kept proportionate to the function of a polis, being neither too small for self-sufficiency nor too large for efficient action (1326a9-b26). In assigning land to the citizens the statesman is told τοῦ μέσου στοχαστέον: the estates must be large enough to support them with dignity, but not so large as to promote luxury (1266b24-28). And the same principle will decide the proper amount of their total wealth (1267a22-28), the site of their city (1330a34-b22), the best natural character for citizenship (1327b18-1328a8), the best physique (1335b6-13), even the type of music to be used in education (1342b-14-34). In all cases the statesman, like the physician and moral guide, must begin with the individual, taking the particular population with its peculiar characteristics and leading it to the most perfect μεσότης of which it is capable. Having stabilized it thus he may hope to bring it in time to the absolutely best μεσότης of the mixed constitution, and perhaps to even a higher form.

For if it has been well-said in the *Ethics* that the happy life is one led without impediment in accordance with virtue, and that virtue is a "middle state" (μεσότης), then a *middle* way of life must be best for a polis, — the "middle state" being that which is possible for each to attain (1295a 36-39).

2. *Norms and Values: the Absolute Good*

Now the physician, moral guide, and statesman, in approaching the particular individual or polis, must first of all have some standard for determining the present condition of his patient or subject. At the same time, in order to evaluate that condition, he must have another standard with which to compare it. What norms or standards, then, does Aristotle provide the expert, that he may correctly diagnose the condition of the individual or the polis, and bring them to a better?

a. *Subjective Norms*

The internal condition of the individual patient or subject will be manifested to the physician and moral guide by his *reactions*, particularly by the responses of pleasure and pain. For Aristotle teaches, as we have seen, that the subject's physical or moral condition is healthy or virtuous if he is pleased and pained by those objects and activities which are known to be pleasing or painful, helpful or harmful, to human nature at its best. Hence, if the subject reacts otherwise, a study of the things that please or pain him will reveal the direction and extent of his deviation. Thus Aristotle provides a criterion by which the expert may judge the subjective state of the individual.

We may be justified in assuming that this also provides the expert with a criterion for judging the effectiveness and suitability of the cure or treatment which he prescribes. For if, eventually at least, the painful reaction subsides and the pleasant increases, he may be confident that it is suitable; while if the opposite occurs, the expert will know that he has failed to "hit the mean" in prescribing for this particular subject.

Likewise, in approaching the particular polis, it seems clear that Aristotle's statesman would be guided by those reactions which correspond to pain and pleasure in the individual, — by the presence of civil strife, faction, and discontent on the one hand, or of friendship and harmonious cooperation on the other. The nature and power of the discontented element in the polis, the cause and intensity of their opposition, will reveal to the statesman the basis of the trouble and the aspects of the present constitution that need adjustment to make it more suitable to the population, and hence more stable. Aristotle warns that in drafting a reform a statesman must watch that it be "one which men can easily be induced, and will readily be able, to graft onto the system they already have" (1289a1-3, reading with Barker). And he will know that he has succeeded in stabilizing the constitution when he finds an increase of friendship, support, and acceptance of the government, at least among the greater or more powerful element in the population (cf. 1270b21-23; 1309b15-18).

b. *Objective Norms:* Μεσότης *and the Absolute Best*

Now the function of the physician, moral guide, and statesman is, according to Aristotle, not only to prescribe the relative good for the subject in its present state but, if possible, to bring it eventually to that state of perfection which is absolutely best, in which the things that are by nature good are also good for the particular individual or polis. To fulfill this function, as we have seen, the expert must be equipped with an *absolute* scale of values, i.e. he must know the absolute best, the perfection of nature, at least upon the physical, moral, and political levels.

(1). *On the Physical Level*: We have seen that Aristotle considers a human being to have attained the perfection of *health* when the opposing elements of his body are habitually balanced in dynamic equilibrium, — ἐν μεσότητι. Then each organ, maintaining the exact blend proper to its nature, functions proportionately, contributing to the well-being of the whole body, to its unified, harmonious operation in subordination to the central governing organ, the heart. This perfection of health can occur in human beings, however, only at the prime of life, when the process of growth is complete and the process of decay not yet begun; for youth and age, growth and decay, are dominated by heat and cold respectively, breaking the equilibrium and interfering with perfectly proportionate response.

Moreover, absolute physical perfection is attainable only by the male of the species. For, as we have seen, Aristotle regards the female as an imperfect male, whose complete development has been arrested from the time of conception by a certain deficiency of vital heat, a deficiency which affects both her physical and intellectual powers, making her by nature inferior to the male (*De Gen. An.* 728a18-22; 737a28; 766a17-23; 775a13-22). And below the perfect male and female, of course, are all those who in various degrees are stunted or distorted in their physical, and psychological, development.

The best condition absolutely on the physical level, then, is the proportionate blend and subordination of parts enjoyed by the healthy male at his prime, enabling his body to function harmoniously as a unified whole in all his complex activities. And everything that is good and pleasant for the physical organism in this condition is to be regarded as good and pleasant absolutely.

(2). *On the Moral Level*: Again, man attains the absolute best on the moral level, as we have seen, when through training and conscious choice his basic emotional powers — the opposing faculties of pleasure and pain — are brought to an habitual disposition of balance ἐν μεσότητι. His powers of emotion and action are then free from conflict, integrated, and poised, ready to respond proportionately as reason directs. In this condition of order and tranquillity the emotional powers, and the senses with which they are linked, are subordinated to, and best able to serve, the intellectual powers. Under the direction of his practical intellect man is then enabled to respond justly, bravely, temperately, etc. in external activity with his fellow men.

But the exercise of moral excellence and prudence in external activity is ordered by nature to the highest activity of man, the exercise of his speculative intellect according to the virtue of wisdom in contemplation. The man who arrives at this condition, where he exercises his highest faculty according to its distinctive excellence, regularily and over a long period of time, and with all his other faculties cooperating in proper subordination, — this man attains the perfection of human happiness, the absolute best possible to the individual. The whole man, with all his faculties fully developed, is then most completely integrated and self-sufficient, and his activity most simple, pleasant, and enduring, the noblest and best of which man is capable. And the objects or activities which are good and pleasant for such a man are absolutely so; for they are good and pleasant to human nature as such, whose perfection he embodies.

This state of perfect development and activity, however, while it represents the condition which human nature is constantly striving to produce, nevertheless seems to be realized in comparatively few instances, as we have seen. "It rarely happens that men achieve their final perfection" (1339b27-38). For, as in the case of health, it is possible only in the mature individual. It requires the best physical and intellectual equipment, developed through careful physical, moral, and intellectual training. It must be cultivated in peaceful surroundings, supported by a sufficiency of virtuous friends and the wealth necessary for leisure. These requirements put the absolute best beyond the reach of most human beings: the young and the immature; women, and those men whose physical and intellectual development have been retarded; fully endowed males who lack, however, the necessary emotional control, or moral and intellectual training; and those whose life is spent, through choice or necessity, in the external activities of war, politics, and business. All these are kept back, for one reason or another, from attaining the best condition possible to human nature, and can only approximate it, sharing in it to the limited extent of their abilities and opportunities.

(3). *On the Political Level*: Finally, human nature works to produce an absolute best on the level of the social organism, the polis. Just as in the physical organism nature secures unity and harmony by subordinating all other organs to the heart, the one organ best fitted by structure, position, and blend to control the whole body; and just as on the psychological level man's conscious activities are unified and harmonized when his lower faculties are properly subordinated and controlled by the faculty highest in nature and excellence, his intellect; so the political organism best achieves unity, harmony, and stability when all other parts willingly submit to be ruled permanently under one qualified by superior nature and virtue to direct the activities of the whole. Thus, for Aristotle, the most natural form of constitution, the best absolutely, is kingship.

If we follow out the implications of Aristotle's principles, it seems clear that under the king of surpassing excellence the polis enjoys the highest possible unity, stability, and justice. It enjoys the closest unity because there is one permanent source of authority and order. Being endowed with superior wisdom, the king (like the intellect and the heart) is best able to dispose and control the other members of the polis so that each is assigned the function for which he is fitted by nature and all cooperate harmoniously to the good of the whole. This imparts to the polis a natural organic unity (cf. *De Motu An.* 703a29-b2). Under such a king, moreover, the polis is most stable, because all members recognize his surpassing ability and virtue, regard him with admiration and good will, and submit gladly to his wisdom and justice (1270b21-24; 1284b28-34). Finally, absolute *justice* prevails in such a polis since (a) the ruler receives his just share: Honor and authority are conferred on the natural basis of intrinsic ability and virtue, rather than upon the basis of extrinsic goods (freedom, wealth, armed might). And these are conferred upon the individual of surpassing excellence in the proportion he deserves; surpassing all together in excellence, he is given honor and authority over all (1284a4 ff.; 1288a25-29). And (b) the ruled receive their just due under a monarch who himself embodies law and justice (1284a13-14). Being endowed with the greatest practical wisdom and the power to implement it, he is able (i) to settle conflicting claims between citizens wisely and with a higher degree of justice in cases which, because of exceptional circumstances, escape the necessarily general provisions of a written law (1282b1-6; 1286a8 ff.), and (ii) he can with greatest justice distribute honors and awards to each according to his contribution to the life of the body politic (*N.E.* 1130b30 ff.).

In this polis, then, where excellence is the constitutional as well as the natural basis of honor and authority, the absolutely good man is also the good citizen; virtue therefore is pursued and cultivated before all else in civic life (1288a33-b2); and the activities of the polis as such are products of harmonious cooperation between a king surpassing in the virtues of command and subjects prepared to carry out his orders with ability and virtue (1326b13-14). Such a polis must enjoy the greatest internal harmony and friendship based upon mutual esteem for excellence (cf. *N.E.* 1157b25-29). It will be self-contained and peace-loving (1325b14-28), not overly aggressive through desire to dominate others (1324b-1 ff.), yet most efficient in its external affairs through the unity, wisdom, and moral virtue of its leaders and subjects. Since its life, then, will be peaceful, provided with the resources for self-sufficient existence, and filled with activity according to virtue, such a polis will achieve the greatest possible share of happiness (1325b14-22).

If we look to the parallel between the polis and the individual, as Aristotle insists (1323b40-1324a2; 1325b14 ff.), it may not be too fanciful to suggest that the polis which realizes this best and most natural form completely is like the rare individual who attains the completeness of the intellectual and moral virtues. As the condition of such an individual is far superior to that of most men, who at best may develop a certain μεσότης of moral virtue, so the condition of such a polis is superior to that of most populations, which at best may develop the μεσότης of the mixed constitution. Accordingly, the degree of happiness possible under the mixed constitution would be surpassed by that enjoyed under the "first and most divine" constitution as the merely *human* happiness of morally virtuous activity is surpassed by the happiness of contemplation, which brings men closest to the divine (*N.E.* 1178a5-24).

(4). *On the Universal Level*: Aristotle's scale of value, of absolute and relative goodness, is not confined, however, to *human* nature on its various levels. Aristotle holds, of course, that of all creatures on earth man is undoubtedly superior to plants and to the other animals in his possession of reason, in his upright stature, in the finer blend of his body, etc. (1256b15-23; *De Part. An.* 656a1-

13; *De Gen. An.* 744a27-32). But man is by no means the highest and best in the universe, nor is the science of man the highest and best form of knowledge in the universal scheme (*N.E.* 1141a21-23). For he is still but a combination of soul and body, and his body, constituted of opposing material elements, is subject to decay and eventual dissolution. Throughout the discussion we have found that Aristotle regards each being as possessing greater goodness, excellence, or value the more it is in act rather than in potency, the less involved it is in matter and the closer it approaches pure form, the more its nature and activity are marked by order and symmetry, unity and simplicity, permanence and stability, by freedom from conflict and destructibility, by completeness and self-sufficiency. And looking about the universe, Aristotle discovers other beings enjoying these attributes in a degree unattainable by man at his best.

Thus the heavenly bodies, — stars and planets — surpass man in the unity, simplicity, and eternity of their nature and activity. They are composed not of the opposing material elements which form substances of the sublunary sphere, but of a single refined element, — αἰθήρ — "more divine than, and prior to, all these" (*De Caelo* 269a30-32). Consequently they are free from the possibility of internal conflict, subject neither to destruction nor to qualitative change, hence stable and everlasting (*ibid.* 270a13-b5). Moving locally in ordered regularity through their orbits, they enjoy an activity which is simple, tireless, eternal, and, apparently, conscious (cf. *ibid.* 285a29, 292a20ff.; *Met.* 1072a23-b4).

The activity of a god is immortality; but this is eternal life. Necessarily, therefore, what is divine must have eternal motion. And since the heaven is of this nature (i.e. is a kind of divine body), that is why it has a circular body, which by nature moves forever in a circle (*De Caelo* 286a9-13; and cf. *De Part. An.* 644b22-26).

By comparison, man's nature and activities are, at their absolute best, far inferior in unity, stability, permanence, self-sufficiency, etc.; for "chance and change are more the mark of things that die" (*De Part. An.* 651b20-21). Man may contemplate these attributes in the heavenly bodies (*ibid.* 644b32-35; *Met.* 1063a11-18), and strive to reproduce them in himself. But in the universal scale of values, human nature at its best can only imitate the superior goodness or excellence of the heavenly bodies, participating in their qualities as far as man's material aspect will allow.

For Aristotle, however, man and the heavenly bodies and all the material substances of the universe at their natural best can only reproduce more or less imperfectly the attributes of goodness as these appear in the ultimate absolute best, — the Unmoved Mover. Here is the supreme form without matter, act without potency, the reality utterly simple, unified, stable, unchanging, eternal, complete, self-sufficient, — the ultimate objective Good (*Met.* 1071b3-23; 1075a12-18).[104] This is the supreme Good which keeps all else in motion by the attraction of its goodness (*ibid.* 1072a26-b15), which the universe in all its varied and complex activities seeks and strives to imitate (cf. *De Gen. et Cor.* 336b27-337a8; *De Anima* 415a25-b8).

Aristotle repeatedly invokes the life and activity of this supreme Good as the ultimate model and standard of the best for man, — the happiness of virtuous activity perfected in contemplation (*N.E.* 1154b21-28, 1178b22-33; *Politics* 1323b21-26; 1325b24-30). But neither the individual nor the polis can share, in their own way, the happiness of this divine activity unless they first reproduce in themselves something of the unity, simplicity, order, and harmony of the supreme model. And in their compound nature individual and polis can only achieve this, first, through proper subordination of parts where there is inequality of nature, — by subordination of the other organs to the heart, the other faculties to the intellect, the other citizens to the king. While where equal and opposite constituents must be combined, the unity, stability, and harmony of the Unmoved Mover is reproduced, as far as possible, by proportionate blending to equilibrium, — ἐν μεσότητι.

[104] Cf. H. Cherniss, *Aristotle's Criticism of Plato and the Academy*, Vol. I (Baltimore, 1944), 361-364, 458-459; P. Aubenque, *Le Problème de l'etre chez Aristote* (Paris, 1962), 499-505.

V

CONCLUSION

The basic notion associated with Aristotle's term μεσότης has now been studied in the context of the medical writers, Plato's *Timaeus*, and a wide selection of Aristotle's writings. Since conclusions have already been formulated at the end of each central chapter, it will suffice to close with a brief recapitulation.

We may recall that several reasons were proposed for studying Aristotle's physiological theory so as to understand better the notion of μεσότης prominent in the *Nicomachean Ethics*. The suggestion of some earlier studies that the notion was drawn from the Greek medical writers, the recommendation by others of the "biological approach" to Aristotle, the frequent occurrence of physiological analogy in the *Ethics* and *Politics*, and particularly Aristotle's use of health as a prime analogue for moral excellence, — all kindled the hope that an investigation of Aristotle's physiological theory, gathered from his own writings in the context of contemporary medical literature, might supply some new insight into his ethical and political thought. At the same time, a survey of available Aristotelian scholarship revealed no previous study attempting just such an investigation.

Accordingly, we set out to study the background of Aristotle's physiological theory in earlier or contemporary medical literature. First, Alcmaeon's definition of health and disease was analyzed as a classic expression of these basic notions. His conception of health as a dynamic equilibrium of opposing constituents blended in proportion was found to be a characteristic doctrine of one group of medical authorities cited *nominatim* by the Menon papyrus. And this group was linked particularly to Aristotle by the fact that the papyrus quotes Plato first and most prominently among them. The deference accorded here to *Plato* as a medical authority confirmed the presumption that his influence was important in the formation of Aristotle's thought even in this area, and suggested that the *Timaeus*, quoted at length in the papyrus, deserved special treatment in a separate chapter. Since no primary sources survive for the study of the other authorities named with Plato, the investigation then turned to some typical elaborations of this basic theory in the anonymous works of the Hippocratic corpus.

Like the authorities cited by the Menon papyrus, the authors of four representative Hippocratic treatises were found to regard the organism as a complex of opposing elements. Health prevails when these elements are properly proportioned so as to balance or temper each other in equilibrium (μετριότης) and when their union is stabilized by mixture and blend (κρῆσις). This internal balance must be maintained through proportionate response to diet and environment; when it is disrupted, so that one element grows excessive and its opposite deficient, the pain and malfunction of disease ensue. To restore the equilibrium of health, the physician must know and reckon with a great variety of factors — diet, regimen, temperature, locale, seasons of the year, age, sex, etc. — so as to reduce the excessive element and restore the deficient in *this particular* individual. In all details of his treatment he must "aim at the mean", applying general principles to the individual case, and judging the effectiveness of his treatment by the subject's responses of pleasure and pain, which he compares to the reactions of the healthy organism, his absolute norm. All things are evaluated according as they maintain, or restore, the unity, stability, and self-sufficiency of the healthy organism. Hence the "good" is marked by measure, proportion, moderation, blending, whatever establishes a "mean condition" of equilibrium in the body; while all that tends to disrupt this (the disproportionate, excessive, deficient, sudden, or violent) are "bad".

It is significant that the medical writers, in discussing the factors affecting the organism as a whole, sometimes verge upon a cosmology, or probe into the psycho-physical basis of moral conduct and the influence of custom and law. Thus they tend to relate medical treatment through its underlying physiological theory to a total consistent view of man and the universe. This tendency Plato praises in Hippocrates (*Phaedrus* 270B-D). And he brings it to fulfillment in his *Timaeus*.

In studying the *Timaeus*, we found that Plato agrees with the medical writers in conceiving the organism basically as a proportionate blend of opposing elements poised in dynamic equilibrium. However, in applying this concept not merely to man's body but to the body and soul of the cosmos, the perfect reproduction of the ideal Living Being, Plato gives it a universal extension nowhere evident in the medical writers. The good and the beautiful in things that come to be are analyzed in terms of measure, proportion, and symmetry. For these join opposing extremes into new unities which operate harmoniously as an equilibrium of opposites. Thus the soul of the cosmos is a mixture of created and non-created Being, blended in proportion, divided according to fixed ratios to form the circles of the Same and the Different, and poised to revolve in perfect equilibrium. The body of the cosmos likewise is formed of opposites, fire and earth, joined by a double μεσότης through air and water, mixed and blended in proportion according to number and bulk and "power" so that none is excessive or deficient. In shape the cosmos is a perfect sphere, its larger components equally balanced about the earth as its center. Embracing all created Being and undisturbed by surrounding environment, it lives on endlessly in inner harmony, subject neither to sickness nor old age.

Since man, the microcosmos, is fashioned according to the same Model as the cosmos, the principle of symmetry, balance, equilibrium of opposites dominates, or should dominate, every aspect of his being also. This, then, is for Plato the characteristic of human excellence. For when all opposing constituents and parts are balanced and blended in dynamic equilibrium, each responds proportionately, performing its proper function in the whole. When his opposing constituents — earth, water, air, and fire — are properly proportioned and blended in each part, man's body grows symmetrically, responds properly to environment, resists dissolution, renews itself through nutrition, and enjoys health, strength, and beauty. When the elements of the sense organs are balanced properly, they respond proportionately and function well. When opposing emotions like fear and anger are moderated and balanced, they leave the mortal soul at peace. When all these elements of mortal soul and body are brought to order in balance and harmony, the circles of the intellectual soul are free to operate in equilibrium, giving knowledge and wisdom. When the three parts of the soul are proportionate, each functions properly, the intellectual controlling its opposite, the nutritive, through the mediation of the affective soul, resulting in the inner possession of justice, prudence, fortitude, and temperance. Finally, the proper balance between body and soul in the total human compound renders it in every respect καλὸς κ'αγαθός. Man is then in the best possible condition, imitating as far as his mortal nature will allow the perfect balance and symmetry of the cosmos.

Thus the basic physiological notion of opposites united by proportion in dynamic equilibrium becomes with Plato a universal principle in the formation of things after the likeness of the ideal Living Being. And the more perfect the proportion, blend, and equilibrium of opposites, the more excellent is the condition of the compound, the more closely does it reproduce the unity, stability, and eternity of the Model.

Against this background of Plato and the medical writers, we then took up the study of Aristotle's physiological theory. His notion of health and disease was found to be essentially that of his predecessors, revealing that he too conceives the organism as a compound of opposing constituents blended in proportion. Aristotle sees the proportion of earth, air, fire, and water in each particular organ as related to its function and fixed by the form. Therefore, as long as its proper proportion and blend are maintained, a part functions naturally and well, i.e. in health; but when its proportion is disrupted by excess and deficiency of one or other element, the organ cannot respond or function properly, so that sickness and dissolution result. Health, then, is that excellence of the body arising from a stable and abiding disposition (ἕξις) of its opposing constituents, such that, balancing each other in proper proportion, they blend to a mean (τὸ μέσον), forming a μεσότης, a middle state or mean condition. And within a certain range the μεσότης will differ in each particular organ, depending upon its nature and function in the whole.

Next, we saw how Aristotle applies the notion of μεσότης in describing other aspects of the organism at its best. Thus, the innate heat necessary to all vital functions must be tempered by cold to a middle state in which it is neither excessive nor deficient. This demands a constant balance between the processes of nutrition and refrigeration. And the heart, the center and source of heat, must be balanced by a brain of proportionate size as center of cold in the body. Again, the heart, the controlling organ of the body, the source and center of all vital activity, is located centrally between front and back, upper and lower, right and left, equally distant from the opposing parts of the body grouped symmetrically about it, forming with them a μεσότης, a balanced distribution of opposites joined in a μέσον. And in the largest and most perfect animals the heart is structured as a μεσότης, the large right chamber in which the blood is hottest and the small left chamber in which the blood is coldest having a single common source and center in the middle chamber where the blood is of intermediate quantity and quality, purest, most tranquil, and best. Similarly, in every detail the processes of nutrition, growth, and reproduction are governed by the principle of μεσότης: extremes proportionately moderated by their opposites produce the best condition and function of the organism.

Advancing then to the level of sensation, we saw how Aristotle solves the problem of perception by applying the notion of μεσότης to the particular and central sense powers. Because each of the particular senses is a kind of μεσότης informing appropriate matter, it is capable of responding proportionately to either extreme of its proper sense quality; and because the central sense power is in form a kind of μεσότης, it can register and unite in itself the responses of various particular senses simultaneously. Again, the heart is found to be, through structure and position, the proper organ of the central sense power. And response depends upon the internal disposition. For when its constituents are blended properly ἐν μεσότητι, the sense organ responds proportionately and perception is excellent; when the constituents lose their proper equilibrium, response is disproportionate and perception impaired. In this the *dynamism* implicit in the notion of μεσότης appears more clearly.

After studying the terms τὸ μέσον and ἡ μεσότης in Aristotle's physiology and sense psychology, we arrived at the central problem of their application in the *Ethics*. It was established first that moral excellence, the excellence of man's "affective soul" or emotional faculties, is explained in the *Ethics* through the analogy of health and strength; and Aristotle introduces the notions of συμμετρία and μεσότης as applying to both. The analogy of health and moral excellence was then followed through in detail. Analysis revealed that for Aristotle moral excellence, like health, is a certain ἕξις of the opposing "powers" (δυνάμεις) which constitute the moral organism (ἦθος). These are the "powers" or faculties by which man experiences pleasure and pain, the basic drives of pursuit and avoidance underlying all the various emotions leading to action. When these "powers" — like the hot, cold, moist, and dry in the body — are habitually balanced ἐν μεσότητι, the moral organism attains its proper state of excellence, ready to respond proportionately, i.e. to "hit the mean" in emotion and action, as right reason directs. Furthermore, the μεσότης of moral excellence, like that of health, is relative to the individual (πρὸς ἡμᾶς), depending upon age, sex, nature, and function in the community. Like health, the μεσότης of moral excellence gives unity, stability, and strength to the moral organism, and its exercise in virtuous activity becomes pleasurable. For Aristotle, then, the man of moral excellence (ὁ σπουδαῖος), like the healthy man, becomes the norm in that 1) he embodies nature at its best and 2) his reactions of pleasure and pain manifest what is good and bad for human nature as such.

To understand the μεσότης of moral excellence concretely and in relation to the total human organism, we turned next to the psycho-physical aspect of emotion and moral excellence. The psychic aspect of pleasure and pain was found to be produced by the μεσότης of the central sense power, while the physical increase of heat accompanying pleasure and the increase of cold coming with pain were traced to the faculty controlling the vital heat in the body. Hence, for Aristotle the center of emotion and movement is the heart, the organ of the central sense faculty and of the vital heat. This explains the close interdependence of the nutritive, sense, and emotional faculties, so that deviation from the proper μεσότης of one brings deviation and disorder in the others. Hence both disordered emotions and disease interfere with the operation of the central sense power in producing accurate images and phantasms, and lead to false judgments. The μεσότης of moral excellence is necessary, then, not only to keep the emotions poised ready to respond proportionately as reason directs, but also to maintain the inner tranquillity which the central sense power requires in serving the intellect.

With this established, we went on to explore the connection between moral excellence and the intellectual virtues. In general, since the intellect in its operations depends extrinsically upon the sense power, the intellect cannot operate unimpeded if unruly emotion disturbs the sense power. Hence the emotional faculties must be ordered in the μεσότης of moral excellence if sense and intellect are to operate properly. Thus, habitual emotional balance enables the *practical* intellect to choose the true good (τὸ καλόν) without the interference of antecedent emotion, which impels man toward the merely pleasurable (τὸ ἡδύ). Without interference of antecedent emotion, perception and judgment are true and the practical intellect indicates the proper means to the end, enjoying the virtue of prudence. Furthermore, the emotions poised ἐν μεσότητι respond proportionately as prudence directs, issuing in right action. Secondly, the emotional stability of moral excellence is necessary if the *speculative* intellect is to enjoy its characteristic virtue of wisdom. For the constituents of wisdom are intelligence and knowledge of the highest realities; and psychologically these are a *settling down* of the mind in the possession of its object. The phantasm presenting the object, then, must be steady and tranquil, produced by a sense power which operates in the stability of health and moral excellence, undisturbed by disease and disordered emotion. Thus the μεσότης of moral excellence contributes to man's happiness in that it orders and perfects his emotional powers, assisting his sense power and intellect to operate unimpeded in action and contemplation, so that he enjoys a unity, stability, and tranquil pleasure in the exercise of his highest faculties at their best. Man's condition and activity are then most like those of the Unmoved Mover.

After studying the application of μεσότης in Aristotle's physiology, psychology, and ethics, we turned finally to his political theory. Here, underlying the details of the *Politics*, we detected two natural principles of order basic to Aristotle's physiological theory. First, in any complex organism nature operates according to a principle of subordination in that lower parts and functions serve the higher for the good of the whole. In the political organism, then, the natural order is that one or more men of superior virtue rule, and that their inferiors obey them willingly, to the good of the whole. Thus monarchy and aristocracy are the most natural forms of polis, and monarchy the absolute best according to nature's rule of economy and efficiency. However, Aristotle considers these natural forms inapplicable to most existing populations because a single leader, or group, of outstanding virtue cannot be found among them. Instead, existing populations are generally split into two opposing factions of free citizens, the numerous poor and the fewer rich, each contributing something equally necessary for the polis, the one its numbers, the other its wealth. The only natural order possible for most existing populations, then, is that which nature uses elsewhere in uniting equal but opposing parts, namely, the order of μεσότης.

This second principle is embodied in *polity*, a form of government which mixes and blends the opposing extremes of democracy and oligarchy. It can be realized in a population where the extremes of rich and poor are balanced and blended in a large middle class. Hence it is the natural form to which most existing populations may aspire. *De facto*, however, many populations are so imperfect that they can sustain only some unnatural or diseased form like democracy, oligarchy, or even tyranny. Each of these violates in progressively greater degree the natural order of subordination and of μεσότης and consequently is marked by greater injustice, discontent, civil strife, and instability. However, one of these deviation forms, though bad by absolute standards, may be the best a particular deficient population can attain. It then becomes the *relative* best, the "best in relation to actual conditions".

Finally, we considered the parallel function of the physician, moral guide, and statesman, their objectives and norms of value. We saw that Aristotle expects these experts to know and, where possible, to produce the absolutely best state of the physical, moral, or political organism. However, in each case this requires the best material, which is rare. The expert will generally have to deal with inferior material, and must bring it as close to the absolutely best condition as its limitations will allow. He must begin with individuals, producing in them the best they can attain, the relative best, though this may fall far short of the absolute best. In this enterprise the reactions of pleasure and pain afford the expert some norms for judging the subjective state of the individual person or polis. But in addition he needs an absolute set of norms to evaluate the condition of the individual subject and bring it to a better. This Aristotle provides, proposing as the absolute best in physical condition the μεσότης of constituents in the healthy male organism at its prime; in moral disposition, the μεσότης of his emotional faculties which enables the virtuous man to live by reason in action and contemplation; in political organization, that monarchy in which the citizens dwell together in unity, stability, and justice under a king of surpassing excellence. For in these conditions the physical, moral, and political life of men most closely approximates that of the universal absolute Good, the Unmoved Mover.

In conclusion, then, we have found that when Aristotle applies the term μεσότης to the form, condition, or disposition of the living organism, it implies a dynamic equilibrium of opposing factors blended proportionately in a mean. This condition of the compound is "best" in that it brings unity, internal harmony, stability, and the capacity to respond proportionately. *Absolutely* "best" is that state of equilibrium maintained by the organism when perfectly formed and operating under optimum conditions. *Relatively* "best" is that state of equilibrium, approximating the absolutely best, which is possible here and now to a particular organism with less than perfect internal composition and external conditions. Aristotle employs this principle of μεσότης in explaining the formation and operation of the organism on the physiological, psychological, ethical, and political levels, so that it becomes one of his most important fundamental notions. If we must trace the antecedents of the notion to any one level or area, the evidence from Plato and the medical writers, as well as Aristotle's own analogies, suggest that it grew out of a *physiological* context.

APPENDIX I

ON THE NOTION OF ΜΕΣΌΤΗΣ IN GREEK MATHEMATICS

In the Greek mathematical writers the term μεσότης is ambiguous. It signifies, first, the *middle* or *middle space* in general, as in the dictum ascribed to the Pythagorean Occelus (Diels-Kranz 48, 8): ἡ τριὰς πρώτη συνέστησεν ἀρχήν, μεσότητα καὶ τελευτήν (and cf. Aristotle, *De Caelo* 268a10, cited by Diels-Kranz 58, 17). Then, more specifically, μεσότης can signify a *mean* in the commonly accepted sense of that term in English, i.e. the middle term in a continuous three-term progression, the mid-point between extremes. Thus, in English we say: "If three numbers are members of an arithmetic progression, the middle one is the arithmetic mean" (*Van Nostrand's Scientific Encyclopedia*, 3rd ed. [Princeton, N. J., 1958], p. 1310, s.v. "Progression"). Likewise the Greek mathematician can say that eight is the μεσότης between six and twelve in an harmonic progression (Diels-Kranz 44A24, line 32). This is obviously the sense in which Plato uses the term in *Timaeus* 36A. However, the notion of *mean* or *middle term* is more often expressed by the Greek terms τὸ μέσον and ὁ μέσος ὅρος (e.g. Diels-Kranz 47B2 [I, 436, lines 10-11] and 58, 2 lines 28-30). Thus Plato uses τὸ μέσον for the *mean* in describing the nature of proportion (ἀναλογία) at *Timaeus* 31C-32A.

The second, and perhaps more common, significance of μεσότης is "a *proportion*" or "*progression*". Burnet remarks that "the Greek word does not mean only or even primarily the arithmetical mean: it is the oldest word for a proportion of any kind and however determined" (*The Ethics of Aristotle* [London, 1900], pp. 69-70). In this sense μεσότης becomes closely identified with ἀναλογία, though the latter term expressed primarily the notion of a *geomet-*

ric proportion, as Iamblichus mentions in his commentary on Nicomachus' *Introduction to Arithmetic* (ed. H. Pistelli [Leipzig, 1894], p. 100, lines 15-18): "One should mention first that the ancients applied the term ἀναλογία primarily to the geometric, but now it is applied more broadly to all the other μεσότητες generically". Modern commentators, following the lead of G. H. F. Nesselmann (*Geschichte der Algebra*, Vol. I [Berlin, 1842], 210-212) accept this as historically accurate. Cf. J. Gow, *A Short History of Greek Mathematics* (Cambridge, Eng., 1884; reprint N.Y., 1923), p. 93, n. 1; T. L. Heath, *The Thirteen Books of Euclid's Elements*, 2nd. ed. revised (N.Y., 1956), II, 292-293. Thus M. L. D'Ooge, in a note on Nicomachus' *Introduction to Arithmetic* II, 21.2 (cited according to the edition of R. Hoche [Leipzig, 1866]), remarks that:

> Two Greek words, ἀναλογία and μεσότης ... may be translated "proportion", and Nicomachus points out that an ἀναλογία is, strictly speaking, a combination of ratios (like 1, 2, 4, i.e. in his classification only the "geometrical proportions", γεωμετρικαὶ ἀναλογίαι or μεσότητες). Properly then an arithmetical progression of three or four terms (e.g. 1, 2, 3, or 1, 2, 3, 4) should not be called an ἀναλογία, but in practice it is so called. Nesselmann states that originally ἀναλογία was applied only to geometrical proportions and μεσότης to the other two, the harmonic and arithmetic, but that in later usage the distinction of terms vanished. It is certain that Nicomachus uses them indiscriminately of all three types ... (*Nicomachus of Gerasa: Introduction to Arithmetic* [N.Y., 1926], p. 264, n. 2).

This close identification of μεσότης and ἀναλογία must have taken place early, for according to Theon of Smyrna (*Expositio rerum mathematicarum ad legendum Platonem utilium*, ed. E. Hiller [Leipzig, 1878], p. 106, lines 15-20) the Peripatetic Adrastus at the beginning of the second century B.C. found it necessary to explain (as Iamblichus did later) that ἀναλογία properly refers to geometric proportion, though "more broadly the other μεσότητες are also called ἀναλογίαι by some people". Cf. Heath, *op. cit.*, II, 292. The first three types of μεσότης — the arithmetic, geometric, and harmonic — are commonly traced back to Pythagoras himself, though Archytas and Hippasus are credited with changing the name of the last from "sub-contrary" to "harmonic" (Iamblichus, ed. Pistelli, p. 100, lines 15-25); several other types of μεσότης were known to Plato and his contemporaries (*ibid.*, pp. 101, lines 1-5; 116, 1-4; 118, 19 ff.); and the number of μεσότητες was finally brought to ten by two later Pythagoreans, Myonides and Euphranor (*ibid.*, 116, 4-6).

It is clear, then, that as practically synonymous with ἀναλογία, μεσότης can only signify a "*proportion*" (i.e. a three-term progression) and not a "*mean*"; and when it has this sense of a three-term progression or "proportion", μεσότης is distinguished from "the mean" as the whole from the part. In fact Nicomachus calls attention to this distinction when he remarks (*Intro. to Arith.* II. 27.1; see D'Ooge, p. 278, n. 2) that μεσότης (in the sense of a three-term progression) is well named because each type of μεσότης is determined by the specific nature of its μέσος ὅρος or mean, not by its extremes (ἄκρα). And Philoponus, commenting on this remark (ed. Hoche [Berlin, 1867], Book II, p. 32, XXVII, 1, lines 5-7), explains clearly: "And for this reason the *whole relationship of the three terms* (ἡ ὅλη τῶν γ̄ ὅρων σχέσις) *is called* μεσότης, from this fact that the μέσος [ὅρος] (mean or middle term) has the determining influence over the whole relationship of the terms."

Since μεσότης is ambiguous, then, its significance in each case must be gathered from the particular context, so that it cannot be translated uncritically as "the mean" at every occurrence. For μεσότης may signify in mathematics the whole relationship of two extremes joined by a mean, i.e. a "proportion". And applied analogously to physical realities, μεσότης will obviously suggest a disposition in which extremes of any kind are balanced in a mean, i.e. a state of equilibrium among opposing factors.

APPENDIX II

CONCERNING THE THEORY OF FRANCOIS NUYENS ON THE DEVELOPMENT OF ARISTOTLE'S PSYCHOLOGY

(*L'Evolution de la psychologie d'Aristote*, Louvain, 1948)

That Aristotle conceives the nutritive-sensitive soul as the form or "first activation" of the living body while *at the same time* regarding it as centered in the heart is a position which will not be acceptable, I realize, to those who agree with Francois Nuyens that "l'idée que l'âme siège dans une partie bien définie du corps est absolument incompatible avec le concept de l'ἐντελέχεια" (p. 165). This incompatibility is assumed throughout Nuyen's study (cf. pp. 45, 47, 55-58, 119, 159-171, 173-180, etc.) and serves as his chief norm for distinguishing the works of Aristotle's "transitional period" (the ethico-political treatises and the bulk of the biological works) from those of the "terminal period" of hylemorphism (*De Anima*, *De Partibus I*, *De Generatione Animalium*, the first six of the *Parva Naturalia*, etc.). It would be out of place here to attempt an extended critique of Nuyen's study, but the acceptance of his hypothesis by scholars of eminent rank (e.g. W. D. Ross, *Aristotle: Parva Naturalia*, [Oxford, 1955], p. 3 ff.) demands some justification for disregarding it in the present context.

Nuyens' main criterion for distinguishing the works of the "transitional period" from those of the "terminal period" is, as we have said, the conviction that "il y a contradiction entre cette idée d'une union de l'âme avec un organe déterminé et la conception de l'âme entéléchie du *De Anima* suivant laquelle le corps tout entier doit être considéré comme le substrat de l'âme" (p. 45). Other commentators also have considered these notions incompatible or contradictory, as Nuyens points out (pp. 44-47, 54-56).

However, before denying the authenticity of any work or distinguishing different periods of development on this basis, it seems that, as historians of Aristotle's thought, we should first ask whether Aristotle himself found these notions contradictory or incompatible. I do not find in Nuyens' study convincing evidence that he did. Rather, one general criticism of Nuyens' approach might be that it tends to read back into *Aristotle's* mind what appears as contradictory to the commentator. "Au sujet du problème des rapports entre l'âme et la corps, on se trouve placé en face de déclarations qu'il est impossible d'accorder entre elles; cela, pour la raison très simple que pour Aristotle lui-même elles sont incompatibles" (pp. 164-165, note 49). Whether (1) the commentator finds two notions irreconcilable, and whether (2) *Aristotle* regards them as such, — these appear to me to be separate problems. Only the latter is of primary importance to the historian, and should be clearly established independently of the former, on the basis of evidence that Aristotle explicitly rejects one notion in favor of the other, or at least that the two notions never occur simultaneously in the same context or work. It is just this kind of evidence that I miss in Nuyens' discussion.

On the other hand, there seems to me to be evidence that Aristotle does speak of the soul in the *same* context or work as *both* the form of the body *and* as centered in a particular organ, the heart, governing the life processes through the innate heat. A few instances must suffice:

1. *Metaphysics* 1035b14ff. is cited by Nuyens (pp. 176-177), and quite rightly, as illustrating the *hylemorphic* concept of the body-soul relationship, and he uses the passage as evidence (*contra* Werner Jaeger) that this book of the *Metaphysics* must be at least contemporary with, if not later than, the *De Anima*. The passage treats the relation of parts to the whole, introducing without previous explanation the statement that "the soul of animals, for this is the substance (οὐσία) of a living being, is their substance according to the formula (ἡ κατὰ τὸν λόγον οὐσία), i.e. the form and the essence of a body of a certain kind ... " (1035b14-16). Aristotle explains that the soul and its parts, as essential substance, are *prior* to the animal and to the body and its parts, though in a sense they are *not*, since they cannot exist apart from the concrete whole,

the composite of body and soul. Then, speaking of the relationship of parts of the *composite* to the whole, he goes on to remark, again without qualification: "Some parts are neither prior nor posterior to the whole, i.e. *those which are dominant* (κύρια) and *in which the formula*, i.e. *the essential substance* (ὁ λόγος καὶ ἡ οὐσία) *is immediately present* (ἐν ᾧ πρώτῳ), e.g. the *heart*, or perhaps the *brain*, — for it does not matter which of the two is of such a nature" (1035b25-28). Thus, immediately after stating as evident the notion that the soul is the substance, form, or essence of the animal Aristotle assumes as equally evident that there is a *dominant* part of the animal body *in which* this exists *primarily*, though he does not care to argue at present whether the organ be the heart or, as Plato and others held, the brain. (On this last, see the remarks of R. D. Hicks in his commentary on the *De Anima*, p. 274 *ad* 408b9. Cf. *De Part. An.* 667b22ff. where the *heart* is said to be the part [μόριον] possessing the sensitive soul *primarily* [πρώτως]).

This is brought out clearly in Alexander's comment (*ad Met.* 1035b25ff., *Commentaria in Aristotelem Graeca* I, 508, lines 26-35):

> And when certain of the parts are separated from it, the whole can exist without them; but certain other parts on being separated destroy the whole at the same time, i.e. such parts as are dominant (κύρια) and in which first of all the λόγος and οὐσία exist, that is, the soul; for the soul of living things is their form (εἶδος). Therefore, whether the *heart* is the source and center (ἀρχή) of animals and in this the soul exists primarily (as Aristotle thinks) or whether it be the brain (as Hippocrates and Plato think), these parts simultaneously with their separation from it destroy the whole. But in saying "... the heart, or perhaps the brain ..." Aristotle adds, "for it does not matter which of the two is of such a nature", that is, whether it is the heart in which the soul happens to exist or whether it is the brain makes no difference to the argument ...

Here, then, Aristotle espouses in the same context the two aspects of soul which Nuyens considers "absolument incompatible". Apparently, therefore, Aristotle was capable of entertaining both notions simultaneously, and if they are contradictory he does not seem to be aware of it. Nuyens offers no comment on *Metaphysics* 1035b25ff.

2. In the *De Anima*, as Nuyens rightly maintains, the notion of soul as form and "entelechy" of the organism is most explicitly and fully developed. Yet even in this work there are clear indications that Aristotle still considers the heart as the center and source of the vital functions, as we have seen it described in the biological works. In the *De Anima* he is not, of course, primarily concerned with the *material* aspect of the organism, i.e. with the heart or any other particular organ. The references to the heart are consequently few, but perhaps not quite so insignificant as Nuyens (p. 247, note 77) and Ross (pp. 7, 8, 12) suggest. For example:

A. Nuyens (p. 159-160, 165-167) and, later, Ross (p. 5-13) assign the *De Partibus Animalium* and the last of the *Parva Naturalia* to Aristotle's "transitional period" on the principle that these works portray the soul as "closely associated with heat and with the hottest organ of the body, the heart" (Ross, p. 6), that is, the soul uses heat as a tool or instrument in nutrition and in imparting motion (cf. *De Part. An.* 652b7-16), both processes being centered in the heart.

However, turning to Book II of *De Anima* (which Ross considers, — p. 17 — "in all probability the latest of Aristotle's writings on psychology") we find Aristotle denying that "the nature of fire" is the sole cause of nutrition and growth, but asserting that: "A co-cause (συναίτιον) in a sense it *certainly* is; but not a cause *absolutely* (ἁπλῶς αἴτιον); *that* is rather the *soul*" (416a13-15). A little later (416b20 ff.) Aristotle explains that in the nutritive process a) it is the *body* that is nourished, b) the nutritive *soul* that nourishes, c) while *that by which* (ᾧ) the body is nourished must be understood to include (1) the *food*, which must be concocted (i.e. changed without changing another, like a ship's rudder) and (2) *the vital heat*, which effects concoction (i.e. is moved by the soul, and itself changes the food, like the helmsman's hand on the rudder, — implying that the soul, like the helmsman, effects change while remaining unchanged: cf. Hicks, pp. 348-349 *ad* 416b26 and 27). Clearly, then, in the *De Anima* as in the biological works, the soul is closely associated with the vital heat, using it as its co-cause or instrument to effect nutrition and growth.

Furthermore, to complete the parallel with the biological works, in *De Anima* II Aristotle evidently implies that the heart is the *center* of this vital heat, i.e. "the hottest organ of the body", when he explains that nature uses the breath as a necessary means of regulating the *internal heat* (420b20-21), and that the area which needs cooling *before all others* (πρῶτος) is the one surrounding the heart (420b25-26).

Therefore the *De Anima* contains *both* the hylemorphic concept of soul *and* the notion that soul operates by means of the vital heat centered in the heart. And the same may be said of the *De Generatione Animalium*, which Nuyens also considers a work of the "period of hylemorphism" (pp. 256-263). Cf. *De Gen. An.* 740a2-23, 740b26 ff.

B. Again Ross, following Nuyens, contends that the biological writings of Aristotle's "mechanistic period" locate the seat of *sensation* and of *movement* in the heart, while "in the *De Anima* no such significance is attributed to the heart" (p. 12). However, examining the references to the heart in the *De Anima* (403a29-b1, 408b1-9, 432b33) we find it invariably singled out as the specific organ in which psychic states are immediately manifested in the body. It is the *heart* which is said to move with the basic drives of anger and fear, and even with the effort of deliberation (διανοεῖσθαι).

The passage at 408b1-19 is especially interesting. Aristotle here emphasizes the *bodily aspect* of psychic states, the aspect which, he says, is often overlooked when we speak of the *soul* as being pained, pleased, bold, fearful, etc. Each of these states is a movement (which implies matter) having its origin in the soul. "For example, to be *angry* or to *fear* is for *the heart to be changed* (moved) in this or that way, and to *reason* (διανοεῖσθαι) involves *this organ* or some other." The nature and manner of the change does not concern him here. (*Sic* 408b1-11). The point is, Aristotle continues, that to say the *soul* is moved, e.g. feels anger, is like saying it weaves or builds. Hence it is better perhaps not to speak of the soul as feeling angry, but rather to say that the composite, the *man*, does so, understanding that the change involved in sensation and emotion does not occur in the soul, but reaches to or starts from the soul. (*Ibid.* 11-19).

The latter half of the passage (lines 11-19) is employed often (17 times) by Nuyens as illustrative of the hylemorphic concept, and properly so. But it would be helpful to have his comment on the first half also (lines 1-11), in which the *heart* is singled out as the organ immediately involved in the movements of anger, fear, etc. which are said to have their origin in the soul. For Aristotle's reference to the heart here, as in the other passages cited above, seems to imply that even in the *De Anima* he regards it as the center of sensation and emotion or movement.

Finally, unless this is granted, it is difficult to see how we can explain the cryptic passage (433b19-27) where Aristotle speaks of the "bodily part" which the appetitive faculty uses as its instrument to move the animal. This part is found "at the point where beginning and end coincide". "For everything is moved by pushing and pulling, so that something must remain fixed, like the center point of the moving circle, from which motion has its origin." (For the mechanics involved here, cf. Pseudo-A. *Mechanica* 847b-16ff.). This becomes intelligible only by relating it to Aristotle's remarks elsewhere on the heart as the center from which the soul moves the body (e.g. *De Part. An.* 647a25-34, 665a10ff., b18ff., 666b14-17, 667a4-7; *De Motu An.* 698a14ff.). Thus Hicks, p. 565 *ad* 433b26, and the Greek commentators cited by him there. It is surprising to find Ross himself agreeing with this in his own commentary on the *De Anima* (*Aristotle: De Anima*, Oxford, 1961, p. 317), identifying the unnamed source of movement as the heart, "which is for A. the central seat of life".

Thus the references to the heart in the *De Anima*, though they are few and incidental, nevertheless reveal clearly enough that Aristotle even in this work regards the heart as the center of nutrition and sensation, the source of emotion and movement in the rest of the organism, the organ in which the vital heat is centered. And the vital heat, operating in both nutrition and emotion (cf. 403a33), is a co-cause with the soul, used by the soul as its instrument in these life functions.

In the *De Generatione Animalium*, as we have indicated above, the soul is regarded as operating through the vital heat centered in the heart. Accordingly the heart is singled out in that work also as the source and center from which soul principally operates in nutrition, growth, sensation, and motion. Cf. 735a1-26; 742a33-b3; 742b33-743a1; 743b26-30; 744b2-6, 12-14, etc. Yet Nuyens quite rightly points out (p. 257) that the *De Generatione Animalium* contains numerous passages expressing the substantial unity of soul and body in the same terms as the *De Anima*.

This must suffice at present to indicate that Aristotle could entertain *simultaneously* in the same context or work the notions of soul which Nuyens regards as contradictory. In view of this fact, I do no think we can safely say that *Aristotle* regarded them as contradictory also; rather the opposite is suggested. So that, despite the difficulties of understanding *how* he reconciled these notions, it seems to me best not to separate the *De Anima* and *De Generatione Animalium* from the rest of the biological works on this basis.[1]

[1] Views similarly critical of Nuyens' position have appeared in print since this appendix was first written in 1958. See I. Block, "The Order of Aristotle's Psychological Writings", *AJP*, LXXII (1961), 50-77, especially 50-61; J. J. Walsh, *Aristotle's Conception of Moral Weakness* (New York, 1963), 66-83; W. F. R. Hardie, "Aristotle's Treatment of the Relation Between the Soul and the Body", *Philosophical Quarterly* XIV (1964), 53-72. W. Theiler, *Aristoteles: Über die Seele* (Berlin, 1959), 73-77, indicates some of the positive reasons for regarding the *De Anima* and *Parva Naturalia* as closely integrated, and perhaps contemporaneous in composition, with the other biological works.

APPENDIX III

THE HEART AS SOURCE AND CENTER OF EXTERNAL ACTION

We have seen that the material aspect of emotion is primarily a *change in the normal balance of hot and cold in the heart*, the pleasurable reaction increasing the heat, the painful the cold. The changes induced in the heart by this increase of heat or cold are described in detail at the end of the *Parva Naturalia* (479b17 ff.). The first, *palpitation* (πήδησις), is a *contraction* caused by an increase of *cold* in the heart. This may be due to disease or to *fear*. For "when one is afraid the upper parts become cold and the hot substance, fleeing away, by its concentration in the heart produces palpitation" (479b22-25; and cf. *De Part. An.* 669a18 ff.). Thus palpitation, associated with fear and the *painful* emotions, is "a contraction of the heart related to the compression of cold" (480a13-14). The second, and opposite, phenomenon is *pulsation* (σφυγμός, σφύξις), associated with an increase of *heat*. This is a throbbing or expansion, compared by Aristotle to *boiling*; "for boiling occurs when the *moist* is *vaporized* (πνευματουμένου τοῦ ὑγροῦ) by the *hot*" and expands as its mass increases (479b31-33). The normal heat of the heart, in constantly vaporizing the moisture from nutriment, causes the regular throb of heart and veins (480a-3 ff.). We may infer that the *increased* θερμότης of the pleasurable emotions *speeds up* this boiling by which the moist is converted into vapor or "air" (πνεῦμα: *De Gen. et Cor.* 330b5), thereby enlarging the heart and the blood vessels with its expansion. Hence *pulsation*, associated with anger and the *pleasurable* emotions, is "the vaporization of the moist when heated" (480a14-15), resulting in the expansion of the heart. Thus an increase of heat or cold moves the heart locally in one direction or the other, causing it to expand and push out or contract and pull in. The general physical principle connecting thermal with local changes is given by Aristotle as follows (*Physics* 260b8-13):

Again all affections have their origin in condensation and rarefaction; thus heavy and light, hard and soft, *hot and cold*, seem to be types of *density* and *rarity*. But condensation and rarefaction are nothing more than *combination* and *separation* ... and in being combined and separated things must *change in respect of place* (Oxford translation).

The beginning of action, briefly, is this: In the heart, the center of sense perception and of motion, an object of thought or sensation is presented by the phantasm as good or bad, helpful or harmful; this evokes the reaction of pleasure or pain, materially manifest as an increase of hot or cold in the heart; this brings an expansion or contraction of the central organ which, by pushing (in expanding) or pulling (in contracting), inaugurates the motions of pursuit or avoidance in the rest of the body.

Through several distinctive features of structure, function, and position the heart is able to communicate motion to the limbs and other organs. We have seen that the heart is unique in being at the same time *homoeomerous* and *non-homoeomerous* or structurally differentiated, "the former in virtue of receiving all that is perceived by the senses, the latter because it has to do with *motion* and *action*" (*De Part. An.* 647a25-30). As source of motion and action "the heart is abundantly supplied with *sinews* ... For the motions of the body commence from the heart, and are brought about by *contraction* (ἕλκειν) and *relaxation* (ἀνιέναι). The heart therefore ... requires some such subservient and strengthening parts" (*ibid.* 666b14-18). Thus, as the heart expands with heat or contracts with cold it pushes or pulls, relaxes or tightens, and motion is communicated mechanically to the rest of the body through the sinews, whose point of *origin* is the heart, whence they extend to the joints and flexures of the bones (*Hist. An.* 515a27-b7).

But the heart communicates motion to the rest of the body not only mechanically but also through a system of *thermodynamics*. For the heart's structure and function as center of *vital heat* and of the nutritive-vascular system enables it to convey *temperature* changes to all parts of the body. As the increasing or decreasing heat associated with emotion causes the vapor (πνεῦμα) *within* the heart to expand or contract, to become tense or to relax, so apparently an increase or decrease of heat communicated *from* the heart operates analogously in the other organs, and in the organism as a whole. Of particular importance here is Aristotle's notion that the heart causes and controls the movement of the *lungs*, and thereby controls the *breath* (πνεῦμα) whose chief function is to cool the entire organism (*De Part. An.* 669a14 ff.; *Parva Nat.* 456a4 ff., 478a12 ff.).[1] For when the heart arrests the breathing process, the vapor or air contained within the organism quickly becomes hot (*Parva Nat.* 478b16-18; and cf. 472b2-3), so that it expands and provides strength or tension throughout the body. "It is impossible to make any movement or perform any action without strength, and the holding of the breath (πνεῦμα) produces strength, — breath from outside in the case of animals that respire, and in those that do not, the innate breath" (*ibid.* 456a16-18; and cf. *Politics* 1336a38-39). It is just this expanding action of heat from within that gives man, of all animals, the power to hold himself upright (*De Part. An.* 669b4-7; 653a30 ff.). And when the vital heat is cooled he sinks down in sleep (*Parva Nat.* 457b24 ff.). Accordingly we may infer, in general, that the heat accompanying pleasurable emotion moves the body to expansive, vigorous, positive action, while the cold of painful emotion causes it to contract, becoming heavy and inert. Thus the movement of the body and its parts is explained also through the expansion or contraction of air or breath (πνεῦμα)[2] due to the increase or decrease of the

[1] This makes it possible also for the heart to produce voice and language i.e. sound made meaningful by conforming to a phantasm. Cf. *De Anima* 420a 20ff., 420b 5ff. (and Hicks' note *ad* 420b 28, p. 389); *De Gen. An.* 776b 9-18, 786b 7-788a16.

[2] Breath (πνεῦμα) or air is well adapted to serve as the medium by which changes of temperature are conveyed from the heart to the rest of the body and converted into local movement through expansion and contraction. For being intermediate between the heat of fire and the cold of water, air or vapor is both most fluid and most capable of being heated or cooled (cf. *De Gen. et Cor.* 330b 3-6, 331a 4-6, 16-20). "Since that which is most fluid boils up and cools down most easily" (*De Part. An.* 658b 9-10). And being of a much more rarified nature than *water* air can pass through the parts quickly (*Parva Nat.*

vital heat centered in the heart. This is a further reason why Aristotle holds that "the hot is of all material substances the most suitable for serving the functions of the soul. For nutrition and *the imparting of motion* are the function of soul, and it is by heat that these are most readily effected" (*De Part. An.* 652b11-13).

Finally, the position of the heart at the *center* of the organism helps Aristotle to explain how a very *slight* change in the heart can result in the ample movements of the outer limbs. The mechanics of motion involved are indicated in the *De Anima* where, discussing the origin of animal movement, Aristotle dismisses the question of the physical organ thus (433b22-27):

To state the matter summarily at present, that which is the instrument in the production of movement is to be found *where a beginning and end coincide* ... separate in definition but not separate spatially. For everything is moved by *pushing* (ὦσις) and *pulling* (ἕλξις). Hence just as in a *circle* so here there must be a point which *remains at rest*, and from that point movement must originate. (Adapted from the Oxford translation.)

That Aristotle has the *heart* in mind here seems clear from the following: The heart, as we have seen, being central in the organism, marks off right and left, above and below, before and behind (*Parva Nat.* 467b28 ff., *De Part. An.* 669b18-22). Recalling the principle that the mid-point (μέσον) is spatially one but logically multiple (cf. *Physics* 262a19-20), we see that the heart above all other organs occupies the position in the body where "beginning and end coincide", being at one and the same time the beginning and end of opposing halves of the body. Moreover, the heart is said to be the physical source and center (ἀρχή) of the organism, roughly equidistant from the extremities, and the point where all movements involving sense and emotion begin and end (*De Part. An.* 666a11-17). Finally, the blood in the *central* chamber of the heart is said to be intermediate in quantity and heat because "the source and center should be *as calm as possible*" (*ibid.* 667a4). Evidently, then, Aristotle detects the perfect geometrical figure,

---

478a18-21). The mechanical and thermodynamic principles suggested above are applied in detail in the *De Motu Animalium*, a work whose authenticity I find no serious reason to doubt.

the *circle* or *sphere*, embodied in the organism as the basis of its structure and dynamics. And the center is fixed in the heart, and in the central chamber of the heart. "Hence just as in a *circle* so here there must be a point which *remains at rest*, and *from that point* movement must originate."

Now Aristotle seems to have had a system of mechanics based upon the circle or sphere, applying, therefore, to the organism with its circular or spherical structure. The principles of his circular or spherical mechanics are found applied to the *universe* and its motion in the *Physics* (e.g. 265a14-b15, 267a22-b8; and cf. Ross' notes *ad loc.*) and in the *De Caelo* (e.g. 286a3-32, 286b10 ff.). But perhaps the clearest commentary appears in the pseudo-Aristotelian work on *Mechanics*, which deals specifically with the problem of "how a great weight can be moved with but little force" (847b1 ff.). Obviously this is but the abstract form of the question: How can the movements of the outer limbs result from a slight change in the heart? The solution is found in the principle of the lever, and that is ultimately explained through the *balance* of a diameter upon the fixed *mid-point* of a circle or sphere (848a11-16):

There is nothing strange in the circle being the origin of any and every marvel. The phenomena observed in the balance can be referred to the circle, and those observed in the lever to the balance; while practically all the other phenomena of mechanical motion are connected with the lever. (Loeb Classical Library translation.)

Briefly, the fixed center of the symmetrical body is conceived as the fulcrum, the unmoving point from which opposite motions arise. And as the extremities of the balance arm move in a greater arc than do points closer to the center, so a small change at the center of the symmetrical organism initiates motion which becomes quite ample in the extremities. These mechanical principles are assumed in the *De Motu Animalium*, which comments (701b25-33):

And it is not hard to see that a small change occurring at the center makes great and numerous changes at the circumference, just as by shifting the rudder a hair's breadth you get a wide deviation at the prow. And further, when by reason of heat or cold or some kindred affection a change is set up in the region of the heart, — even in an imperceptibly small part of the heart — it produces a vast difference in the periphery of the body ... (Oxford translation.)

This must suffice to indicate the broad lines of Aristotle's solution. It is clear that, in order to explain how external action is produced by a slight mechanical or thermal change in the heart, Aristotle relies once more upon the concept of μεσότης, — a symmetrical disposition centered in a point — expressed most perfectly in the circle or sphere. He sees the basic structure of the organism as an embodied sphere centered in the heart, and the heart itself as an embodied μεσότης formed about its central chamber. This chamber, where the blood is preserved pure, moderate in temperature and quantity, and calm, is therefore most perfectly adapted to serve as the source and center of the organism, in which images are converted, through heat and cold reactions of pleasure and pain, into movements of the entire body.

BIBLIOGRAPHY

(The bibliography is not intended to be comprehensive but represents only works used in the preparation of this study and referred to in the text or footnotes).

I. ANCIENT WRITERS: EDITIONS, TRANSLATIONS, COLLECTIONS

A. *Medical Writers*

Diels, H., *Anonymi Londinensis ex Aristotelis Iatricis Menoniis et aliis medicis eclogae, Supplementum Aristotelicum*, III, Pt. I (Berlin, 1893).

Festugière, A. J., *Hippocrate, l'Ancienne Médecine*, Introduction, traduction, et commentaire (Paris, 1948).

Jones, W. H. S., *Hippocrates.* With an English Translation. 4 vols. Loeb Classical Library (London and Cambridge, Mass.), (Vols. I-II) 1923, (Vol. IV) 1931.

——, *The Medical Writings of Anonymous Londinensis* (Cambridge, Engl., 1947).

——, *Philosophy and Medicine in Ancient Greece*, With an Edition of περὶ ἀρχαίης ἰητρικῆς(Baltimore, 1946).

Kühlewein (sometimes "Kuehlewein"), H., *Hippocratis opera quae feruntur omnia*, 2 vols. (Leipzig, 1894, 1902).

Kühn, C. G., *Claudi Galeni opera omnia*, 20 vols. (Leipzig, 1821-1833).

Littré, E., *Œuvres completes d'Hippocrate*, Traduction nouvelle avec texte grec en regard, 10 vols. (Paris, 1839-1861).

Wellmann, M., *Die Fragmente der sikelischen Ärzte Akron, Philistion und des Diokles von Karystos* (Berlin, 1901).

B. Plato

Adam, J., *The Republic of Plato*, Edited with Critical Notes, Commentary, and Appendices, 2 vols. (Cambridge, Engl., 1920).

Archer-Hind, R. D., *The Timaeus of Plato*, Edited with Introduction and Notes (London, 1888).
Burnet, J., *Platonis opera*, Recognivit brevique adnotatione critica instruxit, Tomus IV (*Clitopho, Res Publica, Timaeus, Critias*), Scriptorum Classicorum Bibliotheca Oxoniensis (Oxford, 1902).
Bury, R. G., *Plato*, With an English Translation, Vol. VII (*Timaeus, Critias, Cleitophon, Menexenus, Epistles*), Loeb Classical Library (Cambridge, Mass. and London, 1942).
Cornford, F. M., *Plato's Cosmology*, The *Timaeus* of Plato Translated with a Running Commentary (London, 1937).
Fowler, H. N., *Plato*, With an English Translation, Vol. I. (*Euthyphro, Apology, Crito, Phaedo, Phaedrus*), Loeb Classical Library (London and Cambridge, Mass., 1953).
——, *Plato*, With an English Translation, Vol. III (*The Statesman, Philebus, Ion*), Loeb Classical Library (London and New York, 1925).
Jowett, B., and L. Campbell, *Plato's Republic*, The Greek Text Edited, with Notes and Essays, 3 vols. (Oxford, 1894).
Lamb, W. R. M., *Plato*, With an English Translation, Vol. V (*Lysis, Symposium, Gorgias*), Loeb Classical Library (Cambridge, Mass. and London, 1946).

C. *Aristotle*

Aubonnet, J., *Aristote: Politique*, Livres I et II. Texte etabli et traduit (Paris, 1960).
Barker, E., *The Politics of Aristotle*, Translated with an Introduction, Notes, and Appendixes (Oxford, 1946 [corrected reprint, 1948]).
Burnet, J., *The Ethics of Aristotle*, Edited with an Introduction and Notes (London, 1900).
Dirlmeier, F., *Aristoteles: Nikomachische Ethik*, übersetzt (Berlin, 1956).
Düring, I., *Aristotle's Chemical Treatise*: *Meterologica, Book IV*, With Introduction and Commentary (Göteborg, 1944).
Gauthier, R.-A., O.P. and Jolif, J. Y., O.P., *L'Éthique à Nicomaque*, Tome I: Introduction et Traduction; Tome II: Commentaire (Louvain, 1958-1959).
Grant, A., *The Ethics of Aristotle*, Illustrated with Essays and Notes, 2 vols., 3rd ed. (London, 1874).
Guthrie, W. K. C., *Aristotle: On the Heavens*, With an English Translation, Loeb Classical Library (Cambridge, Mass. and London, 1945).
Hicks, R. D., *Aristotle: De Anima*, With Translation, Introduction, and Notes (Cambridge, Engl., 1907).
Joachim, H. H., *Aristotle: On Coming-to-Be and Passing Away* (*De Generatione et Corruptione*), A Revised Text with Introduction and Commentary (Oxford, 1922).
LeBlond, J. M., *Aristote: Philosophe de la vie. Le livre premier du traité sur les Parties des Animaux*, Texte et traduction, avec introduction et commentaire (Paris, 1945).
Lee, H. D. P., *Aristotle: Meteorologica*, With an English Translation, Loeb Classical Library (London and Cambridge, Mass., 1952).

Newman, W. L., *The Politics of Aristotle*, With an Introduction, Two Prefatory Essays, and Notes Critical and Explanatory, 4 vols. (Oxford, 1887-1902).

Peck, A. L., *Aristotle: Generation of Animals*, With an English Translation, Loeb Classical Library (Cambridge, Mass. and London, 1953).

——, *Aristotle: Parts of Animals*, With an English Translation and a Forward by F. H. A. Marshall. Also: *Movement of Animals*, *Progression of Animals*, With an English Translation by E. S. Forster, Loeb Classical Library (Cambridge, Mass. and London, 1945).

Rackham, H., *Aristotle: Politics*, With an English Translation, Loeb Classical Library (Cambridge, Mass. and London, 1950).

——, *Aristotle: The Athenian Constitution, The Eudemian Ethics, On Virtues and Vices*, With an English Translation, Loeb Classical Library (Cambridge, Mass. and London, 1952).

——, *Aristotle: The Nicomachean Ethics*, With an English Translation, Loeb Classical Library (Cambridge, Mass. and London, 1947).

Ross, W. D., *Aristotle: De Anima*, Edited with introduction and commentary (Oxford, 1961).

——, *Aristotelis Politica*, Recognovit brevique adnotatione critica instruxit, Scriptorum Classicorum Bibliotheca Oxoniensis (Oxford, 1957).

——, *Aristotle: Parva Naturalia*, A Revised Text with Introduction and Commentary (Oxford, 1955).

——, *Aristotle's Physics*, A Revised Text with Introduction and Commentary (Oxford, 1936 [corrected reprint, 1955]).

——, (Oxford Translation) *The Works of Aristotle*, Translated into English under the Editorship of (J. A. Smith and) W. D. Ross, 11 vols. (Oxford, 1908-1931).

Volumes used in the present study:

I. *Categoriae* and *De Interpretatione* by E. M. Edghill; *Analytica Priora* by A. J. Jenkinson; *Analytica Posteriora* by G. R. G. Mure: *Topica* and *De Sophisticis Elenchis* by W. A. Pickard-Cambridge (1928).

II. *Physica* by R. P. Hardie and R. K. Gaye; *De Caelo* by J. L. Stocks; *De Generatione et Corruptione* by H. H. Joachim (1930).

III. *Meteorologica* by E. W. Webster; *De Mundo* by E. S. Forster; *De Anima* by J. A. Smith; *Parva Naturalia* by J. I. Beare and G. R. T. Ross; *De Spiritu* by J. F. Dobson (1931).

IV. *Historia Animalium* by D'Arcy W. Thompson (1910).

V. *De Partibus Animalium* by W. Ogle; *De Motu* and *De Incessu Animalium* by A. S. L. Farquharson; *De Generatione Animalium* by A. Platt (1912).

VIII. *Metaphysica*[2] by W. D. Ross (1928).

IX. *Ethica Nicomachea* by W. D. Ross; *Magna Moralia* by St. George Stock; *Ethica Eudemia, De Virtutibus et Vitiis* by J. Solomon (1925).

XI. *Rhetorica* by W. R. Roberts; *De Rhetorica ad Alexandrum* by E. S. Forster; *De Poetica* by I. Bywater, 1924.

Theiler, W., *Aristoteles: Über die Seele*, übersetzt (Berlin, 1959).

D. *Greek Commentators on Plato and Aristotle*

Alexander Aphrodisiensis, *In Aristotelis Metaphysica commentaria*, M. Hayduck (ed.), *Commentaria in Aristotelem Graeca*, I (Berlin, 1891).

——, *Scripta minora: questiones, de fato, de mixtione*, I. Bruns (ed.), *Supplementum Aristotelicum*, Vol. II, Pt. II (Berlin, 1892).

Aspasius, *In Ethica Nicomachea quae supersunt commentaria*, G. Heylbut (ed.), *Commentaria in Aristotelem Graeca*, Vol. XIX, Pt. I (Berlin, 1889).

Philoponus, Joannes, *In Aristotelis De Anima libros commentaria*, M. Hayduck (ed.), *Commentaria in Aristotelem Graeca*, XV (Berlin, 1897).

Proclus Diadochus, *In Platonis Timaeum commentarii*, E. Diehl (ed.), 3 vols. (Leipzig, 1903, 1904, 1906).

Simplicius, *In libros Aristotelis De Anima commentaria*, M. Hayduck (ed.), *Commentaria in Aristotelem Graeca*, XI (Berlin, 1882).

E. *Mathematical Writers*

Euclid. *The Thirteen Books of Euclid's Elements*, Translated from the Text of Heiberg, with Introduction and Commentary, by T. L. Heath, 2nd ed. revised with additions, 3 vols. (New York, 1956).

Iamblichus, *In Nicomachi arithmeticam introductionem*, H. Pistelli (ed.), (Leipzig, 1894).

Nichomachus Gerasenus, *Introductionis arithmeticae libri II*, R. Hoche (ed.), (Leipzig, 1866).

——, *Nicomachus of Gerasa: Introduction to Arithmetic*, Translated into English by M. L. D'Ooge, with Studies in Greek Arithmetic by F. Robbins and L. C. Karpinski (New York, 1926).

Philoponus, Joannes, 'Εις τὸ δεύτερον τῆς Νιχομάχου ἀριθμητικῆς εἰσαγωγῆς, R. Hoche (ed.), (Berlin, 1867).

Theon of Smyrna, *Expositio rerum mathematicarum ad legendum Platonem utilium*, E. Hiller (ed.), (Leipzig, 1878).

F. *Collections*

Diels, H., *Die Fragmente der Vorsokratiker*, Siebente Auflage herausgegeben von W. Kranz (Berlin, 1954).

——, *Doxographi Graeci, Collegit recensuit prolegomenis indicibusque instruxit*, 3rd ed. (photographic reproduction of 1879 ed.), (Berlin, 1958).

II. MODERN COMMENTARIES, STUDIES, MONOGRAPHS, ETC.

Adam, J., *The Republic of Plato*. Includes commentary. See above, I.B.

Allan, D. J., *The Philosophy of Aristotle* (London, 1952).

Anton, J. P., *Aristotle's Theory of Contrariety* (London, 1957).

Aubenque, P., *Le Problème de L'Etre chez Aristote*, Essai sur la Problematique Aristotelicienne (Paris, 1962).

Aubonnet, J., *Aristote: Politique*, Livres I et II. Contains important introduction. See above, I.C.

Barker, E., *Greek Political Theory: Plato and His Predecessors*, 3rd ed. (London, 1947).

——, *The Political Thought of Plato and Aristotle* (London, 1906).

Beare, J. I., *Greek Theories of Elementary Cognition* (Oxford, 1906).

Block, I., "The Order of Aristotle's Psychological Writings", *American Journal of Philology*, LXXXI (1961), 50-77.

Bourgey, L., *Observation et expérience chez les médecins de la collection hippocratique* (Paris, 1953).

Burnet, J., *Early Greek Philosophy*, 3rd ed. (London, 1920).

——, *The Ethics of Aristotle*. Includes introduction and notes. See above, I.C.

Cherniss, H., *Aristotle's Criticism of Plato and the Academy*, Vol. I (Baltimore, 1944).

——, *Aristotle's Criticism of Presocratic Philosophy* (Baltimore, 1935).

Cornford, F. M., *Plato's Cosmology*. Includes commentary. See above, I.B.

Deichgräber, K., *Die Epidemien und das Corpus Hippocraticum* (Berlin, 1933).

——, "Petronas", Pauly-Wissowa *Real-Encyclopädie* XIX, Pt. I, col. 1191.

——, "Zu Hippokrates' Περὶ 'Αρχαίης 'Ιητρικῆς 9", *Hermes*, LXVIII (1933), 356-358.

Den Dulk, W., ΚΡΑΣΙΣ, *Bijdrage tot de Grieksche Lexicographie* (Leiden, 1934).

Diels, H., "Über die Excerpte von Menons *Iatrika*", *Hermes*, XXVIII (1893), 407ff.

Diller, H., "Hippokratische Medizin und attische Philosophie", *Hermes*, LXXX (1952), 385-409.

——, "Philistion", Pauly-Wissowa *Real-Encyclopädie* XIX, Pt. II, cols. 2405-2408 (Stuttgart, 1938).

Dirlmeier, F., *Aristoteles: Nikomachische Ethik*. Includes introduction and commentary. See above, I.C.

D'Ooge, M. L., *Nicomachus of Gerasa: Introduction to Arithmetic*. Includes notes and essays. See above, I.E.

Düring, I., *Aristotle's Chemical Treatise: Meteorologica Book IV*. Includes introduction and commentary. See above, I.C.

——, "Aristotle's Method in Biology", *Aristote et les problèmes de méthode* (Louvain, 1961), pp. 213-221.

Edelstein, L., "Galen", *The Oxford Classical Dictionary* (Oxford, 1949), p. 377.

——, "Nachträge (Hippokrates)", Pauly-Wissowa *Real-Encyclopädie* Supplementum VI, cols. 1290-1343 (Stuttgart, 1935).

——, ΠΕΡΙ ΑΕΡΩΝ *und die Sammlung der hippokratischen Schriften* (Berlin, 1931).

——, "The Relation of Ancient Philosophy to Medicine", *Bulletin of the History of Medicine*, XXVI (1952), 301.

——, Review of L. Stella, *Importanza di Alcmeone nella storia del pensiero greco* (Rome, 1939), *American Journal of Philology*, LXIII (1942), 371-372.

——, Review of M. Pohlenz, *Hippokrates* (Berlin, 1938), *American Journal of Philology*, LXI (1940), 221-229.

——, Review of W. Jaeger, *Diokles von Karystos* (Berlin, 1938), *American Journal of Philology*, LXI (1940), 483-489.

——, "The Role of Eryximachus in Plato's *Symposium*", *Transactions of the American Philological Assn.*, LXXVI (1945), 85-103.

Festugière, A. J., *Hippocrate, l'Ancienne Médecine*. Includes commentary. See above, I.A.

Fredrich, C., *Hippokratische Untersuchungen*. Philologische Untersuchungen XV (Berlin, 1899).

Garuti, G., "L'Uso e il significato di μέσος (μεσότης) e μέτρον (μέτριος) nella filosofia greca fino al Aristotele", Accademia di Modena *Atti e Memorie*, Serie V, vol. XV (1957), 263-272.

Gauthier, R.-A., *La Morale d'Aristote* (Paris, 1958).

——, and Jolif, J. Y., *L'Ethique a Nicomaque*, Tome II: Commentaire. See under I.C.

Gilbert, O., *Meteorologische Theorien des griechischen Altertums* (Leipzig, 1907).

Godfrey, F. L., "The Idea of the Mean", *Hermathena*, LXXXI (1953), 14-28.

Gomperz, T., *Greek Thinkers: A History of Ancient Philosophy*, Translated by G. G. Berry, 4 vols. (Vol. IV: *Aristotle and His Successors*, 1912), (London, 1901-1912).

Gow, J., *A Short History of Greek Mathematics* (Cambridge, Engl., 1884), (Photographic reproduction, New York, 1923).

Grant, A., *The Ethics of Aristotle*. Includes essays and notes. See above, I.C.

Grene, Marjorie, *A Portrait of Aristotle* (London, 1963).

Grenet, P., *Les Origines de l'analogie philosophique dans les dialogues de Platon* (Paris, 1948).

Hantz, H. D., *The Biological Motivation in Aristotle* (New York, 1939).

Hardie, W. F. R., "Aristotle's Treatment of the Relation Between the Soul and the Body", *Philosophical Quarterly*, XIV (1964), 53-72.

Heath, T. L., *The Thirteen Books of Euclid's Elements*. Includes introduction and commentary. See above, I.E.

Heidel, W., *Hippocratic Medicine: Its Spirit and Method* (New York, 1941).

Heinimann, F., *Nomos und Physis*, Herkunft und Bedeutung einer Antithese im griechischen Denken des 5. Jahrhunderts (Basel, 1945).

Heitz, E., *Die verlorenen Schriften des Aristoteles* (Leipzig, 1865).

Hicks, R. D., *Aristotle: De Anima*. Includes introduction and notes. See above, I.C.

Jaeger, W., *Aristotle: Fundamentals of the History of His Development*, Translated with the Author's Corrections and Additions by R. Robinson, 2nd ed. (Oxford, 1948).

——, "Aristotle's Use of Medicine as a Model of Method in His Ethics", *Journal of Hellenic Studies*, LXXVII (1957), 54-61.

——, *Diokles von Karystos*, Die griechischen Medizin und die Schule des Aristoteles (Berlin, 1938).

——, *Paideia: The Ideals of Greek Culture*, Translated from the German Manuscript by G. Highet, 3 vols. (Vol. III, ch. I: "Greek Medicine as Paideia"), (New York, 1943).

——, Report on date of Diocles in *Sitzungsberichte der preussischen Akademie der Wissenschaften* (Berlin), Jahrgang 1938, p. 237.

Joachim, H. H., *Aristotle: On Coming-to-Be and Passing Away*. Includes introduction and notes. See above, I.C.

——, *Aristotle: The Nicomachean Ethics*, A Commentary by the Late H. H. Joachim, Edited by D. A. Rees (Oxford, 1951).

——, "Aristotle's Conception of Chemical Combination", *Journal of Philology*, XXIX (1904), 72-86.

Jones, W. H. S., "'Hippocrates' and the *Corpus Hippocraticum*", *Proceedings of the British Academy*, XXI (1945), 103-125.

——, *Philosophy and Medicine in Ancient Greece*. Includes introduction commentary. See above, I.A.

Joseph, H. W. B., "Aristotle's Definition of Moral Virtue and Plato's Account of Justice in the Soul", *Philosophy*, IX (1934), 168-181.

Jowett, B., and L. Campbell, *Plato's Republic*. Includes notes and essays. See above, I.B.

Kalchreuter, H., *Die Mesotes bei und vor Aristoteles* (Tübingen, 1911).

Keus, A., *Über philosophische Begriffe und Theorien in den hippokratischen Schriften* (Cologne, 1914).

Kirk, G. S., *Heraclitus: The Cosmic Fragments* (Cambridge, Engl., 1594).

Klippel, M., *La Médecine grecque dans ses rapports avec la philosophie* (Paris, 1937).

LeBlond, J.-M., *Aristote: Philosophe de la vie* (*Parts of Animals*, Bk. I, with introduction and notes). See above, I.C.

——, *Logique et Méthode chez Aristote*, Étude sur la recherche des principes dans la physique Aristotelicienne (Paris, 1939).

Lee, H. D. P., "Place Names and the Date of Aristotle's Biological Works", *Classical Quarterly*, XLII (1948), 61-67.

Leonard, J., *Le Bonheur chez Aristote* (Brussels, 1948).

Lloyd, G. E. R., "Who Is Attacked in *On Ancient Medicine*?" *Phronesis*, VIII (1963), 108-126.

Longrigg, J., "Philosophy and Medicine: Some Early Interactions", *Harvard Studies in Classical Philology*, LXVII (1963), 147-175.

Miller, H. W., "The Aetiology of Disease in Plato's *Timaeus*", *Transactions of the American Philological Association*, XCIII (1962), 175-187.

——, "Philosophy and Medicine", *Classical Journal*, XLIV (1949), 309-318.

Mure, G. R. G., *Aristotle* (New York, 1932).

Muri, W., *Arzt und Patient bei Hippokrates* (Bern, 1936).

——, "Περὶ Ἀρχαίης Ἰητρικῆς IX", *Hermes*, LXXI (1936), 467-469.

Nesselmann, G. H. F., *Geschichte der Algebra*, Vol. I (Berlin, 1842).

Nestle, W., "Hippocratica", *Hermes*, LXXIII (1938), 35.

Newman, W. L., *The Politics of Aristotle*. Includes introduction, essays, and notes. See above, I.C.

Nuyens, F., *L'Evolution de la psychologie d'Aristote*, Preface par A. Mansion (Louvain, 1948).

Oates, W. J., *Aristotle and the Problem of Value* (Princeton, 1963).

——, "The Doctrine of the Mean", *Philosophical Review*, XLV (1936), 382-398.

Ohmsted, E. H., "The 'Moral Sense' Aspect of Aristotle's Ethical Theory", *American Journal of Philology*, LXIX (1948), 42-61.

Olerud, A., *L'Idee de macrocosmos et de microcosmos dans le Timee de Platon* (Upsala, 1951).

Pauly, A., *Paulys Real-Encyclopädie der Classischen Altertumswissenschaft*. Neue Bearbeitung, Unter Mitwirkung zahlreicher Fachgenossen herausgegeben von G. Wissowa (Stuttgart, 1893).

Peck, A. L., "The Connate Pneuma", *Science, Medicine, and History*, edited by E. A. Underwood (Oxford, 1953), vol. I. pp. 111-121.

Pohlenz, M., *Hippokrates und die Begründung der wissenschaftlichen Medizin* (Berlin, 1938).

Poschenrieder, F., *Die naturwissenschaftlichen Schriften des Aristoteles in ihrem Verhältnis zu den Büchern der hippokratischen Sammlung* (Bamberg, 1887).

Raeder, H., "Menekrates", Pauly-Wissowa *Real-Encyclopädie* XV, Pt. I, col. 802, No. 29 (Stuttgart, 1931).

Randall, J. H., Jr., *Aristotle* (New York, 1960).

Reiche, Harald A. T., *Empedocles' Mixture, Eudoxan Astronomy and Aristotle's Connate Pneuma* (Amsterdam, 1960).

Ross, W. D., *Aristotle* (London, 1923; 5th ed., revised, 1949).

——, *Aristotle: De Anima*. Includes introduction and commentary. See above, I.C.

——, *Aristotle: Parva Naturalia*. Includes introduction and commentary. See above, I.C.

——, *Aristotle's Physics*. Includes introduction and commentary. See above, I.C.

——, "The Development of Aristotle's Thought", *Proceedings of the British Academy*, XLVI (1957), 68-76.

Schilling, H., *Das Ethos der Mesotes: eine Studie zur nikomachischen Ethik*, Heidelberger Abhandlungen zur Philosophie und ihrer Geschichte, No. 22 (Tübingen, 1930).

Schumacher, J., *Antike Medizin: Die Naturphilosophischen Grundlagen der Medizin in der Griechischen Antike*, 2nd ed. (Berlin, 1963).

Siwek, P., *La psychophysique humaine d'après Aristote* (Paris, 1930).

Solmsen, F., *Aristotle's System of the Physical World: A Comparison With His Predecessors* (Ithaca, N.Y., 1960).

Souilhé, T., *Étude sur le terme* ΔΥΝΑΜΙΣ (Paris, 1919).

——, *La Notion platonicienne d'intermédiaire dans la philosophie des dialogues* (Paris, 1919).

Stella, L., *Importanza di Alcmeone nella storia del pensiero greco* (Rome, 1939).

Stewart, J. A., *Notes on the Nicomachean Ethics of Aristotle*, 2 vols. (Oxford, 1892).

Stocks, J. L., "ΛΟΓΟΣ and ΜΕΣΟΤΗΣ in the *De Anima* of Aristotle", *Journal of Philology*, XXXIII (1910), 182-194.

Taylor, A. E., *A Commentary on Plato's Timaeus* (Oxford, 1928).

——, *Plato, the Man and His Work*, 3rd ed. (New York, 1929).

Theiler, W., *Aristoteles: Über die Seele*. Includes introduction and commentary. See above, I.C.

Überweg, F., *Grundriss der Geschichte der Philosophie*, Erster Teil: Die Philosophie des Altertums. Elfte auflage, herausgegeben von K. Praechter (Berlin, 1920).

Van der Meulen, T., *Aristoteles, die Mitte in seinem Denken* (Meisenheim, 1951).

Vlastos, G., "Isonomia", *American Journal of Philology*, LXXIV (1953), 337-366.

Wachtler, J., *De Alcmaeone Crotoniate* (Leipzig, 1896).

Walsh, J., *Aristotle's Conception of Moral Weakness* (New York, 1963).

Wehrli, F., "Ethik und Medizin: Zur Vorgeschichte der aristotelischen Mesonlehre", *Museum Helveticum*, VIII (1951), 36-62.

Weidauer, K., *Thukydides und die hippokratischen Schriften* (Heidelberg, 1954).

Wellmann, M., "Diokles", Pauly-Wissowa *Real-Encyclopädie* V, Pt. I, cols. 802-812 (Stuttgart, 1903).

Zeller, E., *Aristotle and the Earlier Peripatetics*: Being a Translation from Zeller's "Philosophy of the Greeks" by B. F. C. Costelloe, M.A., and J. H. Miurhead, M.A., 2 vols. (London, 1897).

GENERAL INDEX

All numbers refer to pages; items in footnotes are indicated by the letter "n" following the page on wich they appear. A. = Aristotle; med. wr. = medical writers; P. = Plato.

accuracy, difficulty of in medicine, 39, 40, 66, 71
Adam, J., 131n
Adrastus, 345
Aetius, 22, 109n
age, effect on body, 49; on emotions, 253n
Alcmaeon of Croton, 22-24, 28, 31, 34, 43, 59, 67, 138, 334
Alexander of Aphrodisias, 349
Allan, D. J., 17
alteration (*alloiôsis*) in A., 166n
analogia, *see* proportion
Anaxagoras, 34
animals, lower, in P., 96, 97
Anton, J., 18n
apeiron, in P., 128, 129, 133, 134, 148, 149
appetite (*epithymia*), in P., 112, 114
appetition, *see* pursuit and avoidance
Archer-Hind, R., 79n
Archytas, 345
aristocracy, 290-293, 341
Aristotle, 15-20, 22, 24, 25, 26, 34n, 35n, 36n, 60, 63, 75, 77n, 81n, 94n, 129n, 130n, 131n, 134n, 150n, 156, 157ff. *passim*
Aspasius, 259n
art, in A., 169
Athenaeus, 127
Aubenque, P., 333
Aubonnet, J., 285n

balance, *see* equilibrium
balance = symmetry, 191, 195n, 196n
Barker, E., 17, 279, 285n, 299n, 305n, 311n
Beare, J., 200n, 210n, 247n
beautiful, the, in P., 137-138, 336-337
beauty, physical, 58, 238
bile, in P., 122, 123, 125-132
biological approach, in A., 18
biological analogy, 150n
biological works, of A., 18
blend, *see krasis*
blend, in sensible qualities, 204
Block, I., 353n
blood, in A., 186, 187, 189, 192, 217-219, 255-256
  in P., 109, 116, 117, 122
body, cosmic, in P., 78ff.
body, human, purpose of, in P., 102
bone, in A., 169ff.
  in P., 104, 122
Bourgey, L., 146n
brain, in A., 190-195, 196, 338
  in P., 104, 126
Burnet, J., 17, 29n, 78n, 223n, 344

chararacter, moral, 58, 61, 74, 143, 229

Cherniss, H., 7, 29n, 83n, 333
childhood, in P., 117,141n,
choise, in A., 266-269, 271
chronology of A.'s works, 20, 347-353
climate, influence of, *see* seasons
cold, center of, in body, 190-193
cold, in emotional states, 253, 255-257 340, 354-355
combination, chemical, in A., 167-172
compound, body-soul, in A., 197-200, 248
  in P., 136-138, 145, 337
conflict, in polis, 302-304
constitutions, abs. and rel. best, 285, 292-293, 296, 308-311, 322-324
  unnatural, 301-307, 322-324
contemplation, in A., 278-279, 281-282, 333, 341
  in P., 142
Cornford, F., 23n, 78n, 80n, 84n, 85n, 86, 90n, 91n, 92n, 94n, 98, 100n, 105n, 106n, 109n, 110n, 128, 130n
cosmos, as living organism, 77 ff.

death, in A., 181, 185
  in P., 112
decay, in A., 176
Deichgräber, K., 27n. 34n, 39n
deliberation, in A., 266
democrcy, 294, 301 ff., 305-306, 309-310, 324-325, 341
Den Dulk, W., 37n
desire (*erôs*, in P., 112, 113
diaphragm, in A., 218n
  in P., 114
Diehl, E., 79n, 84n
Diels, H., 25n, 26n
diet, 28, 30, 33-42, 143; *see* food
digestion, 36, 175
dikê, righteousness, 97, 98, 101, 113
Diller, H., 16, 28, 31n, 33n, 39n
Diocles of Carystos, 32n, 34n, 35n, 130n
Dirlmeier, F., 17
disease, in A., 173-174, 187, 194, 226-227, 232-237, 257, 304-305, 337, 340
  in med. wr., 19, 22, 24, 27, 29, 36, 43, 68, 143, 335-337
  in P., 137-138, 143; of cosmic body, 83, 84; of human body, 100, 101, 119, 120-123; of human soul, 123-136
disposition, *see hexis*
D'Ooge, M., 345
dreams, in A., 218-219
Düring, I., 21n, 169n, 188n
dynameis, "powers", qualities, in A., 164-173, 175-176, 202, 209, 231 ff., 294, 339
  in med. wr., 22, 23, 29, 35, 36, 40-42, 67
  in P., 81, 83, 84, 105, 110, 111, 121, 133, 147, 152
dynamics, of *mesotês*, 221-222

Edelstein, L., 16, 22n, 32n, 34n, 35n, 53n, 77n, 151n
education, in P., 100, 101, 136, 140
emotions, in A., 197, 232-237, 246, 247-260, 276, 277, 281, 328, 339-341, 351-352, 354
  in med. wr., 55
  in P., 97, 103, 106, 107, 112, 113, 133-134, 152, 337
Empedocles, 28n, 29, 30, 34, 35, 82, 146n
entelichy, soul as, in A., 347-353
environment, influence of, 30, 34, 47, 51 ff., 62, 63, 69, 83, 105, 111, 116 ff., 151-153, 161, 174-178
equilibrium, balance, 24, 27, 28, 29, 31, 43, 44, 46, 47, 51, 52, 55, 57, 64, 66, 69, 83, 85, 89, 90, 95, 97, 101, 105, 108, 109, 111, 112, 114, 117, 118, 119, 121, 138, 139, 144, 145, 151, 153, 163, 167, 171-172, 176, 178, 181, 188, 194, 196, 206, 212, 215-216, 220-221, 231 ff., 236, 246, 254, 258, 280-283, 296-301, 309, 323-325, 334-343, 346
error, in A., 256-257
Eryximachus, 118, 144, 150, 151n
essence, in A., 169, 198
Euripides, 75n
evil, in med. wr., 74, 335
  in P., 137-138
excellence, in gen., 56,73, 158-160, 336

of body
in A., 159-162, 231 ff.: in P., 150
moral, in A., *254, 271(def.), 320-322, 339-340
of soul, in P., 150
exercise, in P., 137-138, 140-142

Festugière, A., 29n, 32n, 33n, 37n, 39n, 43n
fibrin, in P., 123
flesh, 104, 105, 122, 169 ff.
food, 28, 35-41, 116, 117, 186-187; *see* diet
form, in A., 164n, 169, 197-200, 201-202, 206-209, 217, 221, 252, 314-315, 347-353
Fowler, H., 128n, 146n
Fredrich, C., 27n, 34n
free-birth, basis of right, 289, 293-294

Calen, 25n, 34n, 77, 130n, 146, 156, 188n
Gauthier, R., 17
generation, animal, 185
generation-destruction, in A., 165-167
Gilbert, O., 30n
Godfrey, F., 17
gods, in A., 282, 332
in P., 78-79, 94, 95, 96, 98, 102, 103, 113, 142, 145, 147, 150, 151
Gomperz, T., 16
good, in A., 269, 277-283, 332, 340; absolute and rel., 243-245, 285, 292-293, 310-311, 314-324, 327-333, 341-342
in med. wr., 73, 335
in P., 137, 138, 153, 336
Gow, J., 345
Grant, A., 17, 149n
Grene, M., 18
Grenet, P., 81n
Guthrie, W., 319n

habit, *see hexis*
habituation (*ethos*), 228-229, 240, 268
Hantz, H., 18
happiness, in A., 277-283, 329, 331, 333, 341
Hardie, W., 353n
head, in P., 102, 114, 115
health, theory of, 19
in Alcmaeon, 22-24, 31, 334
in A., 32, 158-174, 195, 220, 226-227, 232-237, 238, 304, 314-315, 318-319, 327-328, 337
in med. wr., 26 ff., 42, 44, 47, 52, 64, 73, 143, 334-335
in P., 77 ff., 100, 101, 119 ff., 121, 135, 137-138, 143, 144, 149, 150
healthy man, as norm, 72, 155, 243-245, 327-328, 335
hearing, in P., 103, 109
heart, in A., 183-189, 195-196, 199-200, 217-220, 255-258, 338, 340, 347-359; three-chambered, 188-189, 219, 338, 357
in P., 113
heat, innate or vital, 44n, 176, 177, 179-182, 184,185- 190-191, 195, 218-219, 253, 255, 338, 340, 350-357
Heat, T., 79n, 80n, 345
heavenly bodies, stars and planets, 89, 92, 94, 95, 96, 103, 147, 175n, 332-333
Heidel, W., 22n
Heinimann, F., 34n, 73, 75
Heitz, E., 158n
Heraclitus, 34, 35n
Herodotus, 75n
hexis, lasting disposition, habit, 15, 57, 64, 69, 98, 124, 158-160, 176, 223, 226, 229 ff., 231, 234-237, 248, 269, 337, 339
Hicks, R., 197n, 225n, 349, 350, 352
Hiller, E., 345
Hippasus, 345
Hippocrates, 34n, 77n, 146n, 336, 349
Hippocratic collection, 16, 26, 32 ff., 34n, 67, 142 ff., 146n, 335
Hoche, R., 83n, 345, 346
homoeomerous parts, in A., 168-174, 195, 220
humors, 28, 31, 33, 42, 46
hylemorphism, 347-353

Iamblichus, 346
ideal state, as norm, in P., 155

image, *see* phantasm
imagination, in A., 251
individual, in A., 311, 316-318, 325, 339, 342
individual traetment in medicine, 38, 40, 49, 50, 66, 70, 71, 335
intellect, in A., 261-262, 278-279, 340; practical, 263-269, 340; speculative, 272-276, 340
in P., 115
intelligence, in med, wr., 55, 65, 74
in P., 94, 95, 97, 103, 143, Intelligence, supreme, 148, 150

Jaeger, W., 16, 19, 20, 31n, 32n, 33n, 34n, 43n, 162n, 163n, 225n, 226n, 285n, 314n, 348
Joachim, H., 17, 167n, 168n, 170n, 173n
Jolif, J., 17
Jones, W. H. S., 23n, 25n, 31n, 32n, 33n, 34n, 36n, 39n, 43n, 53
Joseph, H., 17
Jowett, B., and L. Campbell, 131n
Justice, 289; absolute and rel., 302, 330

Kaibel, G., 127n
kairos, 40-41, 49, 73, 123
Kalchreuter, H., 15
Keus, A., 16, 23n
kingship, in A., 290-292, 330-331, 341
in med. wr., 61
Kirk, G. S., 35n
Klippel, M., 16
knowledge (*epistêmê*), in A., 273-276, 340
in P., 93, 95, 152, 153, 337
Krämer, H., 21n
krasis, blend, 23, 24, 35-38, 42, 43, 44, 46, 59, 64, 68, 69, 73, 91, 96, 105, 119, 143n, 145, 159, 161, 167-172, 175-176, 191, 192, 193, 227, 239-242, 295-300, 324-325, 335-339
Kühlewein, H., 33n

Lamb, W., 119n
LeBlond, J., 18
Lee, H., 20n, 169n
Leonard, J., 225n
Littré, E., 33n
liver, in P., 114
Living Being, ideal, in P., 78, 155, 157, 336-337
Lloyd, G., 33n
logos, in A., 163, 169-171
in P., 95, 98n, 144
Longrigg, J., 16
lungs, in P., 114

male, superiority of, 288
man, vs. animals, plants, 286-288, 331-332
marrow, in A., 193
in P., 103, 104, 122
mathematical analysis, in P., 86-87, 91-93, 146-147
mathematics, 79n, 81-83, 214n, 344-346
matter, in A., 165-166, 209-211, 217 ff., 221, 248, 252-253, 298-299, 308, 314-316, 321-322, 342
mean, 15-20, 36n, 38, 65, 73, 82, 90, 105, 123, 134-135, 145, 157 ff., 171-172, 178 ff., 182, 186, 188-189, 190-191, 193-194, 195, 200, 206, 226-227, 235, 297, 323-325, 335, 337-339, 344-346; *see meson*
mean state, condition, 107, 181, 187
measure, standard, 38, 39, 40, 119, 134, 149, 150; *see also metron*, norm
medical theory, 16, 33, 35, 44, 51, 73n, 75, 105, 150n, 156
medical writers, 16, 18, 19, 68, 78, 83, 87, 88, 119, 120, 138, 141, 142 ff., 159, 178, 188n, 312, 334-335, 343
medicine, art and science, 35, 37, 38, 39n, 41, 63, 66, 119, 148, 225, 311 ff.
medium of sensation, 210
melancholy, 127, 226-227, 256
memory, in A., 262
Menecrates, 26, 31, 34n, 67, 68
Menon papyrus, 25-32, 34n, 67, 77, 130n, 142 ff., 334-335
meson, 59, 79, 81, 163, 183-184, 215-216, 219, 300

mesotês, 15, 18, 19, 20, 82, 93, 95, 97, 123, 147, 163, 171, 173-174, 200, *205-206, 207-209, 212, 215-216, 219, *220-222, 226-227, 227ff., 232-237, 249, *254, 270, 280-282, 295-301, 318-319, 319-324, 327-328, 333-334, 336, 338-346, 359
methodology, 20, 334ff.
in A., 188n, 217n, 223-227, 283, 312-314, 324
in P., 77-78, 90, 95
metriotês, *see* moderation
metron, measure, 49,71,73,163
microcosm, vs. macrocosm, 48, 74, 97, 102, 116, 139, *144n, 145, 153, 336
middle class, 295, 298-300, 341
Miller, H., 16, 77n
mixed constitution, 296-301, 323-324, 341
mixture, in polis, 298-299
moderation (*metriotês*), 42, 43n, 54, 59, 64, 68, 69, 73
moral excellence, in A., 222-283, 320-322, 339-340
in med, wr,. 15, 19, 60, 62
in P., 98, 135, 138, 156
moral guide, 19, 155-156, 233, 284, 311ff., 320-322, 342
more and less, in A., 204
in P., 128, 133
Mure, G., 17
Muri, W., 38n, 39n, 71, 72

nature, natural structure (*physis*), in A., 169, 284-285, 314
in med, wr., 60, 75
Nesselmann, G., 345
Nestle, W., 34n
Newman, W., 17
Nicomachus of Gerasa, 345-346
nomos, law, civic institution, 60, 62, 75, 135-136, 141
norm, measure, 243-245; absolute and rel., 72, 73n, 155, 311ff., 326-333; *see* healthy, virtuous man as norm
nourishment, in A., 181-182
nutrition, in A., 179, 183, 185, 186-188, 350
nutritive process, in P., 99, 116ff.
Nuyens, F., 20, 200n, 218n, 225n, 347-353

Oates, W., 7, 17, 148n, 149n, 207n, 208n, 249n, 319n
Occelus, 344
Ogle, W., 186n
Ohmsted, E., 17, 249n
old age, 117, 132n, 176, 318-319
oligarchy, 294, 301ff., 306-307, 310, 324-325, 341
ontological basis of value, in P., 154-155
opinion, in P., 93, 95, 152, 153
opposites, in P., 128, 138
and sensation, A., 203-204
opposing constituents, in A.: of body, 160, 164-167, 190, 194, 209-211, 226-227, 231ff., 337ff.; of polis 294-295, 324-325
in med. wr., 22, 23, 26, 27, 28, 31, 43, 44, 46, 64, 67
in P., 144, 145, 149, 150, 151, 336; of cosmic body, 79, 83, 85, 87-88; of cosmic soul, 91, 101; of human soul, 106, 133, 134
organism, moral, in A., 229ff., 232ff., 314, 339
organism, political, in A., 283-314, 322, 329-330, 341-342
organism, total human, in P., 102, 138
organs, in A., 182ff., 199, 217-220
in P., 102

pain, in A., 232-237, 248-250, 326-327, 339, 354, 356
in med. wr., 36, 38, 40, 47, 72, 335
in P., 97, 106, 107, 109-113, 124, 133, 152
palpitation, of heart, 354
Peck, A., 18, 164n, 165n, 173n, 174n, 180n, 181n, 190n, 196n
perception, in A., 245, 247, 249-252, 281, 338
Preton of Aegina, 26, 27, 31, 34n, 67, 68
phantasm, 251- 252, 256, 263, 265, 273-275, 340
Philistion, 26, 28-31, 34n, 67, 68, 122,

130n, 131n, 146n
Philolaus, 33n, 130n, 131n
Philoponus, 83n, 161, 209n, 346
phlegm, in P., 122, 123, 125-132
physician, 19, 48, 51, 63, 65, 70, 73, 75, 146, 155-156, 233, 311 ff., 318-320, 342
physiological analogy, 20, 157 ff., 224ff. 228, 230, 241, 245, 302-305, 312-314, 334
physological theory, 19, 22, 25, 32, 41, 42, 144, 148, 157 ff., 178 ff., 223 ff., 298 ff., 334, 336-337, 241
physiology, in A., 234 ff.; of emotion, 247-260; and knowledge, 276
physique, physical types, 49, 61, 62
physis, *see* nature
Pistelli, H., 345
Plato, 18-20, 25, 26, 28, 32, 34n, 35n, 36n, 60, 63, 75, 77-156, *passim*, 157, 159, 161, 166n, 178, 188n, 192, 218n, 233, 334-337, 343, 344, 346, 349
pleasure, in A., 232-237, 242-245, 248-250, 259, 283, 317, 326-328, 339, 354, 356
in P., 36n, 97, 106, 107, 109-113, 124, 133, 152
index of disposition of body, 40, 60, 326-327, 335
pneuma, 354, 356
Pohlens, M., 34n
point, analogue of sense, 213-215, 357
polis, in A., 284-285, 289
political science, 284, 311 ff.
political theory, in A., 341 ff.
polity, *see* mixed constitution
Polybus, 34n
Poschenrieder, F., 16
"powers" or qualities, *see dynameis*
primary bodies, in A., 164-173, 201-202
in P., 87, 116, 117, 120
Prime Mover, 278, 333, 341, 342
Proclus, 79n, 80n, 83n, 84n, 85n, 91n, 92n, 93n, 94, 98n, 100n, 107n, 149n
production (poiêsis), theory of, in A., 314-315
prognosis, 53n
proportion = *analogia*, 79, 82, 89, 121, 344-346
= *symmetria*, 66, 89, 111, 123, 137, 139, 159, 161-163, 175n, 226-227, 231, 235, 295-299
proportion = extremes joined in a mean, 38, 81n, 86, 87, 88, 91, 97, 101, 104, 108, 119, 134, 138, 146-147, 155, 178 ff., 186, 191, 220, 324-325
proportion, in treating patient, 24, 28, 35-41
proportions, geometric, harmonic, arithmetical, 93, 344-346
proportionate (*symmetros*), 23, 24, 28, 68, 138
proportionate response, 69, 105, 109, 111, 118, 143, 151-153, 177, 207, 220-222, 236, 245, 281, 337, 339-340
prudence (*phronêsis*), 270, 272, 328, 340
psychological organism, in P., 95, 133, 135
psychology, in A., 197, 223 ff., 234 ff., 274-276, 282, 340, 347-353
in P., 90, 106, 107, 135
psycho-physical interdependence, in A., 197 ff., 253
in P., 123, 124, 133
pulsation, of heart, 354
pursuit and avoidance, 233, 236, 246, 249-250, 253, 255, 257, 265, 339, 355
Pythagoras, 345

Rackham, H., 285n
Raeder, H., 26n
Randall, J., 17
refrigeration, in A., 181-182
Reiche, H., 181n
Reincarnation, in P., 97
relative mean, in A., 237-239, 319-324, 339
respiration, 30, 177, 356
respiratory system, in P., 116, 117, 122
response, reaction, 39, 40, 72; *see also* proportionate response

Ross, W. D., 17, 160n, 163n, 180n, 217n, 218n, 285n, 293, 313, 347, 350, 351, 352

Schilling, H., 17
Schumacher, J., 16
sciences, division of, in A., 223n
seasons, climate, influence of, 47, 51, 57, 58, 118, 119, 143
self-control, vs. virtue, 240, 266-267
self-sufficiency, 65, 73, 89, 90
sensation, sence perception, 39n
 in A., 192, 197ff., 201ff., 203, 338
 in P., 97, 99, 106, 107, 110-111, 145, 152
sense central, in A., 183, 212-220, 251-255, 258, 338-339
sense organs, in A., 192, 202, 205
 in P., 107, 109, 337
senses, in A., 182, 191-193, 205-212, 250, 273-275, 338
 in P., 103
sensibles, proper, in A., 203-204
sex, physical characteristics of, 49
sight, in P., 103, 108
Simplicius, 162, 209n
sinews, 105, 122, 355
situation, of polis, effect, of, 52, 53-56
Siwek, P., 200n, 247n
sleep, in A., 192, 218
Smyth, H. W., 181n
soil, effect on health, 52, 58
Solmsen, F., 164n
Souilhé, T., 15, 23n
soul, in A., *164n, 179-180, 183, 184, 197-200, 202, 222, 224n, 227-228, 233ff., 260, 277, 339, 347-353
 in med. wr., 58, 61, 65, 75
 in P.; cosmic, 78, 90-96, 336, human, non-rational, 103, 106ff., 113-115, 126, 134-135; rational, 96-101, 124, 126, 128, 152; dependence on body, 100, 101; tripartite, 145, 337; diseases of, 129-132; exercise of, 140-142
speech, in P., 103
spherical structure, 89, 102, 336, 357-359
spleen, in P., 115
stability, 87, 88, 90, 189, 303-304;
 of health, 56, 64, 69, 73, 239-242
statesman, 19, 63, 75, 155-156, 284, 308-311, 322-326, 342
Stella, L., 22n
Stocks, J., 17
stomach, intestines, in P., 115
subordination, principle of, in A., 286ff., 301-302, 341
symmetria, see proportion
symmetros, see proportionate

taste, in P., 108
Taylor, A, E., 23n, 31n, 77n, 79n, 80n, 81n, 82n, 85, 92n, 95n, 98n, 100n, 107n, 109n, 111n, 126, 127, 130n, 131n, 150n
teleogy, in A., 182, 250n, 279-280, 287
 in P., 147
Theiler, W., 353n
Theon of Smyrna, 345
Theophrastus, 109n
theory of knowledge, in P., 94
therapy, techniques of, 50, 71
Thompson, D'Arcy, 18, 188n
topography, influence of, 52, 58
touch, in A., 204n, 208
 in P., 108, 109
tyranny, 301ff., 307-308, 341

Überweg, F., and K. Praechter, 16
uncontrolled, the (akratês), 266-267
universal, formation of, in A., 261-262

values, in A., 279-280, 286ff., 311-334, 342
 in med. wr., 73, 154
 in P., 138-139, 153, 154
Van der Meulen, T., 17
vascular system, in A., 183
 in P., 107, 113
vice, 19, 134, 158-160, 229, 234, 241; *see also* soul, diseases of
virtue, basis of rights, 289
virtue, moral, *see* moral exellence
virtue, natural (vs. moral), 269
virtues, in med, wr.. 75

virtuous man, description, 268
as norm, in A., 155, 243-245, 329, 339
vital activity, center of, in A., 185
Vlastos, G., 23n
void, 116, 174

Wachtler, J., 22n
Walsh, J., 353n
waters, influence on health, 52, 56
wealth, basis of rights, 289, 293-294
Wehrli, F., 16
Weidauer, K., 75n
Wellmann, M., 28n, 31n, 34n, 130n
winds, influence on health, 52
wisdom, 112, 137, 148, 272, 278, 337, 340
woman, in A., 318-319, 321-322, 329
in P., 97
women, types of, 54

Xenophon, 151n

youth, in P., 132n; in A., 253n; *see also* childhood

Zeller, E., 16

INDEX OF PASSAGES CITED

AETIUS
Placita, ed. Diels
IV, 16, 4: 109n
V, 22, 1: 29n
V, 30. 1: 22
ALCMAEON
Fr, 4 (Diels-Kranz 24B4): 22
ALEXANDER ARPHRODESIUS
In Metaphys.,
C.A.G. I, 508, 26-35: 349
Questiones III ,9,
Supplementum Aristotelicum, vol. II, pt. II, p. 96, 1.14ff.: 216n
ANDOCIDES
I, 18 and 35: 151n
ANONYMUS LONDINENSIS (Menon papyrus) W. H. S. Jones' ed.: 25, 34, 334
IV-XX: 25
IV-XX: 25
IV, 26, 28: 25
IV, 28: 142
VII, 15-24: 34n
XIV, 7, 9: 25
XIV, 9-11: 142
XVIII, 41-46: 130n
XIX, 2-18: 34n
XIX, 18-33: 26
XX, 1-14: 27
XX, 16-23: 28n
XX, 25-49: 28
XX, 33-37: 29
ARISTOTLE
Categories
8b27: 158n
8b27-9a4: 158
9a28ff.: 232
9a28-b9: 202
11a20ff.: 159n
11b3-6: 233
13a24-31: 320
De Anima
403a3, 16: 248
403a5-7: 197
403a16-19: 197, 225n
403a19-22: 254
403a23-24: 254
403a24ff.: 248
403a29ff.: 252
403a29-33: 197
403a29-b1: 351
403a31: 248
403a31-32: 255
403a33: 352
403b7-8: 197, 221
403b17: 225n, 248
403b18-19: 197, 221
404b16ff.: 94n
407a32-34: 274
407b33-408a2: 161
408a1-5: 225n
408a14ff.: 169
408b1ff.: 248
408b1-9: 351
408b1-11: 351
408b1-19: 202, 351
408b1-30: 272
408b7ff. 248
408b7-8: 255
408b11-19: 352
408b13-18: 199
408b15-18: 251

408b23-25: 276
409b15: 225n
411a26-28: 198, 225n
411a26-30: 248
411b27-30: 201
412a16: 225n
412a19-20: 197
412a21-28: 198
412a28-b6: 198
412b10-413a6: 198, 205
413a2-3: 210
413b4-9: 204n
413b11 ff.: 248
413b11-32: 198
413b22-24: 250
414a16-27 198
414a29 ff.: 198
414b4-5: 250
415a1-2: 201
415a25-b8: 333
415b15-22: 182n, 199
415b21 ff. 248
415b22-27: 199
415b24: 202
416a8-18: 179n, 205
416a13-15: 350
416a15-18: 169
416a18 ff.: 186n
416b8-15: 179
416b20 ff.: 350
416b32 ff.: 203
416b35 ff.: 207n
417b3-418a6: 202n
417b18-23: 275
418a31 ff.: 211
419a17 ff.: 203
420a4-11: 210
420a16-b3: 211n
420a20 ff.: 356n
420a33 ff.: 211
420b5 ff.: 356n
420b20-21: 351
420b25-26: 351
421a25: 212
422a20 ff.: 205
422b5-6: 212
422b8-10: 212
422b23-31: 204
422b24-423a17: 211
423b25-424a2: 211
423b27-424a10: 208
423b27-30: 202
424a4: 206n, 208n
424a5-7: 206, 207
424a7-10: 206, 208n
424a10-15: 205
424a17-19: 201
424a17-24: 202
424a24 ff.: 205
424a24-28: 217, 218n
424a25-28: 209
424a30-33: 211
424b1: 208n
424b1-3: 201, 210
424b31 ff.: 211
425a2-7: 211
425a3-9: 209n
425b23: 201
425b24: 251
426a8-12: 203
426a27 ff.: 249
426a27-b7: 208n, 211
426b5: 204n
426b8 ff.: 213
426b8-11: 204, 212
426b14-15: 213
426b17 ff.: 213
426b20-21: 213
426b28-29: 213
426b29 ff.: 213
427a9-14: 213
427a11-14: 214, 216
427a14: 214n
429a1-2: 262
429a4-8: 267
429a7-8: 276
429a10-b23: 261
429b13-16: 202
431a1-8: 202n
431a8-10: 249
431a10-14: 249
431a14-17: 263
431a17-19: 215
431a18-19: 220
431a21-22: 216
431b2 ff.: 251
431b2-9: 265
431b6-7: 225n

431b6-10: 266
431b15-16: 214n
432a3-7: 262
432a8-10: 264
432a13-14: 262
432a22-b7: 225n
432a31-32: 224n
432b31-32: 255
432b33: 351
433a1-3: 267
433a7-8: 267
433a10-12: 267
433a20ff.: 251
433b6: 266
433b19-27: 352
433b22-27: 357
433b27-30: 265
433b28-39: 251
434a5-10: 266
434a16-21: 274
434a29: 201
434a30-b2: 182
434b10ff.: 250n
434b11-24: 204n
435a14-15: 209n
435a21: 206n, 208n
435a21-b3: 210
435b3-25: 204n
435b7ff.: 249
435b8-18: 211

De Cealo
268a10: 344
269a18-19 164
269a30-32: 332
269a31ff.: 181
269b30ff.: 164
270a13-b5: 332
276a33ff.: 164
281a2-28: 168n
285a29: 332
286a3-32: 358
286a9-13: 332
286b10ff.: 358
292a20ff.: 332
292a22-b20: 319
293b: 89n
304b23-24: 166n
307b6: 168n
308a14-33: 164
311a15ff.: 164
312b1-3: 173n
321b18-322a27: 186n

De Generatione Animalium
728a18-22: 328
731a24: 184n
734b20-735a8: 218n
734b28ff.: 164n
734b31-735a5: 169
734b33: 169
735a1-26: 353
736a1: 180n
736b30-737a8: 180
736b33ff.: 176n
737a1-8: 180
737a28: 328
738b25-27: 218n
739b34ff.: 183, 186
739b34-745b23: 164n
740a2-23: 351
740b26ff.: 351
740b26-741a3: 179n, 284
740b33: 169
741b15-18: 185
741b18-24: 185
741b25ff.: 185
742a14: 176n
742a15: 180n
742a17ff. 183
742a25-744b28: 179n
742a33-b3: 353
742b33-743a1: 353
743a1ff.: 186
743a27-33: 175n
743a27-36: 181n
743a29-31: 172n
743a36-b6: 169
743a37: 190n
743b26-30: 217n, 218n, 353
743b27-33: 190
744a27-32: 332
744a31: 191
744b2-6: 353
744b12-14: 353
744b12-28: 189
762a20ff.: 181n
765b7-766b27: 175n
766a17-23: 328
766b15-16: 175n

766b35-767a35: 175n
767a14ff.: 308
767a14-24 163n
767a14-28: 175n
767a16-20: 172, 227
767a17: 190
767a17-18: 169
767a24-28: 175n
767a36-b18: 308
768n29-33: 187
769b4ff.: 308
770b9-18: 308
770b10-18: 308
775a13-22: 328
776 b9-18: 356n
777b7-8: 177
777b24ff.: 175n
777b28-30: 174
778b2-11: 164n
778b11-13: 169n
779a26-780a25: 212
779b26-780a25: 209n
783a33-b2: 179n
784a31-b13: 176n
784a34ff.: 176n
784a34-b2: 187
786b7-788a16: 356n

De Generatione et Corruptione
304b23-305a33: 165
319a17-320a6: 165
319b5-15: 166n
319b33-34: 166n
321b18-322a27: 186n
322b18-19: 165
322b23-25: 166
323b1ff.: 207n
323b30-324a9: 210
323b30-324a14: 165
324a12-13: 165
324b5-10: 165
324b5-23: 165
324b7-10: 166
327a-b33: 168n
327b22-31: 168n
327b24-26: 168
327b26-33: 168
328a8-9: 168
328a9-10: 168
328a11-12: 168
328a23-27: 166
328a27-28: 167
328a28-31: 168n
328a28-32: 167, 169
328a30-31: 167n
328a33ff.: 186n
328b18-23: 168n
328b20-23: 167
329a24-26: 166n
329a24-b3: 165
329b24-34: 168
329b27ff.: 164n
330a30: 162n
330b1-6: 165
330b3-6: 356n
330b5: 180n, 354
330b5-6: 191
330b22ff.: 209n
330b31ff.: 210
330b31-331a6: 165
331a4-6: 356n
331a13ff.: 166n
331a15ff.: 162n
331a16-20: 356n
331a25-b2: 166n
332a28: 166n
332a36-b1: 166n
333a21ff.: 166n
333a23-27: 166n
333b8-20: 169
333b17-20: 170
334a22ff.: 170
334a23-25: 170
334b5-8: 170
334b8-14: 170
334b8-30: 168n
334b14-17: 169, 171
334b17-18: 171
334b22-25: 166n
334b23-26: 171
334b26-30: 171
334b28: 173n
334b31ff.: 186n
334b35-335a3: 165n
335b6-7: 169, 170
335b8: 172n
336b27-337a8: 333

De Interpretatione
16b19-21: 274

De Motu Animalium
698a14ff.: 352
701b25-33: 358
702a16-19: 265
703a29-b2: 330
De Partibus Animalium
639b4ff.: 164n, 170
639b12-19: 314
639b12-21: 170
639b27-31: 316
640a10-b4: 170n
640a34-36: 164n
640b1-4: 164n
641a18-33: 223n
642a18-23: 169
644b22-26: 332
644b32-35: 332
645a10: 184n
645b19:20: 182
646a13-18: 168, 231
646a13-24: 173n
646a15ff.: 168n
646a15-20: 202
646a25-b2: 316
646a25-b4: 314
646b15ff.: 168n
647a3-24: 211
647a6-35: 220
647a25-30: 355
647a25-32: 217n
647a25-34: 352
647a25-b8: 186
647b29-36: 189
648a3-4: 192
648a37ff.: 168
648b2-6: 174
649a19: 190n
649b9ff. 164n
650a2ff.: 176n
650a3-8: 179
650a3-b13: 186
650a32ff.: 186
650a37-b3: 187
650b19-23: 192
650b27-30: 256
650b33-35: 256
650b35: 253, 255
651a13-14: 256
651b20-21: 332
652a24ff.: 192
652a27-33: 193
652a28: 191
652a31-33: 195n
652b7-16: 350
652a7-23: 190
652b8-17: 179, 180n
652b11-13: 357
652b17-20: 163n, 172, 227
652b18: 191
652b19-20: 200
652b26-27: 191
652b27: 191
652b27-33: 193
652b33-653a10: 174n
652b36: 194
653a2: 130n
653a12-18: 192
653a14: 191
653a29-34: 191
653a30ff.:356
653b3-5: 194
653b5-8: 191
653b23-29: 211
654b29ff.: 184n
655b3: 179
655b29ff.: 184
656a1-13: 331
656a3-13: 195n
656a14ff.: 192n
656a15: 192
656a24: 192n
656a28ff.: 217n
656b3-8: 193
656b4-7: 219
656b5-6: 257
656b33-34: 196n
658b1-10: 193n
658b9-10: 356n
665a10ff.: 352
665a11-13: 217n, 255
665a11-25: 184
665b10ff.: 183, 186
665b18ff.: 352
665b18-22: 184
666a3: 218
666a6ff.: 186
666a11-17: 218, 357
666a12-16: 255

666a13-16: 184n
666a20-22: 185
666a26-28: 184n
666a28: 195n
666b7-11: 189n
666b14-18: 352, 355
666b22 ff.: 188
666b22-667a7: 188n
666b25 ff.: 188
667a1-4: 189, 219
667a2-3: 189n
667a4: 357
667a4-7: 189, 219, 352
667a14-15: 220
667a15-17: 256
667a17-20: 256
667a23-25: 256
667b22 ff.: 349
667b27-32: 186, 218
668b12-16: 186
669a14 ff.: 356
669a18 ff.: 354
669b4-7: 356
669b13 ff.: 188n
669b18 ff.: 195n
669b18-22: 184, 357
670a3-7: 196n
670a24-26: 185
670b5-8: 187
670b19-23: 196n
672b10 ff.: 218n
673b11-12: 182
673b12: 194
673b26: 173n
683a20-26: 287
686a7-12: 192
686a25 ff.: 195n
687a12: 184n

Eudemian Ethics
1216b11 ff.: 224n
1218a21-24: 260
1219a1 ff.: 277
1220a2-4: 228
1220a18-21: 228
1220a23-33: 230
1220a38-64: 228
1220b5 ff.: 232
1221b33-39: 233
1222a1-3: 234
1222a3-5: 260
1222a6-13: 236
1228b30-38: 241
1228b34-35: 260
1229a33 ff.: 252
1235b31-1236a1: 244
1239b30-38: 226
1248b8 ff.: 283
1248b27-34: 244
1249b12-23: 261

Eudemus: 173n
fr. 45 Rose: 161

Historia Animalium
492a7, b1: 212
496a4: 188
496a17 ff.: 188
496a20-22: 189
511b10: 28n
512b13 ff.: 34n
513a27 ff.: 188
513a32-35: 189
513a35: 188n
515a27-b7: 355
520b22-23: 187
520b29-32: 187
521a32-34: 187
590a12-18: 176
596b20 ff.: 177n
608a10 ff.: 232

[Magna Moralia]
1208a9-21: 261

[Mechanica]
847b1 ff.: 358
847b14 ff.: 217n
847b16 ff.: 352
848a11-16: 358

Metaphysics
980b27 ff.: 262
981a3-4: 273
981b2- 983a11: 278
982a20-983a11: 274
986a21 ff.: 196n
986a27-b1: 22n
993a17-22: 169
993a17-23: 29n
993b21-24: 223n
1013a4-7: 185
1015a10-11: 170n
1021b20-23: 159n

1022b1-3: 158
1022b10-14: 158n, 232
1023a6-7: 320
1023a23-25: 271
1025b19-29: 223n
1025b34-1026a7: 223n
1026a15ff. 274
1026a18-20: 223n
1029b2-13: 317
1032b1-12: 315
1032b1-32: 314
1035b14ff.: 200, 348
1035b14-16: 348
1035b25ff.: 349
1035b25-28: 349
1036b21-32 198
1043a10: 204n
1043a30: 225n
1043a35-36: 218n
1044b15-20: 218n
1057b8-10: 211n
1063a11-18: 332
1066b27-31: 166n
1068a11-13: 159n
1070a14: 225n
1071b3-23: 333
1072a23-b4: 332
1072a26-b15: 333
1072b22-25: 278
1072b23-27: 283
1074a31: 274
1075a12-18: 333
1092b17ff.: 169
Meteorology
328a22ff.: 179n
339a19-21: 164n
347b2-7: 190n
348b2-8: 190n
378a10ff.: 175
378b10-388a9: 168
378b27- 379a11: 174
378b32-379a1: 169
379a12ff.: 176
379a22-26: 176
380a1: 170
381b24-382a11: 169
382a3-4: 165n
382a19: 206n, 208n
382b5ff.: 190n
384b28-385a4: 202
388a10ff.: 169n
389b28ff.: 169n, 170n
Nicomachean Ethics, 16, 17, 19, 20, 223, 334
1094a-b12: 311n
1094a26-27: 223
1094a27-b12: 285
1095a4-10: 276
1095a4-12: 267
1095a5: 223
1095a8-10: 267
1095b2-4: 224
1097a11-14: 311
1097a15-20: 280
1097b1ff.: 277
1097b1-7: 280
1097b2-4: 280
1097b22ff.: 277
1097b22-1098a20: 277
1097b33ff.: 227
1097b34-1098a3: 277
1098a18-19: 277
1098a26-34: 224
1099b3-4: 277
1099b21-23: 314
1099b29-33: 311n
1102a16: 163n
1102a16-17: 225n
1102a23-26: 224
1102a26-28: 224n
1102a27-32: 224n
1102a28ff.: 227
1102a30-34: 224n
1102a33-b12: 228
1102b13-29: 228
1102b25: 224n
1102b27-28: 267
1102b28-29: 240, 269
1102b29-1103a3: 265
1102b31: 228
1103a1-3: 224n, 288
1103a2-4: 228
1103a4ff.: 227
1103a14-26: 284
1103a17ff.: 228
1103a24-26: 229
1103a27-33: 229
1103b1-3: 229

1103b8-9: 229
1103b14-18: 229
1103b23: 229
1103b26ff.: 224n
1103b26-32: 224n
1103b27-28: 223
1104a4-10: 311
1104a11-19: 163n
1104a11-27: 230
1104a12-14: 163n
1104a13-14: 224, 228
1104a18: 174n
1104a18-24: 235
1104a25-26: 235
1104a27ff.: 269
1104a27-30: 230
1104a27-63: 230
1104a32-63: 234, 235
1104a35-63: 241
1104b4-8: 242
1104b8ff.: 233
1104b14-15: 234
1104b24-26: 260
1104b27-28: 234
1105a10-13: 233
1105a13-16: 234
1105a29-b1: 271
1105b12ff.: 224n
1105b19ff.: 232
1105b21-24: 225n, 232
1105b24-26: 232
1105b26-27: 232
1106a15-23: 238n
1106a22-24: 238n
1106a27ff.: 237
1106a36-b8: 237
1106b16-28: 235
1106b36-1107a1: 15
1106b36-1107a8: 271
1107b30ff.: 134n
1109a20-24: 39
1111b7-9: 264
1111b14-16: 267
1112a1-17: 264
1112a2-5: 269
1112a15-18: 266
1113a9-12: 266
1113a23-34: 243
1113a24-b2: 259
1113b3ff.: 264
1114a31ff.: 259
1114b6ff.: 264
1114b27-29: 269
1115b18-24: 271
1116a9: 259
1116a17-b2: 293
1117a6-9: 252
1117a20-22: 268
1117a31: 259
1119a5ff.: 269
1119b13-18: 260
1119b15-18: 240, 269
1120a27-28: 242
1120b7-9: 239
1120b24-25: 239
1121b8-10: 305
1122b23-26: 239
1128a11-12: 229
1128b12: 252
1129b4-6: 318
1130b30ff.: 330
1138b35ff.: 229
1139a7-13: 263
1139a23-26: 270
1139a26-29: 263
1139a32ff.: 264
1139b18-36: 273
1140b6-7: 270
1140b12ff.: 259
1140b31ff.: 273
1140b31-33: 273
1141a17-20: 273
1141a21-23: 332
1141b1-4: 274
1141b3: 274
1143b2-3: 273
1143b15-17: 272
1144a2-3: 272
1144a8-9: 270
1144a15ff.: 270
1144a23-36: 270
1144b1ff.: 269
1144b1-12: 270
1144b13-14: 269
1144b16-17: 270
1144b26-28: 270
1145a3-4: 280
1145a4-10: 272

1145b30-32: 267
1146a10ff.: 268
1146a10-13: 269
1147a ff.: 267
1147a10ff.: 259, 267, 276
1147a18-23: 267
1147a24: 224n
1147b6-9: 276
1147b9: 224n
1149a31-32: 252
1149a33-34: 252
1150b19-22: 267
1150b19-29: 259
1150b22-25: 268
1150b25-29: 256
1151a20ff.: 267
1151a25-27: 267
1151b23-32: 259
1151b33ff.: 240
1152a2-3: 269
1152a25-27: 241n, 293
1152b1ff.: 242
1152b33ff.: 242
1153a3ff.: 242
1153a14-16: 248
1153a31: 260
1153b16-18: 277
1154b3-16: 276
1154b7: 260
1154b10: 253n
1154b12-14: 227
1154b12-16: 256
1154b21-28: 333
1154b23-24: 259
1154b26-29: 283
1155b2-11: 224n
1156a32-34: 267
1157b25-29: 331
1158b12ff.: 239
1158b18-20: 239
1159b20-24: 226
1160a34: 297
1160b10-23: 302n
1160b13-14: 299n
1160b18-22: 305
1160b20: 305
1161a31-33: 303
1162a22ff.: 239
1165b18-20: 320
1166a27-30: 241
1166b7-22: 259
1166b18-24: 241
1167b28-1168a10: 224n
1168a13-17: 234
1170a13: 224n
1172a19ff.: 242
1173a18-23: 320
1173a24-28: 163n, 226, 238
1173a26-28: 318
1174b14ff.: 242
1174b33-35: 242
1176a8-29: 317
1176a11-22: 244
1176b17-18: 225n
1177a12ff.: 278
1177a13-18: 279
1177a25-28: 278
1177b1-25: 280
1177b19-25: 279
1177b26-29: 280
1177b27ff.: 283
1178a1: 81n
1178a5-24: 331
1178a8-22: 280
1178a9ff.: 247
1178a9-24: 225n
1178a14-16: 247
1178a17-23: 272
1178a20-23: 248
1178a22-23: 225n
1178a24ff.: 282
1178b22-23: 283
1178b22-33: 333
1178b32ff.: 282
1178b33-35: 277
1179b10-16: 293
1179b13-16: 259
1180a1: 268
1180b8-16: 311
1180b13-15: 311
1180b25-28: 312
1181b3-5: 312
1181b4-5: 316
1181b15-16: 284
1181b18-19: 317
1181b18-23: 312
Parva Naturalia, 157
436a8ff.: 248

436a17-19: 158n
436b6: 202
436b10-22: 201
436b16-17: 250n
436b19-437a2: 182
436b20-22: 250n
437a1-3: 250n, 262
438b25 ff.: 183
439a3: 190n
439b27-440a6: 249
440a33 ff. 204n
440b10-22: 202
440b13-20: 204n
440b20: 249
442a12 ff.: 249
442a12-25: 204n
445b4 ff.: 202
445b21 ff.: 204
449a5 ff.: 216
449a9-11: 216
449a17-20: 216
449a20: 215n
449b30-450a25: 263
449b31: 263
449b31 ff.: 273
450a4-13: 274
450a10: 251
450b: 129n
450b1-12: 263
450b2-9: 276
450b8-11: 256
450b9 ff.: 220
452a30: 240
453a9-10: 263
453a15 ff.: 220
453a15-23: 256
453a15-b8: 276
453a15-b11: 263
453a18-26: 227
453b25 ff.: 218
454a7-12: 205
454a9-10: 215
455a14 ff.: 217
455a33: 217
455b17: 277
455b28 ff.: 192
456a4 ff.: 356
456a4-6: 217
456a16-18: 356
456a30 ff.: 186
456b18 ff.: 276
457a27 ff.: 227
457a27-34: 256
456b24 ff.: 356
547b30: 191
458a3: 130
458a10 ff.: 188, 189
458a19: 188
459a15-17: 251
459a5 ff.: 251
460b3-10: 257
460b11-16: 257
460b20: 251
460b27-461b12: 276
461a7-b5: 219
461a8 ff.: 257
461a24-27: 219
461b12 ff.: 218
463b26-33: 177n
464a6 ff.: 204
464a33 ff.: 227
464b32-34: 158n
465b1-33: 175n
465b18-22: 176
465b27-30: 177
466b15-28: 177
466b31-32: 184
467b13-16: 183
467b18-34: 218
467b26-28: 183
467b28 ff.: 357
467b28-34: 218
467b28-468a1: 183
468a13 ff.: 183, 188n
468b28-32: 183
468b31 ff.: 183
469a10-13: 217
469a10-17: 183
469a17-19: 218
469a18-19: 183
469a23-26: 183
469a26 ff.: 183
469b6-11: 185
469b7-20: 176n
469b12-13: 186
469b14-16: 185
469b20 ff.: 181
469b20-470a4: 176

470a5-7: 181
470a20-21: 180
472a31-b3: 177
472b2-3: 356
474a25ff.: 186
474a25-b3: 184
474b10-13: 180
474b10-24: 176
474b12: 185
474b13-23: 181
474b25ff.: 177
476a16: 181
477a27-31: 176
477b14ff.: 176
477b17-19: 176
477b30-478a3: 177
478a12ff.: 356
478a18-21: 357n
478a29: 185
478b16-18: 356
478b32-33: 185
478b32-479b6: 186
479a8-9: 174n
479a8-14: 169n
479a28-30: 180
479b17ff.: 354
479b22-25: 354
479b31-33: 354
480a3ff.: 354
480a13-14: 354
480a14-15: 354
480b21ff.: 158n

Physics

Book VII, 173n
189b6: 165
192b9ff.: 284
193b24-25: 214n
193b31-32: 214n
193b34-35: 214n
194a12-b15: 224
194a22-b8: 315
194b32-195a3: 314
198b34ff.: 284
199a15-16: 314
204b12-22: 166n
210a18-20: 271
210a20: 162n
217b20-23: 175
219b12-220a26: 214n
219b13-32: 214n
220a10-11: 214
222a12ff.: 214n
224b28-34: 215
226b1-9: 215
229b15-20: 215
244b5-245a4: 202
245b8: 158n
246a10-248a9: 158
246a11-246b2: 159
246a11-12: 159
246b3ff.: 227
246b3-7: 174, 177n, 178
246b3-10: 159
246b4: 174n
246b4ff.: 163n
246b6: 173
246b9: 174
246b10-20: 160
246b17-20: 241n
246b20ff.: 160
247a8ff.: 249
247b1ff.: 274
247b1-13: 275
247b13-248a9: 275
253a7ff.: 214n
253a7-20: 175n
259b1ff.: 175n, 214n
260a29ff.: 186n
260b8-13: 355
261b19: 177n
262a19-20: 214, 357
262a19-23: 215
262a21-26: 214n
262b19: 214n
256a14-b15: 358
267a22-b8: 358

[Physiognomonica]

813b7ff.: 218n

Politics, 16, 18, 20, 223, 284, 334

1252b1-15: 291
1252b2: 287n
1254a22-39: 288
1254a34-b3: 243
1254b3-16: 288
1256b16-17: 287
1259a38-b18: 288
1259b10-18: 292
1259b15-16: 291

1259b23ff.: 239
1260a4-14: 288
1260a5-24: 321
1260a18-20: 239
1261a38-40: 291
1261b1-7: 291
1265b15-23: 331
1265b29: 299n
1266b24-28: 325
1267a22-28: 325
1270b21-23: 303, 327, 330
1273b10-12: 291
1277a6-13: 321
1277b13-30: 321
1277b17-25: 239
1278b9-12: 298
1279a19-22: 302
1279a38-39: 297
1279a40-b5: 293
1279b1-5: 299n
1279b4-5: 299n
1279b37-1280a7: 294
1280a2-7: 294
1280a8-15: 301
1280a23-25: 301
1281a4-8: 289
1282b1-6: 330
1282b8-13: 302
1282b14-17: 285
1283a14-16: 289
1283a14-19: 294
1283a34-41: 293
1284a1-3: 289
1284a3-10: 292
1284a3-12: 290
1284a4ff.: 330
1284a13-14: 330
1284b9: 163n
1284b23-33: 290
1284b28-34: 330
1284b33-34: 290
1284b35ff.: 307
1285a25-30: 292n
1285b20ff.: 305n
1285b30-34: 292n
1286a8ff.: 330
1286b8ff.: 302n
1287b40-42: 301
1288a1-2: 293
1288a1-3: 291
1288a8-10: 292
1288a10-13: 292
1288a13: 293
1288a13-15: 299n
1288a15-29: 292n
1288a25-29: 330
1288a28-29: 292
1288a33-37: 290
1288a33-b2: 331
1288a38-b3: 293
1288b10ff.: 312
1288b10-20: 312
1288b12-13: 316
1288b15-16: 317
1288b22-39: 313
1288b24-28: 322
1288b28-33: 317, 323
1288b34-35: 317
1289a1-3: 327
1289a2-8: 323
1289a6-8: 317
1289a32-34: 290
1289a40: 291, 307
1289b2-5: 305
1289b9-10: 305
1289b18-20: 308
1289b18-23: 317
1289b31-32: 299n
1289b33-40: 299n
1290b3-4: 294
1291b2-13: 294
1291b14ff.: 305n
1292a18-24: 307
1292a31-38: 306
1292a39ff.: 305n
1292b23ff.: 305n
1293a13ff.: 305n
1293b3-8: 293
1293b33-35: 297
1293b41-42: 290, 293
1294a9-11: 290, 293
1294a15-17: 297
1294a23-24: 297
1294a30ff.: 297
1294a41-b1: 297
1294b13-14: 297
1294b14-40: 298
1294b39-40: 304

1295a1 ff.: 305n, 307
1295a19-21: 301
1295a19-23: 303
1295a25 ff.: 296
1295a25-32: 300
1295a35-b2: 297
1295a36-39: 326
1295b4: 177n
1295b4-6: 300
1295b6 ff.: 300
1295b35-40: 300
1295b38-40: 300
1296a7-9: 300
1296a13-16: 305
1296a22-34: 295
1296a23-33: 323
1296a24-33: 303
1296b3-9: 304
1296b7-13: 310
1296b13 ff.: 308
1296b15-17: 305, 308
1296b17-24: 309
1296b25-31: 310
1296b25-33: 309
1296b31-35: 310
1296b35-38: 324
1296b38-40: 309
1296b39-1297a8: 299
1297a6-7: 323
1297a7-8: 304
1297a14 ff.: 305n
1297a29-36: 306
1297a36 ff.: 305n
1297a38-b1: 297
1297b1-8: 299n
1297b16-28: 299n
1298a10 ff.: 305n
1298a33-34: 307
1298a34 ff.: 305n
1298a39-40: 307
1298b13 ff.: 305n
1298b14-38: 324
1298b27 ff.: 305n
1299a35-36: 291
1299b10: 287n
1301a26-39: 303
1301a26-b29: 303
1301a36-37: 302
1301b29-37: 309
1301b40-1302a2: 293, 294
1302a4: 305
1302a4-7: 304
1302a8-11: 306
1302a8-13: 305
1302a8-16: 304
1302a14-15: 300
1302a29-31: 303
1302b15-18: 324
1302b33 ff.: 324
1302b33-1303a3: 304
1302b35-37: 163n
1304a38-b4: 305
1307a6 ff.: 302n
1307a6-8: 323
1307a7-20: 304
1307a17-18: 305
1307a17-20: 306
1308b11-20: 324
1308b25-31: 325
1309a15-27: 324
1309a28-33: 324
1039b7-9: 294
1309b15-18: 327
1309b16-18: 305
1309b18-20: 324
1309b20-35: 304, 324
1309b22-35: 307
1309b23-30: 163n
1310a2-11: 303
1310a2-13: 324
1310b3-8: 307
1310b33-34: 290
1310b40-1311a2: 292n
1311a3-7: 303
1311a5-8: 292n
1312b33-38: 307
1312b38-40: 292n
1313a7-8: 292
1313a10-11: 307
1313a13-16: 303
1313a34 ff.: 305n
1313a35-1314a29: 303
1314a30-1315b17: 325
1315a31-33: 294
1315b12-13: 306
1317a19 ff.: 305n
1317a23-29: 310
1317a40-b10: 302

1318a19-23: 302
1318a31-32: 294
1318b7ff.: 310
1318b7-38: 325
1319b2-4: 307
1319b13-17: 324
1319b25-28: 325
1319b32-33: 293
1320b18ff.: 305n
1320b18-33: 310
1320b19-22: 325
1320b21-23: 307
1320b32-39: 307
1320b34-39: 319
1321a1-2: 305
1321a3-27: 306
1323a18-19: 322
1323b21-26: 333
1323b40-1324a2: 331
1324b1ff.: 331
1325a8-11: 322
1325b7-10: 291
1325b8-10: 293
1325b10-13: 289
1325b14ff.: 331
1325b14-22: 331
1325b14-28: 331
1325b24-30: 333
1325b33ff.: 316
1326a9-b26: 325
1326a38: 163n
1326b13-14: 331
1327b18-1328a8: 325
1328a37-38: 322
1328a37-b2: 316
1328a38-40: 322
1330a34-b22: 325
1331b38-39: 315
1331b39-1332a2: 316
1332a4-11: 315
1332a28-b11: 316
1332b12-23: 291, 292
1332b17-18: 292
1332b23-25: 292
1333a17ff.: 279
133a17-37: 287
133a22-24: 279, 287
1333a29-30: 311, 318
1333a31-b2: 279
1334b10-28: 287
1334b13-29: 279
1335b6-13: 325
1336a38-39: 356
1337a2-3: 284, 314
1337a19-22: 284
1339b27-38: 329
1342b14-34: 325

Posterior Analytics
72a25ff.: 273
78b18-20: 162
79a8-10: 214n
99b20ff.: 273
99b36ff.: 273
99b36-100a8: 262
100a3ff.: 274
100a6-7: 274
100a12-b2: 274
100a17-b1: 262

[Problemata]
862b27: 130n

Rhetoric
1361a3: 163n
1361b3: 163n
1361b7-14: 238
1362b14-18: 163n
1369b16-20: 268
1369b19-27: 266
1370a5ff.: 242
1370a6-7: 240, 268
1370a13: 242
1370a27-35: 234, 250
1370a28ff.: 251
1370b1-4: 268
1371a25: 242
1372b12ff.: 266
1378a32-34: 252
1378b8-9: 252
1382a19-22: 252
1383a14-16: 252
1383a15-18: 252
1384a23-25: 252
1389a2ff.: 253
1389a18-b32: 253n
1389b32: 253, 255

Topics
107b27-32: 211n
116b17: 163n
116b17ff.: 161

119a30-32: 211n
123b29: 177n
139b21: 162n, 163n
145b7ff.: 160n, 161n, 162
145b8: 163n
ASPASIUS
In N.E., C.A.G., XIX, pt. I, p. 157: 259n
ATHENAEUS
6,262A. ed. Kaibel, 127n
DIOCLES OF CARYSTOS
Fr. 55, 57, ed. Wellmann, 130n
DIOGENES LAERTIUS
VIII, 83: 22n
EMPEDOCLES
Diels-Kranz, ed.
31A74: 30n
31A78: 29n
B17.27-29: 29n
B21.9-14: 29n
B23: 29n
B96: 29n
B98: 29n
GALEN
ed. Kühn, II, 130: 130n
V, 667-668: 146n
X, 5: 28n
XV, 25: 25n
HERODOTUS
III, 80, 142: 23n
HIPPOCRATIC CORPUS, 18, 32ff., 57n, 335
Airs, Waters, Places Jones, Loeb. Class. Libr. ed.: 32, 33n, 34, 51-63, 67, 69, 70, 74, 119, 135
ch. 1: 51
ch. 2: 52
ch. 3: 54
ch. 3-6: 53
ch. 4: 54
ch. 5: 55, 75n
ch. 5. 6-28: 54
ch. 6: 55
ch. 7. 53-57: 56
ch. 7. 79-90: 56
ch. 7. 80-82: 56
ch. 7. 82-93: 57
ch. 7-9: 56, 57
ch. 10, 7: 71n
ch. 10, 11: 57
ch. 11: 58
ch. 12. 7-13: 58
ch. 12. 13-19: 59
ch. 12. 16-18: 74n
ch. 12. 35-38: 59
ch. 12. 40ff.: 72n
ch. 12. 40-44: 60
ch. 12, 16, 23: 75n
ch. 12-24: 58
ch. 14: 60
ch. 15: 60
ch. 16. 3-14: 61
ch. 16. 12-14: 75
ch. 16. 14ff.: 62
ch. 16. 31-33: 63
ch. 19. 33-35: 61
ch. 23: 61
ch. 23. 9-12: 61
ch. 23. 19-30: 61
ch. 23. 25-30: 75
ch. 23. 30-41: 62
ch. 24. 19-22: 63
Ancient Medicine
Jones' ed. (*Phil. and Med. in Ancient Creece*), 32, 33, 35-45, 55, 58, 59, 67, 68, 69
ch. 1: 35
ch. 2: 35
ch. 3: 35
ch. 3, p. 51. 26-36: 36
ch. 5, p. 52. 19-31: 37
ch. 6: 38, 72n
ch. 6, p. 53. 1-2: 38
ch. 7, p. 53. 14-15: 38
ch. 8: 38
ch. 9: 39, 187
ch. 9, p. 54. 6-8: 39
ch. 9, 54. 8-11: 39
ch. 10: 39, 72n, 127
ch. 10-12: 40
ch. 11: 39
ch. 12, p. 55. 30-33: 56
ch. 12, p. 55. 31-33: 40
ch. 14: 41
ch. 14, p. 56. 33: 41
ch. 14, p. 57. 14-31: 41
ch. 14, p. 57. 16-17: 42
ch. 14, p. 57. 16-23: 72n

ch. 14, p. 57. 20-21: 43
ch. 15: 39
ch. 15, p. 58. 9-12: 42
ch. 16, p. 58. 17-20: 42
ch. 18, p. 59. 31-32: 44
ch. 19, p. 60. 10-12: 44
ch. 19. 1/ 60. 24-31: 44, 72n
ch. 19, p. 60. 33-34: 44
ch. 19, p. 61. 5-7: 43
ch. 20: 35, 45n
ch. 21: 45n
ch. 22, p. 62. 24-25: 42
ch. 22-23: 45n
ch. 23: 45n
Aphorisms: V, 62: 70n
VII, 66, 67: 72n
Breaths: ch. 1. 3-7: 71n
ch. 14: 67n
Epidemics: I, ch. 2, 11, 14, 17, 19: 73n
I, ch. 23: 71n
I, III: 58n
III, ch. 15: 58n
Humors: ch. 15: 74n
Joints: ch. 56, 57: 73n
Nature of Man, 32, 33, 45-51, 52, 67, 68
Jones, Loeb. Class. Lib. ed.
ch. 2: 45
ch. 3: 45, 70
ch. 3. 24-28: 46
ch. 4: 72n
ch. 4. 1-4: 46
ch. 4. 4-9: 46
ch. 4. 10-20: 47
ch. 5. 1-17: 46
ch. 5, 6, 7: 45
ch. 7: 27n, 28n, 46, 47, 70n
ch. 7. 48-66: 48
ch. 8: 48
ch. 8. 11-14: 48
ch. 9: 71n
ch. 9. 2-6: 49
ch. 9. 6-8: 49
ch. 12. 34-44: 49n
ch. 15. 21-37: 50n
ch. 15. 30-37: 49n
Nutriment: 34: 57n
Places in Man: 42ff.: 72n
Prognostic: ch. 2, 3: 72n
Regimen I, 32, 33n, 34, 35n, 64-67
Jones, Loeb. Class. Libr. ed.
ch. 2: 66
ch. 2. 41-48: 66
ch. 2. 61-67: 57n
ch. 3: 64
ch. 4: 64
ch. 7. 18: 71n, 74n
ch. 8, 9: 70n
ch. 9, 10: 144n
ch. 10: 70n
ch. 25: 49n
ch. 32: 64
ch. 32. 3: 57n
ch. 33: 49n
ch. 34: 49n
ch. 35: 65, 75n, 129n
ch. 35. 15-17: 65
ch. 35. 25-27: 67n
Regimen II-IV, 66
II, ch. 66. 61-62: 71n, 72
III, ch. 67. 1-13: 71n
III, ch. 67, 69: 66n
Regimen in Health, 33n, 50-51
Jones, Loeb. Class. Libr. ed.
ch. 1: 50
ch. 1. 22-26: 50
ch. 2: 50
ch. 2. 1-8: 49n
ch. 2. 8-14: 49n
ch. 2. 14-18: 49
ch. 3, 4: 49n, 50
ch. 6. 7-11: 49n
Sacred Disease
ch. 9, 10: 131n
ch. 17: 67n, 75n
Sevens, ch. 6: 144n
IAMBLICHUS
In Nicomach., ed. Pistelli
p. 100. 15-18: 345
p. 100. 15-25: 345
p. 101. 1-5: 346
p. 116. 1-4: 346
p. 116. 4-6: 346
p. 118. 19ff.: 346
NICOMACHUS GERASENUS
Arithm. Intro. ed. Hoche, II. 27. 1, 346
OCCELUS
48, 8 Diels-Kranz: 344

PHILOPONUS
In De An., *C.A.G.*, XV, 144-145, 161n
435-436: 209n
In Nicomach., ed. Hoche
Bk. II, p. 28, XXIV, 11- 4-9, 83n
p. 32, XXVII, 1. 5-7: 346
PLATO
Charmides 159E: 128
Cratylus, 127
Critias 111E4, 118n
112C: 81n, 35
Epinomis 976A1, 118n
Gorgias, 163n
508A: 79n, 149n
Laws, 136, 140
649E: 126n
666B: 127, 132n
757E: 126n
777B: 126n
790C-791C: 141n
791C, D: 126n
791D: 127
792A: 126n
792C, D: 135
935A: 126n
Phaedo 69C, D, 100n
Phaedrus 268A ff., 151n
270B-D: 336
270C, E: 146n
Philebus, 36n, 106, 135, 148
23C ff.: 128
24A ff.: 149
25B, C: 149
25C: 128
25D, E: 149
25E: 149
26A6, B1: 118n
26B: 150
26B, C: 148
29A-30C: 144n
30A-C: 148
30C: 148, 150
31D-32B: 112
32B, C: 112
33C: 112
33C ff.: 129
33D: 107n
34D ff.: 112
45C-E: 153
47D ff.: 112
52C, D: 112
53C-55C: 111n
63E-64A: 154
64C-E: 154
66A ff. :154
Protagoras 315C, 151n
Republic, 136, 140, 142
373E: 155
411A ff.: 133n
411C: 126n
443A-444E: 155
444C: 153
444D, E: 156
449A: 155
556E: 143n
564B: 131n
586C: 126n
590A: 126n
618D: 128
Statesman, 136
283E-284C: 154n
Symposium, 113, 150
186A-188D: 144n
186B: 150n
188A, B: 119
Theatetus, 106, 135
147D ff.: 80n
152D-153C: 141n, 145n
161A: 126n
174D: 126n
191C ff.: 129
194C: 129
Timaeus, 18-20, 26, 30, 31, 32, 77, 334, 335, 336
24C, D: 118
30A: 99
30B: 78
30C: 78
30D: 78, 140, 155
31B-32C: 91
31B, C: 79, 90n
31C-32A: 344
31C-32B: 93
32A: 80
32A ff.: 92
32A-C: 82
32B: 87
32C: 83

32D-33A: 83
33A, B: 84
33B: 89
33C, D: 89
33D-34A: 89
34A, B: 90
35A: 91
35A-36D: 90
35B, C: 91
35C-36B: 93
36A: 344
36A2: 92n
36B-D: 93
36B5: 92n
36D: 92
36D, E: 94
36E: 95
37A: 95, 155
37A6: 152
37A, B: 94
37A-C: 152
37B, C: 95
37D: 155
38C: 92
38C, D: 89, 94, 95
40A: 96
40A, B: 95
41C: 78, 96
41D: 90, 96
41D, E: 97
41Eff.: 97
41E-42A: 96
42A: 106, 134
42A, B: 97
42B: 101, 113
42B, C: 101
42C: 97
42C, D: 98, 101
42E: 78
42E-43A: 98
42E-44D: 123
43A: 102
43Aff.: 98
43B2: 98n
43B, C: 107
43C5-7: 127
43C-E: 99
43D: 97
43E3: 98n
43E-44A: 99
44A2: 152
44B, C: 100
44D: 78, 102
44D-45A: 102
45A, B: 103
46Bff.: 147
45B, C: 121
45B-D: 108
46C, D: 124
46D, E: 147
47B, C: 103
47C, D: 103
49A-50A: 145n
50A: 85
50A-C: 86
50C-51B: 86
51D: 86
52D-53A: 85, 119, 143n
52D-53B: 86
53A: 98n, 99
53C: 86
53C-55B: 86
53C-68E: 86
53E: 86
54A: 86
54B: 81n
55D-56A: 79
55D-56C: 87
55E: 87
56A, B: 87
56C: 81n, 87
56C-57C: 87
57A: 88
57A, B, E: 121
57C: 81n
57C, D: 120
57D-58C: 88
57E: 28n
58A: 116, 121
58C: 120, 121
58D: 121
58D, E: 88
58E-59A: 132n
60A: 110, 111
60C: 88
61C: 109n
61D-62B: 108
61E-62B: 132n

62B, C: 109
64A: 110
64A, B: 109
64A-C: 107
64C: 108
64C-65B: 110
65A: 109n
65C: 109
65C-66C: 108
66C: 110
67A: 110
67A-C: 109
67C: 120
76Cff.: 108, 121
67D: 132n
67D, E: 108, 211n
68A: 131n
68E: 83n
69B: 120, 140
69B-C: 89
69C: 78, 103, 106, 115
69C, D: 133
69D: 106, 109n, 134, 145
69Dff.: 113
70B: 114
70C-D: 31n, 132n
70D: 114
70E: 114
72E-73A: 115
73A: 115
73B: 104
73B, C: 104
73C, D: 104
73E-74A: 104
74B: 105
74B, C: 131n
74C: 105, 130n
74D: 105
75D, E: 103
76E: 116
77A: 116
77B: 114
77C, E: 109n
77C-81B: 116
77E: 113
78E: 132n
79A-E: 31n
80C: 116
81A: 116
81Aff.: 85
81A, B: 117
81B-D: 117
81C-82B: 84
81D, E: 111
81E: 118
82A: 120
82A, B: 121
82B: 84n, 121
82E: 18n
83C: 130n
83C, D: 131n
84B: 132n
84C-E: 31n
84C-85E: 122
84D: 122
84D-86A: 131n
85A: 125, 131n
85A, B: 123
85B: 123, 130n
85B5ff.: 130n
85C, D: 123
85D: 84n, 130n
85E: 130n
86A: 31n, 121
86B: 124, 133
86Bff.: 124
86B, C: 100n, 124
86D, E: 125
86E-87A: 125n
87A, B: 136
87C: 138
87C, D: 137
87E-88B: 137
88B: 112, 137
88C: 138, 140
88D, E: 140
89A: 141
89Dff.: 141
89E-90A: 140
90Aff.: 142
90C: 140
90C,D: 95, 142
91A, B: 104
92C: 78, 120
PROCLUS
In Tim., ed. Diehl
II, 17. 17ff.: 79n
II, 18. 29-19.3: 149n

II, 18-20: 79n
II, 19. 21-22: 98n
II, 20. 26: 79n
II, 21. 14ff.: 80n
II, 22. 23-24: 149n
II, 23.9ff.: 79n, 80n
II, 25.4-8: 80n
II, 25.13: 80n
II, 26.29-27.7: 80n
II, 58.18: 84n
II, 62.31-63.9: 84n
II, 63.12-16: 83n
II, 64.17: 84n
II, 198.2ff.: 94n
II, 198. 29ff.: 92n
II, 198. 29-200.21: 93n
II, 200.21ff.: 92n
II, 212.3-5: 92n
III, 133: 89n
III, 286.7ff.: 107n
III, 329-332: 107n
III, 352.16: 100n
SIMPLICIUS
In De An., C.A.G., XI, 53: 162n
XI, 165: 209n
THEON SMYRNAEUS
Expositie, ed. Hiller, p. 106. 15-20, 345
THEOPHRATUS
De Sensu 6, 85: 109n
XENOPHON
Memorabilia III, 13, 151n

STUDIES IN PHILOSOPHY

1. Nathan Rotenstreich: *Humanism in the Contemporary Era*. 1963. 171 pp. *f* 23.—
2. Troy Wilson Organ: *The Self in the Indian Philosophy*. 1964. 184 pp. *f* 26.—
3. P. Allan Carlsson: *Butler's Ethics*. 1964. 196 pp. *f* 25.—
5. Lowell Nissen: *John Dewey's Theory of Inquiry and Truth*. 1966. 112 pp. *f* 16.—
6. Ronald D. Milo: *Aristotle on Practical Knowledge and Weakness of Will*. 1966. 114 pp. *f* 17.—
7. F. Warren Rempel: *The Role of Value in Karl Mannheim's Sociology of Knowledge*. 1965. 125 pp. *f* 17.—
8. Ralph Philip Joly: *The Human Person in a Philosophy of Education*. 1965. 147 pp. *f* 17.—
9. G. D. Bowne: *The Philosophy of Logic, 1880-1908*. 1966. 157 pp. *f* 19.50
10. Joe K. Fugate: *The Psychological Basis of Herder's Aesthetics*. 1966. 303 pp. *f* 32.50
11. John J. Fitzgerald: *Peirce's Theory of Signs as Foundation for Pragmatism*. 1966. 182 pp. *f* 23.—
12. Allen Lacy: *Migual de Unamo: The Rhetoric of Existence*. 1967. 289 pp. *f* 32.—
14. David L. Perry; *The Concept of Pleasure*. 1967. 224 pp. *f* 25.—
16. Fred D. Newman; *Explanation by Description: An Essay on Historical Methodology*. 1968. 104 pp. *f* 14,—

Bei Fragen zur Produktsicherheit wenden Sie sich bitte an:
If you have any questions regarding product safety,
please contact:

Walter de Gruyter GmbH
Genthiner Straße 13
10785 Berlin
productsafety@degruyterbrill.com